Designing Information Architecture

A practical guide to structuring digital content for findability
and easy navigability

Pabini Gabriel-Petit

Designing Information Architecture

Portfolio Director: Ashwin Nair
Relationship Lead: Nitin Nainani
Program Manager: Ruvika Rao
Content Engineer: Nisha Cleetus
Technical Editor: Simran Ali
Copy Editor: Safi s Editing
Proofreader: Nisha Cleetus
Indexer: Manju Arasan
Primary Production Designer: Alishon Falcon
Production Designer: Alishon Falcon
Growth Lead: Sohini Ghosh

First published: March 2025

Production reference: 3281125

Published by Packt Publishing Ltd.
Grosvenor House
11 St Paul's Square
Birmingham
B3 1RB, UK

ISBN 978-1-83882-719-9

www.packtpub.com

To my husband, Richard. Thank you for making many things possible in my life. Also to the memory of my father and mother.

Pabini Gabriel-Petit

Foreword

My first encounter with the concept of information architecture (IA)—as both a practice and a field of study—took place around 2000, during the bumpy shift from the dotcom boom to the dotcom bust. At that time, I was working at Razorfish—a leading digital agency that helped pioneer the role of the *information architect*—and living in Germany. With my background in library and information science, I was immediately attracted to IA. It felt so natural—helping businesses grow through better information experiences. This also marked the beginning of my journey into a field that was shaping the way we interact with digital information.

In my studies at Rutgers University, I was exposed to many of the core theories and voices that would become foundational to IA. Models for information-seeking behaviors, which were already central in information science, soon found their place in the applied practice of IA. I was fascinated by the faceted-navigation project that Marti Hearst led at Berkeley, which laid the groundwork for many of the navigation frameworks we still see today. This blending of theory and practice started to gain traction around 2000, giving IA its modern contours.

Inspired by the work of people such as Richard Saul Wurman—who is often credited with coining the term *information architecture*—I delved more deeply into the field. The release of the now-iconic *Information Architecture for the World Wide Web*—affectionately called the "Polar Bear Book"—ignited a movement, attracting practitioners from diverse fields. I eagerly absorbed as much content on the subject as I could find, hungry to refine my skills and understanding.

My professional journey provided hands-on experience in designing Web sites and structuring navigation systems for a range of companies. This practical work soon demonstrated the true power of IA: taxonomy, labeling, and navigation are *not* just aesthetic improvements. They are powerful tools that drive meaningful business outcomes.

This dual perspective from combining academic learning and on-the-ground experience eventually shaped my own book *Designing Web Navigation* (2007), in which I distilled the principles I had come to rely on in my everyday work. IA is a practical extension of information science, with both disciplines mutually reinforcing one another. Today, IA has achieved such recognition that formal degree programs now teach it, preparing a new generation for the complexities of digital organization.

The rise of IA reflects a critical response to the digital age's demand for organized, accessible information. As online environments scaled to massive enterprise Web sites, structuring information became a necessity, and IA emerged as the essential toolkit to meet that need—particularly in transactional contexts such as ecommerce sites. After all, people can't buy what they can't find. Findability, it turns out, is a huge factor at the front end of just about every commercial endeavor online.

Fast forward to our modern, data-driven era, where artificial intelligence (AI) and big data are reshaping our digital experiences, raising questions such as the following: *People can just enter a search term or speak into a microphone and computers will do the rest, right? Isn't organizing information manually a thing of the past? Do we still even need AI?*

The answer to that last question is a resounding *yes*. We need IA more than ever.

Far beyond being a dotcom-era Silicon Valley trend, IA has proven itself a foundational discipline that is adaptable and resilient within a changing digital landscape. The fact that new books on IA are now being published underscores its enduring relevance.

In today's world, the volume of information that we encounter is overwhelming. The concept of *information overload* has evolved from a theory to a daily reality for many. Ironically, one of the cures for too much information is more information—namely, more metadata to control and organize it all. So, IA remains indispensable in helping users match their information needs with resources. The advent of AI and advanced data processing haven't diminished IA's relevance but enhanced it. If anything, the critical thinking that IA fosters is more crucial now than ever, providing clarity amid a flood of information.

Structured information will always have its place in information-rich environments—particularly those of high value—even if AI and other tools may do a larger share of the organizing and finding for us. Web sites for companies, online shopping, universities, and governmental departments still need navigation systems, taxonomies, and tagging schemes that are crafted in a human-centered manner. The heart of IA lies in creating clear paths to guide people toward the information they need—something that today's technologies, powerful as they are, cannot wholly replace.

Designing Information Architecture offers a fresh, updated perspective on IA, presented with a thoughtfulness that reflects many years of practice. Pabini brings to this work not only a wealth of personal insight but also the voices of global experts, enriching the reader's understanding of IA's multifaceted nature. And she does this in a very concise manner, making the ideas and concepts readily accessible and creating an indispensable reference for future generations of IAs.

Whether you're new to designing digital information spaces or a seasoned IA practitioner, you'll find this book valuable and packed with goodness. It offers both foundational knowledge and new insights that will deepen your appreciation for the field. Begin exploring these pages, and prepare to strengthen your understanding of the durable, dynamic world of IA.

Jim Kalbach

Chief Evangelist at Mural

Author of Mapping Experiences and Designing Web Navigation

Contributors

About the author

With more than 30 years working in user experience at Silicon Valley companies such as Apple, WebEx, Google, Cisco, and many startups, Pabini is currently the principal consultant at Strategic UX, providing UX strategy and design consulting services to product-development companies. She is also the publisher and editor in chief of *UXmatters*.

Pabini has led UX strategy, design, and user research for Web, mobile, and desktop applications for consumers, small businesses, and enterprises, in diverse product domains. Her past UX leadership roles include head of UX for sales and marketing IT at Intel, senior director of UX and design at Apttus, principal UX architect at BMC Software, VP of user experience at scanR, and manager of user experience at WebEx. As a UX leader, she has facilitated conceptual modeling and ideation sessions; written user stories; defined and prioritized product and usability requirements; designed and established corporate design systems, standards, and guidelines; and integrated lean UX activities into agile-development processes.

Pabini is passionate about creating great user experiences that meet users' needs and get business results. Working collaboratively with business executives, multidisciplinary product teams, and other UX professionals, Pabini has envisioned and realized holistic UX design solutions for innovative, award-winning products that delighted users, achieved success in the marketplace, and delivered business value. A strategic, systems thinker, Pabini has diverse experience that enables her to synthesize innovative solutions for challenging UX strategy and design problems.

A thought leader in the UX community, Pabini was a founding director of the Interaction Design Association, as well as the founder of *UXmatters*, a leading publication for UX professionals.

Thank you to the UXmatters authors who helped make this a better book by providing feedback on chapters and contributing figure images—especially Nate Davis, Michael Morgan, and Jim Ross.

About the reviewers

Kat Kmiotek is a quality engineer at Houseful who works on test automation, IaC, and Continuous Integration/Continuous Delivery pipelines. She sees engineers as users of her work and cares about making things easy to use. While she focuses on testing and quality, Kat is interested in how good design can improve products for everyone who uses them. Her curiosity about usability heuristics helps her create more user-friendly solutions in her technical work.

Rod Marshall has over 30 years of experience working as an information architect. He tends to work on large, content-heavy CMS/data-driven Web sites or large dynamic Web applications. Rod has worked in several business sectors, particularly online marketplaces, consumer product reviews, personal finance and wealth management, and commercial property portfolio management. As an information architect working in these sectors, Rod has specialized in data modeling, data analysis, reporting, and presentation. The IA discipline of creating taxonomies and nomenclatures has been a huge benefit to him.

Table of Contents

Part I: Fundamentals of Information Architecture

1

6

Understanding and Structuring Content   161

7

Classifying Information   215

8

Defining an Information-Architecture Strategy   251

Part III: Designing Information Architectures for Digital Spaces

9

Labeling Information   281

10

Designing and Mapping an Information Architecture   309

11

Foundations of Navigation Design   339

12

Designing Navigation   391

13

Designing Search   461

Index   529

Other Books You May Enjoy   544

Preface

In today's digital world, information architecture (IA) touches the lives of everyone who is online—whether at work or in their personal life. Digital information spaces have proliferated across all the platforms that people use today, from Web sites, intranets, and ecommerce stores to information-centric mobile apps and social-media apps. Well-designed IAs make it easier for people to find what they need online, whether they're at work, looking for entertainment, shopping, or conducting serious research. The goal of every information architect is to remove any obstacles that might prevent people from finding the information they need.

Around the turn of the century, when the design of Web sites predominated, there was a time when some information architects were talking about "big IA." But we now refer to this broader discipline as user experience (UX) design, which comprehends the design of both information spaces and applications. Although IA is a key component of all UX design, it is just one of the relevant disciplines that UX design comprises. Other key disciplines of UX design include interaction design, which focuses predominantly on the design of highly functional applications that help people get things done rather than the design of information spaces, and visual interface design, which is an essential aspect of all software user interface (UI) design. Increasingly, designers have specialized in one aspect of UX design or another, but it is essential that all designers have broad knowledge of all its disciplines.

Over the last few decades, IA has matured both as a discipline and a professional practice. IA has continually evolved as new platforms and design challenges have arisen. Its greatest recent advancements have come from the impacts of artificial intelligence (AI) on search experiences. In the near future, AI will likely transform the practice of IA.

Who this book is for

This book is for information architects and, more generally, UX professionals at all levels, their colleagues on the development teams with whom they work to create digital information spaces, and the businesses for which they create them. In short, this book is for all students and practitioners of IA, as well as the many professionals who need to understand IA and the value it provides to developers of digital information spaces.

What this book covers

Chapter 1, What Is Information Architecture?, defines the practice of IA, discusses its key goals, considers information ecosystems, and describes the value of IA to both users and businesses. The chapter also provides an overview of the IA design process and describes the people responsible for creating an IA and the skills they must have, as well as the key components of an IA.

Chapter 2, How People Seek Information, describes users' information-seeking and sensemaking models, which are helpful in understanding the strategies, tactics, and behaviors that people employ when looking for information. The chapter also discusses some common information-seeking behaviors that are typical when people are looking for specific, known items of information versus those they employ when exploring an information space.

Chapter 3, Design Principles, provides a solid grounding in UX design principles that have their basis in our understanding of and need to support the cognitive and physical capabilities and behaviors of human beings. These broadly applicable design principles offer a firm foundation for user-centered design (UCD). The chapter also explores wayfinding design principles, including placemaking, orientation, navigation, labeling, and search principles.

Chapter 4, Structural Patterns and Organization Schemes, explains the two key components of an organizational model for a digital information space: the structural patterns that define the structural relationships between the groups of related content objects that particular types of digital information spaces comprise and the organization schemes for organizing their content such as exact, ambiguous, and hybrid schemes, as well as their typical applications. The chapter also describes Richard Saul Wurman's early model for organizing information: LATCH.

Chapter 5, UX Research Methods for Information Architecture, considers various UX research methods—such as types of card sorting—that apply specifically to the design of effective IAs, as well as UX research methods for evaluating the findability of information spaces. The chapter also explores usage-data analytics—for example, Web analytics; path, or clickstream, analysis; and search-log analytics—and how to synthesize insights from UX research to inform information-architecture strategy and design.

Chapter 6, Understanding and Structuring Content, looks at some methods of helping teams to better understand and structure the content that an information space comprises. Methods of understanding an information space's content include content-analysis heuristics, content-owner interviews; content mapping, inventories, and audits; and competitive analyses. The chapter also considers *content modeling*—that is, ways of decomposing content into its logical components and elements—which enables the content to support contextual navigation; chunked, personalized, filterable, and sortable content; and content reuse.

Chapter 7, Classifying Information, discusses the principles, key goals, and challenges of classifying information; considers types of metadata and the use of metadata schema to provide the basis for information categorization; and describes an information space's content objects. The chapter covers various types of controlled vocabularies in depth, including synonym rings, authority files, taxonomies, thesauri, ontologies, semantic networks, and faceted classification schemes; considers how to choose preferred, variant, broader, narrower, and related terms; and provides a step-by-step process for developing a controlled vocabulary.

Chapter 8, Defining an Information-Architecture Strategy, considers the concerns of IA strategy, which defines strategic outcomes across four dimensions: business value, user needs, the scope and structure of an information space's content, and leveraging implementation technologies. Strategic concerns include alignment with business strategy, synthesizing UX research findings, understanding an organization's content, learning about implementation technologies, and the big IA-strategy picture. The chapter also covers envisioning, communicating, and validating conceptual models, as well as documenting and presenting IA strategy.

Chapter 9, Labeling Information, describes the attributes of effective labeling systems for information spaces' organizational structure, especially the labels for its global and local navigation systems, but also those that identify and provide structure to its pages. Thus, the chapter focuses primarily on the design of labels and icons for navigation systems, contextual hyperlinks and buttons, and progressive disclosure controls. It also covers the translation of textual labels. Plus, the chapter explores the discovery, definition, and testing of optimal labels for categories, navigation systems, and pages.

Chapter 10, Designing and Mapping an Information Architecture, covers some UX research methods for evaluating an existing information space's organizational structure and findability. It also describes how to choose an information space's organizational model, comprising its structural pattern and organizational scheme. The chapter then provides an overview of the process of categorizing and labeling an information space's content and explores some methods of diagramming and documenting an IA.

Chapter 11, Foundations of Navigation Design, explores the primary concerns of navigation design and the key objectives of designing a navigation system. It also provides some universally applicable navigation design guidelines. The chapter then covers the various types of navigation in depth, including structural navigation, navigation pages, associative links, supplementary navigation, and complementary navigation tools, as well as Web browsers' navigation capabilities.

Chapter 12, Designing Navigation, provides some navigation design patterns for organizing and representing groups of hyperlinks, including fundamental navigation elements, desktop navigation patterns and layouts, mobile navigation patterns and layouts, and progressive disclosure. The chapter also covers creating, presenting, critiquing, and testing navigation design deliverables such as sketches, wireframes, mockups, prototypes, and navigation design specifications.

Chapter 13, *Designing Search*, considers some challenges that users encounter in searching information spaces, as well as the need for and value of implementing an internal search system. The chapter then explores the design of usable internal search systems in depth, including how to optimize the quality of search results and improve the overall search experience through the implementation of specific underlying technologies. The chapter also provides search user-interface design patterns and explores some specialized types of search systems, including those that are driven by generative artificial intelligence, personalized search, semantic search, faceted search, parametric search, and advanced search.

Download the references

You can download the references for this book from GitHub at `https://github.com/PacktPublishing/Designing-Information-Architecture`. If there's an update to the references, it will be updated in the GitHub repository.

Text conventions for this book

Here are some text conventions that have been applied throughout this book.

- *Italics* indicate new terms being defined, emphasis, or book titles.
- **Bold** indicates UI text and run-in heads.
- `Code in text` indicates HTML and CSS code.

Note—Looks like this.

Get in touch

Feedback from our readers is always welcome.

General feedback: If you have questions about any aspect of this book, email us at `customercare@packtpub.com` and mention the book title in the subject of your message.

Errata: Although we have taken every care to ensure the accuracy of our content, mistakes do happen. If you have found a mistake in this book, we would be grateful if you would report this to us. Please visit `www.packtpub.com/support/errata` and fill in the form.

Piracy: If you come across any illegal copies of our works in any form on the Internet, we would be grateful if you would provide us with the Web address or Web site name. Please contact us at `copyright@packt.com` with a link to the material.

If you are interested in becoming an author: If there is a topic that you have expertise in and you are interested in either writing or contributing to a book, please visit `authors.packtpub.com`.

Share Your Thoughts

Once you've read *Designing Information Architecture*, we'd love to hear your thoughts! Scan the QR code below to go straight to the Amazon review page for this book and share your feedback.

https://packt.link/r/1-838-82719-6

Your review is important to us and the tech community and will help us make sure we're delivering excellent quality content.

Free Benefits with Your Book

This book comes with free benefits to support your learning. Activate them now for instant access (see the "*How to Unlock*" section for instructions).

Here's a quick overview of what you can instantly unlock with your purchase:

PDF and ePub Copies   **Next-Gen Web-Based Reader**

Access a DRM-free PDF copy of this book to read anywhere, on any device.

Use a DRM-free ePub version with your favorite e-reader.

Multi-device progress sync: Pick up where you left off, on any device.

Highlighting and notetaking: Capture ideas and turn reading into lasting knowledge.

Bookmarking: Save and revisit key sections whenever you need them.

Dark mode: Reduce eye strain by switching to dark or sepia themes

How to Unlock

Scan the QR code (or go to packtpub.com/unlock). Search for this book by name, confirm the edition, and then follow the steps on the page.

UNLOCK NOW

Note: Keep your invoice handy. Purchases made directly from Packt don't require one

Part I:
Fundamentals of
Information Architecture

This part provides some fundamentals of information architecture, including design principles, structural patterns, and organizational schemes, and includes the following chapters:

- *Chapter 1, What Is Information Architecture?*
- *Chapter 2, How People Seek Information*
- *Chapter 3, Design Principles*
- *Chapter 4, Structural Patterns and Organization Schemes*

1

What Is
Information Architecture?

This chapter introduces and provides an overview of the professional practice of information architecture (IA)—one of the design specialties that make up the broader practice of user experience (UX) design. In this chapter, I'll present some foundational knowledge that you'll find useful whether you're learning about information architecture because you plan to pursue a career as an information architect, you need to take on IA activities because there is no information architect on your design or project team, or you want a better understanding of the role of any information architects on your team.

This chapter conveys basic knowledge regarding the purview of information architecture as a professional practice and considers the following key topics:

- Defining information architecture
- Goals of information architecture
- Information ecosystems
- The value of information architecture
- Who is responsible for information architecture?
- Core IA skills
- Overview of the IA process
- Key components of information architecture

In reading this book, you'll build on this foundation to gain a deeper understanding of the many facets of the discipline of information architecture.

> **Free Benefits with Your Book**
> Your purchase includes a free PDF copy of this book along with other exclusive benefits. Check the *Free Benefits with Your Book* section in the Preface to unlock them instantly and maximize your learning experience.

Defining information architecture

Let's first consider the meanings of the individual words *information* and *architecture*. *Information* is a broad term that encompasses everything from the sorts of raw factual and numeric data that organizations store in relational and object-oriented databases to the more meaningful—sometimes even insightful—content on Web pages, on pages in mobile apps, in documents of various types, and in images and other media.

In relation to a professional practice, the term *architecture* originally referred to the planning and design of physical spaces—particularly buildings. But since the advent of computers and software, *architecture* has also referred to the design and organization of the components of computing and software systems.

Who coined the term *information architecture*? The prolific author and information designer Richard Saul Wurman wrote the following in his 2001 book *Information Anxiety 2*:

"When I came up with the concept and the name *information architecture* in 1975, I thought everybody would join in and call themselves *information architects*. But nobody did—until now. Suddenly, it's become a ubiquitous term. ... Effective information architects make the complex clear; they make the information understandable to other human beings. If they succeed in doing that, they're good information architects." [1]

In 2004, Dirk Knemeyer—a design entrepreneur who has cofounded companies such as GoInvo and Genius Games and currently leads SciStories—interviewed Wurman, asking how he had chosen the term *information architecture*. Wurman responded with the following:

"The common term then was *information design* ... Information design was epitomized by which map looked the best—not which took care of a lot of parallel systemic parts. That is what I thought *architecture* did and was a clearer word that had to do with systems that worked and performed. Thought *architecture* was a better way of describing what I thought was the direction that more people should look into for information, and I thought the explosion of data needed an architecture, needed a series of systems, needed systemic design, a series of performance criteria to measure it." [2]

Nevertheless, it was Lou Rosenfeld and Peter Morville who, in 1998, defined *information architecture* as a practice in their seminal work *Information Architecture for the World Wide Web*—which the IA community affectionately refers to as the *polar bear book* because of the bear on its cover. [3] Both Lou and Peter have backgrounds in library and information science (LIS), which forms the foundation of their thinking. Their view of information architecture—with its focus on organization, clear labeling, navigation, and search—differs fundamentally from that of Wurman, and it established the foundation for the profession upon which all information architects have since built.

What exactly is *information architecture*? Here is a brief definition of the term as I'll use it in this book:

Information architecture is a design practice that focuses on defining the structure of digital information spaces—for example, Web sites, intranets, social-networking communities on the Web, and information-rich digital products such as Web and mobile applications—by organizing information and supporting findability and usability through well-designed labeling, navigation, and search systems.

While some information architects have expanded their view of information architecture to comprehend physical spaces, the focus of this book is on information architecture for digital information spaces, in service of providing a practical book for people working on digital products and services.

Around the time I launched *UXmatters* in 2005—when I was giving a lot of thought to the distinctions between information architecture, interaction design, and UX design—I wrote a detailed definition of *information architecture* for the site's glossary. That definition previews the scope of the information I'll cover in this book:

"A UX design discipline that defines the structure of digital information spaces—including Web sites, intranets, online publications, applications, and other digital products—with the goal of supporting findability and usability. Information architecture encompasses the creation of taxonomies of the hierarchical and associative relationships that exist between content objects; controlled vocabularies that effectively communicate the nature of and relationships between content objects; labeling for navigation systems that makes information browsable; metadata, retrieval algorithms, and query syntaxes that produce useful search results; and the content and format of both individual search results and sets of results. Good information architectures make digital information easier to navigate, search, and manage; balance breadth and depth appropriately; and enable users to readily find the information they need. Information-architecture deliverables include content inventories,* wireframes, site maps, and flow diagrams." [4]

Note—With today's increased levels of specialization, a content analyst, content manager, or content strategist might now be responsible for creating content inventories.

I wrote that detailed definition of *information architecture* for UX professionals, who are knowledgeable about its practices. Let's begin building your understanding to the same level by exploring the basic and advanced concepts that the definition merely highlights. By reading this book, you'll quickly learn everything you need to know to create an effective information architecture.

Goals of information architecture

The goals of information architecture include the following:

- Making information easy to find
- Providing information scent
- Supporting browsing
- Supporting search
- Creating a sense of place
- Making content easy to consume
- Combatting information overload

Making information easy to find

Findability is the efficiency and effectiveness with which users can find specific content within a digital information space and is essential to usability. How do practitioners of information architecture assist people in finding the information they need? By structuring, grouping or categorizing, and labeling information, they help people make sense of information spaces. Organizing information objects into discrete groups or categories that are meaningful to people and assigning clear labels to them are paramount. The design of usable labeling, navigation, and search systems enables people to easily find the information they need.

Providing information scent

In the late 1990s, Peter Pirolli and Stuart Card of Xerox PARC (Palo Alto Research Center) published their "information foraging theory," which frames human information-gathering activities in terms of the ways "adaptive pressures work on users of information that are analogous to ecological pressures on animal food foraging…." In their report, they wrote: "The problems and constraints of [information-foraging] environments can be thought of as forming abstract landscapes of information value and costs, such as the costs of accessing, rendering, and interpreting information-bearing documents." [5]

A key concept of information foraging is *information scent*, which enables people to instinctually follow environmental cues in gathering the valuable information they need, while ignoring irrelevant information. These cues include visual and textual signposts that convey semantic meaning to people as they browse for information—proximal cues that let information seekers know whether they're getting *hotter* or *colder* in relation to the information they want, as in a game of Hot and Cold. Obviously, people pursue the paths that promise to be most fruitful.

The information-foraging environments that information architects create can facilitate the effectiveness and efficiency with which users can gather information. Providing the right environmental cues, or signposts, aids users in successfully accomplishing their information-seeking goals. This is especially important for complex information spaces. At the same time, providing information scent enables organizations to capture the most valuable of resources: users' attention.

Supporting browsing

Information architecture supports browsability by doing the following:

- Creating logical structures for information spaces that match users' mental models and, thus, encourage exploration
- Chunking content onto pages at an optimal level of granularity to prevent what Jared Spool calls *pogo-sticking*—that is, forcing users to navigate back and forth between pages unnecessarily by decomposing content into chunks at too fine-grained a level [6]
- Designing usable global, primary navigation systems, comprising links with clear labels, to enable users to gain access to the information they need and show users where they are now and where they can go next

- Providing global or local supplementary navigation systems such as site maps and alphabetical indexes to provide alternative means of gaining access to information

- Presenting contextual links in a consistent location on pages to support navigation within a page

- Providing useful links to related information as supplementary means of navigation, either by embedding them within content or by grouping them in a consistent location on pages

- Displaying clear page titles to ensure that users know where they are—titles that use language that is similar to that of the links that display the pages

Supporting search

A search system is necessary for any digital information space that is sufficiently extensive that a navigation system alone would not adequately support findability. Information architecture plays an important role in ensuring the usefulness and usability of a search system. To better support search, practitioners of information architecture can do the following:

- **Define descriptive metadata**—Using metadata to describe content of any type gives semantic meaning to that content and provides greater precision in searching, enabling search systems to deliver more useful search results in response to users' queries.

- **Implement a controlled vocabulary and thesaurus**—Supporting a domain-specific vocabulary, including variants of terms, and synonyms of search terms increases the likelihood that searching would deliver useful content to users.

- **Provide a usable search user interface**—Users should be able to easily type or revise a search query. For a large digital information space such as an ecommerce site, it should be possible to limit the scope of a search to a specific facet of the information within that space.

- **Suggest search strings**—Automatically suggesting search strings that would deliver useful results, in response to whatever the user types, helps users to search more efficiently.

- **Display meaningful search results**—Effectively displaying the most salient information on search-results pages helps users to accurately identify the content they need.

Creating a sense of place

Increasingly, users are experiencing digital information environments as places—and designers are consciously creating contexts that are conducive to that perception. As Jorge Arango—an architect of physical spaces by training, but an information architect in practice—says in *Living in Information*, "With the growing pervasiveness of information systems in our daily lives, placemaking has started to emerge as a primary concern in the design of information systems." [7]

Designing navigable digital information spaces that emulate certain characteristics of real-world places and, thus, create a sense of place, lets you take advantage of users' knowledge of familiar places to make digital spaces easier to use, convey what users can do there, set the right user expectations, and elicit desired user behaviors.

Making content easy to consume

Information architecture is about making it easier for people to consume content. In addition to supporting navigation and search, doing the following facilitates information consumption:

- Applying a consistent organizational structure that reflects the user's mental model of the content within a digital information space
- Orienting users to ensure that they know where they are and prevent their becoming lost in hyperspace
- Creating clear labels for navigation links and page titles that make it clear users have arrived at their intended destination
- Appropriately decomposing content into chunks so there is neither too little nor too much content on a page
- Personalizing content, prioritizing content that meets users' needs, and thus, reducing the amount of information users must peruse to find what they need
- Providing convenient links to related information

Combatting information overload

Bertram Gross, in his 1964 book *The Managing of Organizations*, was the first to use the term *information overload*, which Alvin Toffler popularized in his 1970 book *Future Shock*. This term describes a state of mind that arises when people receive more information within a short period of time than they are capable of rationalizing for use in their decision-making. [8]

Data on the Web is accruing at an astounding rate. According to a Bernard Marr article on *Forbes*, "How Much Data Do We Create Every Day? The Mind-Blowing Stats Everyone Should Read," people generated 2.5 quintillion bytes of data every day in 2019. In just two years, we generated 90 percent of the data that currently exists in the world, and the pace of data creation is only accelerating with the growth of the Internet of Things (IoT). [9] This is more information than any human being could possibly keep up with and, as Daniel Tunkelang writes in his book *Faceted Search*, creates a "scarcity of the most valuable resource of all—the user's attention." [10]

Because of this burgeoning volume of information, digital information spaces have become increasingly complex, and the danger exists that they could become unusable. However, by effectively structuring and categorizing information and, thus, improving findability, the practice of information architecture provides a bulwark against information overload. Users can find the information they actually need without having to wade through a morass of irrelevant information. This reduces the *information anxiety*—another term coined by Richard Saul Wurman [11]—and frustration that users would otherwise experience as a result of trying to process too much information, which has become exacerbated as people use more and more devices, information sources, and services.

According to a Pew Research Center report from 2016, information overload is less of a factor for most people than it was a decade earlier. Just 20% felt overloaded, down from 27% in 2006. "The large majority of Americans do not feel that information overload is a problem for them. ... 77% say they like having so much information at their fingertips. Two-thirds (67%) say that having more information at their [disposal] actually helps to simplify their lives." [12]

Information ecosystems

British ecologist Arthur Tansley introduced the concept of an *ecosystem* in his writings in 1935, using the term to describe "the whole system, ... including not only the organism-complex but also the whole complex of physical factors forming what we call the *environment*." [13] While the term *ecosystem* originated in the life sciences, some have adapted it for use in information science. Others, including Morville and Rosenfeld in *Information Architecture for the World Wide Web*, have adopted the term *information ecology*. [14]

People thrive in information-rich environments. We exist within overlapping information ecosystems at various levels of granularity—ranging from the personal to the Web, a massive, worldwide information ecosystem. Depending on users' current context, their larger information ecosystem could comprise *different*, more granular ecosystems. For example, a business information ecosystem might comprise departmental, corporate, organizational, marketplace, and governmental ecosystems.

Despite an information ecosystem potentially having a much greater scope than a typical Web site or mobile app, many of the same approaches to organizing information and designing information architectures still apply. Some ecosystems have emerged organically, while others have been consciously designed—but as they evolve, either the messiness of organic growth or the order of design may predominate at any given moment in time, unless an ecosystem is maintained systematically, with great discipline.

A holistically designed information ecosystem typically comprises *users* of various types, a *business* or other organization that serves their needs, and *content* that the business or the users themselves create to meet those needs. Of course, all information-architecture solutions must serve the unique needs of both their users *and* the business or organization sponsoring the creation of a digital information space—that is, the needs of a specific information ecosystem. Plus, every designed information ecosystem exists within the broader *context* of a marketplace or community. Figure 1.1 depicts the elements of a business's information ecosystem and the information architecture that connects users with that ecosystem, whose elements we'll look at in greater depth next.

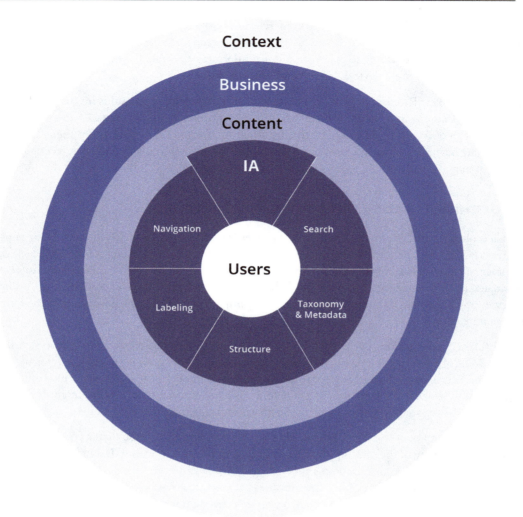

Figure 1.1—The elements of an information ecosystem and information architecture

Users

Practitioners of information architecture follow a user-centered design (UCD) process to ensure that the IA solutions they create meet the needs of the users who engage with them. They do user research to understand those users' information needs, tasks, and information-seeking behaviors, then create personas that represent key user groups, or target audiences. Depending on users' wants and needs, they might simply consume the content within a digital information space or contribute to it—either by creating user-generated content (UGC) or tagging, voting on, sharing, or linking to content.

Each unique audience has a particular mental model of an information space, based on their existing knowledge and prior experience, and uses a specific vocabulary that should factor into the design of an information architecture.

Content

Digital information spaces are dynamic and their content may evolve constantly. There are complex interrelationships and dependencies between the content elements in a designed information space. The information architect must understand its existing structure and both the scope and types of the existing and planned content to create a sustainable information architecture. These might include Web or mobile pages; documents in various formats; different types of digital media such as images, audio, or video; other digital assets; or databases—even the metadata that enables people to find the content within a digital information space. That metadata could consist of information such as the content's title, document format, owner, creator, vendor, or groups or categories to which the content belongs.

The business

Commercial Web sites, intranets, and extranets are designed information spaces that exist within the context of a business or organization. A business or organization that sponsors the design and development of such digital information spaces does so to meet its strategic business goals. According to David A. Aaker, *business strategy* comprises the following:

- A product/market investment decision
- Marketplace in which a business competes
- Level of investment
- Allocation of resources across business units
- Functional strategies, including product strategy, pricing, and market segmentation
- A foundation for sustainable competitive advantage
- Strategic assets and competencies
- Synergy across business units [15]

An information architect who designs a digital information space needs to understand the strategy of the business sponsoring it—particularly its vision for that information space, the available design and development resources, and any technical constraints that exist—and must work toward the business's strategic goals. To do this effectively, the information architect must also understand the culture of the business and build strong relationships with stakeholders and all members of the multidisciplinary team responsible for design and development.

An information architecture is a tangible manifestation of a business's or an organization's strategy, which conveys meaning to its target audience and differentiates the business or organization in the marketplace.

Context

The context in which a designed, digital information space exists includes its intersections with other digital information spaces. For a business, these might include one or more Web sites, intranets, extranets, ecommerce sites, and mobile apps. Together, these information spaces form the business's information ecosystem, which itself exists within the context of a broader information ecosystem comprising the business's marketplace, competitors, sales channels, customers, users, social media, the press, and more.

Creating holistic, cross-channel information ecosystems

As information sources and the devices on which people consume information proliferate, information architects must create holistic, cross-channel information ecosystems that interoperate across devices. For example, if users look for information on their mobile phone, then go to their computer, they should easily be able to resume their information-seeking task there. Working across devices should be a seamless experience.

Any information that is available on one device should be available on *all* of a user's devices. However, the ways in which specific platforms display or communicate that information might differ, depending on the size of a device's screen or its lack of a screen. Nevertheless, the language and organization of the information should be consistent across channels and the information useful regardless of the user's current context.

The value of information architecture

Information architecture provides value to both the people who use a digital information space and the business sponsoring its design and development. Let's consider the value proposition that information architecture offers to users *and* the business.

The value of information architecture to users

Designed information architectures provide many benefits to the people using digital information spaces. An effective information architecture provides value to users in the following ways:

- Improves the user experience (UX)
- Helps people make sense of information spaces
- Makes information easier to find
- Reduces the frustration of users' being unable to find information

- Improves the quality of the information people find
- Minimizes the likelihood that people would find the wrong information or no information
- Improves decision-making
- Facilitates browsing
- Facilitates search
- Saves users' time
- Makes it easier to identify content
- Makes content easier to consume
- Makes information easier to manage
- Prevents users from experiencing information overload
- Reduces the need for documentation, training, and support

The value of information architecture to the business

Designed information architectures can deliver great benefits to the businesses and organizations that sponsor their creation. An effective information architecture provides value to the business in the following ways:

- Improves ease of use
- Increases employee productivity
- Enhances customer service
- Increases customer satisfaction
- Increases customer adoption
- Improves the business's understanding of customers
- Increases sales on ecommerce sites by ensuring customers can find the products they want and by upselling related products to them
- Reduces the overall costs of information seeking to the business
- Minimizes negative business impacts resulting from workers' lacking accurate, useful information
- Improves decision-making
- Reduces duplication of effort
- Communicates and facilitates alignment on business strategy
- Creates competitive differentiation
- Engenders trust

- Builds brand loyalty
- Minimizes employees' needs for documentation, training, and support
- Reduces customers' needs for documentation and support
- Reduces the overall costs of documentation, training, and support
- Prevents unnecessary reimplementation costs
- Makes information spaces easier to manage and maintain
- Reduces maintenance costs [16]

Who is responsible for information architecture?

Ideally, a highly trained and experienced information architect should be responsible for a digital information space's information architecture. A large organization with a high level of design maturity and a complex information ecosystem would likely employ a team of information architects. But with high design maturity, even a small organization might employ an information architect or, more likely, would work with an IA consultant. However, in a very small organization or one with a low level of design maturity, some other professional might have to take responsibility for information architecture. In such a case, it is essential that the person responsible for information architecture has the necessary mindset, IA knowledge and skills, and sufficient experience to be successful.

Likely candidates for other roles taking responsibility for information architecture include UX designers, content strategists, usability professionals, information designers, library scientists, business analysts, product managers, developers, marketers, and technical writers.

In large organizations, people in more specialized UX roles might take on responsibility for specific IA activities—for example, a content analyst. In contrast, within smaller companies, an information architect might also be responsible for content strategy.

Core information-architecture skills

"Effective information architects make the complex clear; they make the information understandable to other human beings."—Richard Saul Wurman [17]

Practitioners of information architecture must develop a deep understanding of both the target audience for a digital information space—including the language they use and their mental models for organizing the information of its domain—and the goals of the business sponsoring its development. They must also comprehend the full scope of the content that information space comprises and be capable of doing very detailed work in organizing that information. Finally, they must have the ability to synthesize all of this information into a holistic IA solution that meets users' needs.

Accomplishing these goals requires a broad skillset. Now, let's consider some essential attributes, knowledge, and skills that an information architect must have to take on the responsibility of creating the information architecture for a digital information space.

Systems thinking

What exactly is *systems thinking*? According to *The Fifth Discipline Fieldbook*, by Peter M. Senge and his coauthors, it is "a way of thinking about, and a language for describing and understanding, the forces and interrelationships that shape the behavior of systems." In describing strategies for systems thinking, they say: "Although systems thinking is seen by many as a powerful problem-solving tool, ... it is more powerful as a language, augmenting and changing the ordinary ways we think and talk about complex issues." [18]

Systems thinkers are able to think abstractly, address complex problems, appreciate the interrelationships between the various elements of a system, and approach problem-solving holistically. Being a systems thinker is especially important for an information architect who is designing complex or cross-channel information ecosystems.

UX research and user-centered design

Of course, all UX professionals should possess UX research and user-centered design (UCD) skills. The practice of information architecture benefits particularly from the following UX research and UCD activities:

- Content analysis
- Stakeholder research
- User research
- Developing personas
- Building mental models
- Card sorting
- Tree testing

Empathy for users

Empathy is an essential quality for information architects—and all UX professionals. Empathy for users enables an information architect to understand their motivations and accurately perceive their needs and emotions, then use that understanding to create better solutions for them. Being empathetic lets an information architect look at things from different people's perspectives and internalize what they learn about the target users of a digital information space. [19]

Information organization

Effective practitioners of information architecture can perceive, understand, and communicate the relationships between the information elements within a digital information space, organize information in ways that make sense to users, and devise holistic IA solutions.

IA design skills

Designing an information architecture requires the following knowledge and skills:

- A deep understanding of design principles that are pertinent to the design of an information architecture
- Labeling design
- Designing and mapping information architectures
- Navigation design
- Search design

Strong language skills

Designing an information architecture requires excellent language skills—in the language of the target audience for a digital information space. Both the hidden underpinnings of an information architecture and the visible elements that users experience rely on the use of language. An information architect must organize content into groups or categories, define metadata for content, choose optimal labels for navigation links, and design search systems. All of these design activities require working with language.

Overview of the information-architecture process

The design of an information architecture occurs within the broader context of the UX design process, which I explored in depth in my *UXmatters* column *On Good Behavior* "Design Is a Process, Not a Methodology." [21] In this section, I'll highlight the aspects of the design process that play the largest role in ensuring the creation of an optimal information architecture.

Stakeholder and user research

Stakeholder research involves interviewing the business leaders who have sponsored and are defining the strategy for an IA project, project team members who are leading the implementation of that strategy, and others who have a stake in the project's outcome, including customers. This research ensures that those responsible for designing an information architecture understand the business needs driving the project, as well as the desired outcomes.

User research consists of a variety of generative-research techniques—including user interviews, contextual inquiries, user observations, field studies, various open card-sorting methods, tree testing, and surveys—and enables practitioners of information architecture to understand the needs, vocabulary, mental models, and information-seeking behaviors of existing or prospective users. [20] On the basis of the findings from user research, the UX team can define personas that represent the various types of users who make up the target audience for a digital information space.

Stakeholder and user research occur during a project's *Discovery* phase. In Chapter 5, *UX Research Methods for Information Architecture*, you'll learn about some specific UX research methods that are useful in designing information architectures.

Information-architecture strategy

A project's information-architecture strategy derives from an organization's business strategy and the stakeholder and user research that the UX team has conducted. Thus, the IA strategy is an output of the project's *Discovery* phase.

Louis Rosenfeld, Peter Morville, and Jorge Arango, the authors of *Information Architecture: For the Web and Beyond*, define information-architecture strategy as follows:

"An *information architecture strategy* is a high-level conceptual framework for structuring and organizing an information environment. It provides the firm sense of direction and scope necessary to proceed with confidence into the design and implementation phases. It also facilitates discussion and helps get people on the same page before moving into the more expensive design phase." [22]

An information-architecture strategy provides preliminary design guidance on which a team can align—regarding the definition of a taxonomy and metadata; the optimal structure for a scalable information space that can accommodate the addition of new content over time; the organization of content; the design of labeling, navigation, and search systems; and the eventual implementation and administration of an information architecture. Chapter 8, *Defining an Information-Architecture Strategy*, explores IA strategy in depth.

Design and usability testing

The goal of the *Design* phase is to realize the IA strategy; iteratively design, test, and deliver an IA solution that provides value to both users and the business. Designing an information architecture comprehends creating the structure of a digital information space and determining the organization of its content, which you'll learn about in Chapter 6, *Understanding and Structuring Content*; the definition of a taxonomy and metadata, in Chapter 7, *Classifying Information*; designing labeling systems, in Chapter 9, *Labeling Information*; designing and mapping information architectures, in Chapter 10, *Designing and Mapping an Information Architecture*; designing navigation systems, in Chapter 11, *Foundations of Navigation Design*, and Chapter 12, *Designing Navigation*; and designing search systems, in Chapter 13, *Designing Search*.

Usability testing enables a team to evaluate the quality of a designed information architecture, identify areas for improvement through iterative design, and ultimately, validate a design solution.

Implementation and administration

The users of an information architecture include the content-management (CM) team, which is responsible for the analysis, collection, management, and publishing of content. In a small organization, there might be a CM team of one; in a large enterprise, different people, or multiple people, might be responsible for each of the roles that Table 1.1 outlines. According to Bob Boiko, the primary goal of content management "is to break information away from its presentation and instead focus on its structure...." [23] Information architects have a similar goal. Achieving this goal requires a well-defined process and clear policies.

Role	Implementation Responsibilities	Maintenance Responsibilities
Content manager	Content management, including the planning and implementation of CM initiatives and technology acquisitions	Content management, including governance
Business analysts	Creating a business strategy or aligning CM initiatives with business strategy, building upon existing efforts, fostering support for CM initiatives, setting goals, and devising a CM strategy	Establishing a governing body and ensuring the CM team remains aligned with the business strategy
Information architects	Content analysis, which includes gathering content requirements; designing a logical, holistic structure, or content model, for digital content; creating a metatorial framework, or categorization scheme, for organizing content; defining metadata, and applying metadata to content	Governing the application of a metatorial framework to the content in a digital information space; ensuring content conversions have correctly identified, chunked, and tagged content components; applying metadata to content; maintaining a metatorial guide, and training others on the guide's use
Software developers	Template, content-management system (CMS), and application development	Template and CMS development
Content creators	Writing, graphic design, media production, and editing	Managing acquisitions, writing, graphic design, media production, and editing

Table 1.1—Key roles on a content-management team*

Note—The information in Table 1.1 derives from Chapter 11 of Bob Boiko's *Content Management Bible*. [24]

Key components of information architecture

Let's briefly consider the following key components of the practice of information architecture:

- Taxonomies
- Metadata
- Labeling systems
- Organizational structure
- Navigation systems
- Search systems

Refer to Figure 1.1, earlier in this chapter, for a diagram depicting these key components.

Taxonomies

A *taxonomy* provides a systematic classification of the content within a digital information space and supports its organization into a hierarchy of predefined, but evolving, categories, or classes, of information. You'll learn how to define an information-classification system in Chapter 7, *Classifying Information*.

Metadata

Metadata is a type of data that describes other data—such as specific Web or mobile pages, documents, content elements, or digital assets. Descriptive metadata facilitates identifying and finding particular instances of information. Chapter 7, *Classifying Information*, also covers how to define metadata.

Labeling systems

Effective labels enable users to seek and identify the information they need. Choosing what domain-specific language to use consistently throughout a digital information space and designing an optimal labeling system are the subjects of Chapter 9, *Labeling Information*.

Organizational structure

The organization of information is at the core of information architecture. Chapter 4, *Structural Patterns and Organization Schemes*, describes some common structural patterns and organization schemes that you can employ in structuring information within a digital information space. Chapter 6, *Understanding and Structuring Content*, explains how to discover what content already exists, what content the team needs to create, and how to structure that content. In Chapter 10, *Designing and Mapping an Information Architecture*, you'll learn how to design and map an information architecture.

Navigation systems

A digital information space's information architecture becomes manifest in its navigation system. The design of a navigation system leverages all the foundational work that goes into creating an information architecture that optimally supports browsing. Chapter 11, *Foundations of Navigation Design*, and Chapter 12, *Designing Navigation*, cover the design of navigation systems in depth.

Search systems

Designing search systems also relies heavily on the foundational work of creating an information architecture—especially the definition of an information-classification system, metadata for specific instances of information, the vocabulary to use within an information space, and its labeling system. In Chapter 13, *Designing Search*, you'll learn about the complexities of designing a search system.

Summary

This chapter defined the practice of information architecture, discussed its key goals, looked at the more expansive concept of information ecosystems, and communicated the value of information architecture to both users and the business. This chapter also provided an overview of the IA process, within the context of the broader UX design process, and described some of the people who might be responsible for creating an information architecture and the skills they must have to practice information architecture successfully. Finally, this chapter previewed the key components of information architecture.

Next, in Chapter 2, *How People Seek Information*, you'll learn about people's common information-seeking needs and behaviors. Understanding these needs and behaviors is foundational to building an effective information architecture.

References

To make it easy for readers to follow links to the references for this chapter, we've made them available on the Web: `https://github.com/PacktPublishing/Designing-Information-Architecture/tree/main/Chapter01`

Further reading

Books

My goal in writing *Designing Information Architecture* is to create an easy-to-use tutorial and reference book, in which you can find the practical guidance you need to learn how to design information architectures for digital information spaces.

Depending on your reading preferences, other books on information architecture that you might read include the following:

- If you want to read the classic, definitive, comprehensive book on information architecture, read the fourth edition of the polar bear book, *Information Architecture: For the Web and Beyond*, by Louis Rosenfeld, Peter Morville, and Jorge Arango. I first learned about information architecture many years ago when I read the second edition, by Morville and Rosenfeld, which was originally titled *Information Architecture for the World Wide Web*.

- Another early book, *Information Architecture: Blueprints for the Web*, which Christina Wodtke wrote back in 2003, is now in its second edition, for which Christina partnered with Austin Govella.

- If you prefer more philosophical books, are fascinated by the parallels between organizing things in the real world and structuring digital information spaces, or just enjoy good storytelling, read Peter Morville's *Ambient Findability* and *Intertwingled*; Andrea Resmini and Luca Rosati's *Pervasive Information Architecture: Designing Cross-Channel User Experiences*; Andrew Hinton's *Understanding Context: Environment, Language, and Information Architecture*; or Jorge Arango's *Living in Information: Responsible Design for Digital Places*. These are all great books by leading thinkers on information architecture.

- If you want to read a brief book about information architecture, read Donna Spencer's *A Practical Guide to Information Architecture*, now in its second edition, or Lisa Maria Martin's more recent book *Everyday Information Architecture*, whose chapters closely integrate design guidance with examples.

- If you want to learn about two design specialties within the discipline of information architecture, the design of navigation and search systems, I recommend James Kalbach's *Designing Web Navigation* and Tony Russell-Rose and Tyler Tate's *Designing the Search Experience: The Information Architecture of Discovery*, respectively.

- I was Contributing Editor on Greg Nudelman's book *Designing Search: UX Strategies for eCommerce Success*. The idea for this book began with his column *Search Matters*, which we published on *UXmatters* from 2009 through 2011. The book includes contributions from many leading thinkers on the design of search systems.

Articles and papers

In addition to the articles and papers among the references, here are a few sources of articles on information architecture:

- *UXmatters*—On my Web magazine *UXmatters*, we've been publishing articles on the full breadth of topics relating to User Experience since 2005, including many articles on information architecture. [https://www.uxmatters.com/topics/design/information-architecture/]

- I particularly recommend Nate Davis's long-running column on *UXmatters*, *Finding Our Way*, which we published from 2011 through 2017. [`https://www.uxmatters.com/columns/finding-our-way/`]

- *Boxes and Arrows*—Christina Wodtke founded the first Web magazine focusing exclusively on information architecture in 2002. [`http://boxesandarrows.com/`]

2

How People Seek Information

To design an effective information architecture for a digital information space—as well as navigation and search systems that serve the needs of your target users—you must first understand the actual goals, motivations, and specific information-seeking needs and behaviors of those users.

The term *information seeking* refers broadly to all of the activities that people pursue when looking for information. People who seek information share some common information-seeking strategies, tactics, and behaviors that they can choose to employ, depending on their current needs: What type of question does a user need to answer? What information would answer that question? How would the user go about finding that information online? Where would the user look for answers to the question? How would the user determine whether he had found sufficient information to meet his needs? By answering these questions, you can build a model of the user's information-seeking strategy, tactics, and behaviors in a particular context.

To learn about how people seek information, we'll explore the following key topics in this chapter:

- Understanding users' information-seeking needs
- Information-seeking models
- Information-seeking strategies and tactics
- Information-seeking behaviors

Understanding users' information-seeking needs

All people have diverse information-seeking needs. In some cases, they might want just a single correct answer to a question. In other cases, they would be satisfied with a few basic facts. In still other cases, they might need in-depth information that helps them either meet their research goals *or* make simple or complex decisions. Plus, users often need to find the same information more than once. Peter Morville and Lou Rosenfeld, the authors of *Information Architecture for the World Wide Web*—the seminal work on the discipline of information architecture—refer to these common information-seeking needs as follows: [1]

- **Known-item seeking**—Users are looking for a specific item of information such as *What country in the European Union has the largest population?* This question has a single correct answer.

- **Exploratory seeking**—Users want to find some information that adequately satisfies an immediate need—for example, to find someplace to go over a long weekend by asking *Where can I travel to view fall foliage nearby?*

- **Exhaustive research**—Users are conducting in-depth, secondary research—as I did when writing about *information-seeking models* for this chapter.

- **Refinding**—Users need to find the same or similar information again—for example, *Where are wildfires currently burning in California?*

Sometimes, users' information needs are highly ambiguous when they're beginning an information-seeking process, then evolve as a user learns about a domain's subject matter and vocabulary and becomes familiar with the information landscape.

Conducting in-depth, generative user research is essential to understanding the goals, motivations, and information-seeking needs of your users. While a detailed discussion of generative user research is beyond the scope of this book, Chapter 5, *UX Research Methods for Information Architecture*, covers data analytics, which can help you understand what information people are seeking on an existing information space.

Information-seeking models

Over the last four decades or so, academic and corporate researchers have identified information-seeking models and posited theories regarding the strategies, tactics, and behaviors that people use when they're looking for information. While some of these models are more useful than others, they've all contributed to our understanding of how people seek information, which is fundamental to designing effective information architectures.

For your reference, I'll briefly describe, in chronological order, some of the specific information-seeking models and theories that researchers have devised, as follows: [2, 3]

- **1981—Thomas D. Wilson**—professor and researcher at the Department of Information Studies, University of Sheffield, and founder of the *International Journal of Information Management* and *Information Research*—first published his model of information behavior and coined the term *information seeking behavior*. [4, 5, 6]

- **1983—Brenda Dervin**—professor of communication and researcher at Ohio State University and first president of the International Communication Association—devised her *sensemaking theory*, which complements research on information seeking. See the section "Sensemaking," later in this chapter. [7, 8]

- **1987—David Ellis**—later a professor and researcher at the Department of Information Studies, Aberystwyth University—submitted his PhD thesis to the Department of Information Studies, University of Sheffield, on a *behavioral model for information retrieval–system design* that addresses eight common information-seeking activities. See the next section "Patterns for information-seeking activities." [9, 10]

- **1989—Marcia J. Bates**—professor and researcher at the Graduate School of Library and Information Science, University of California, Los Angeles—proposed a dynamic, *berrypicking model* of information seeking, based on years of research on information-search tactics. See the section "Berrypicking model," later in this chapter. [11, 12]

- **1991—Carol Kuhlthau**—professor and researcher at the School of Communication and Information at Rutgers University—developed a six-stage *information-search process (ISP) model*, showing how users' emotions change at each stage of an information-seeking process. See the section "Information-search process model," later in this chapter. [13, 14]

- **1992—Vicki L. O'Day and Robin Jeffries**—researchers at the Software and Technology Laboratory at Hewlett-Packard—published research on *orienteering*, focusing on how users digest the results of an information search, then use them to solve the problem that prompted the search. [15]

- **1995—Nicholas J. Belkin**—professor and researcher at the School of Communication, Information, & Library Studies at Rutgers University—and his colleagues Colleen Cool, Adelheit Stein, and Ulrich Thiel posited a theory on *information-seeking strategies*. [16]

- **1995—Peter Pirolli and Stuart K. Card**—researchers at Xerox Palo Alto Research Center (PARC)—presented their in-depth, information-foraging theory, which includes models for information patches, information scent, and information diets. See the section "Information foraging," later in this chapter. [17]

- **1997—Gary Marchionini**—professor and researcher at the University of North Carolina at Chapel Hill—wrote *Information Seeking in Electronic Environments*, based on his research on information-seeking strategies. [18, 19]

- **1998—Alistair G. Sutcliffe and Mark Ennis**—Sutcliffe was later a professor of Systems Engineering at the School of Informatics at the University of Manchester—theorized about cognition and information retrieval. [20, 21]

- **1998—Chun Wei Choo, Brian Detlor, and Don Turnbull** [34, 35]— faculty of Information Studies at the University of Toronto—devised an *integrated model of browsing and searching on the Web*. See the section "Integrated model of browsing and searching," later in this chapter. [22]

- **2004—Kalervo Järvelin and Peter Ingwersen**—professors and researchers at the Department of Information Studies at Tampere University and the Royal School of Librarianship and Information Science at the University of Copenhagen, respectively—studied information seeking from a task-oriented perspective. [23]

- **2005—Allen Edward Foster**—when at Aberystwyth University—published his *nonlinear model of information-seeking behavior*. [24]

- **2005—Anne Aula** presented her dissertation on *user strategies for Web search* at the University of Tampere. [25]

- **2006—Ralf Schlosser**, Oliver Wendt, Suresh Bhavnani, and Barbara Nail-Chiwetalu—Schlosser, who is a professor and researcher in The Department of Communication Sciences & Disorders at Northeastern University, and his colleagues—explored the *pearl-growing model* of information seeking. See the section "Pearl-growing model," later in this chapter. [26]

In the following sections, we'll take a deeper dive into some of this academic and corporate research to learn about the models and theories that are most helpful in understanding how users seek the information they need within digital information spaces. Learning about these information-seeking models can help you understand the behaviors that you observe in the course of conducting generative user research with your target users. Some of these models look at similar behaviors from different perspectives. Use or adapt whatever model works best for you within a given context.

Patterns for information-seeking activities

David Ellis's behavioral model for information retrieval–system design identifies eight patterns for information-seeking activities that are useful in different situations, which he defines as follows:

1. **Starting**—Conducting an initial search for relevant sources of information

2. **Chaining**—Following a series of linked references such as citations of source materials in books or papers or information on the Web

3. **Browsing**—Searching for information on topics of potential interest

4. **Differentiating**—Filtering information sources based on the type and quality of their information

5. **Monitoring**—Periodically checking specific sources of information to keep up with new developments in a particular area of interest

6. **Extracting**—Methodically perusing a specific source of information to find pertinent information

7. **Verifying**—Validating that information is accurate

8. **Ending**—Conducting information-seeking activities that are necessary to finish researching a topic or to conclude a project [27]

Berrypicking model

In her paper "The Design of Browsing and Berrypicking Techniques for the Online Search Interface," Marcia Bates describes the behaviors of people conducting "real-life searches," as follows:

"Users may begin with just one feature of a broader topic or just one relevant reference, and move through a variety of sources. Each new piece of information they encounter gives them new ideas and directions to follow and, consequently, a new conception of the query. At each stage they are not just modifying the search terms used in order to get a better match for a single query. Rather the query itself—as well as the search terms used—is continually shifting, in part or whole. This type of search is … called an *evolving search*.

"Furthermore, at each stage, with each different conception of the query, the user may identify useful information and references. In other words, the query is satisfied not by a single final retrieved set, but by a series of selections of individual references and bits of information at each stage of the ever-modifying search. A bit-at-a-time retrieval of this sort is … called *berrypicking*." Bates chose this term because the activity of "picking huckleberries or blueberries in the forest" is analogous to this information retrieval activity. "The berries are scattered on the bushes; they do not come in bunches," so the person gathering berries has to pick them one by one. An information seeker's search need might or might not evolve. Figure 2.1 depicts a berrypicking, or evolving, search.

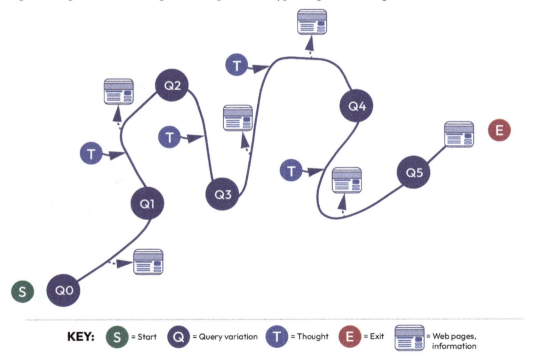

Figure 2.1—Marcia Bates's berrypicking model

Bates's berrypicking model acknowledges the reality that people's information-seeking needs and goals are dynamic and evolve as they discover new information and learn more about the information domain they're exploring. Her model shows that information seeking is a highly iterative process. In fact, people may have ongoing information needs that they pursue across multiple search sessions. [28]

Information-search process model

Carol Kuhlthau's six-stage *information-search process (ISP)* models the holistic process of information seeking, considering the user's tasks, as well as the specific *feelings*, thoughts, and actions that are characteristic of the user's experience at each of the following stages in the search process:

1. **Initiation**—The user "first becomes aware of a lack of knowledge or understanding and feelings of *uncertainty* and *apprehension* are common."

2. **Selection**—The user identifies "a general area, topic, or problem … and initial uncertainty often gives way to a brief sense of *optimism* and a readiness to begin the search."

3. **Exploration**—When the user encounters information that is "inconsistent [or] incompatible…, [*frustration*], *uncertainty*, *confusion*, and *doubt* frequently increase and people find themselves [losing] confidence." Many information seekers never progress beyond this stage.

4. **Formulation**—Once the user's information need comes into focus, she achieves clarity, and "uncertainty diminishes as *confidence* begins to increase."

5. **Collection**—Once the user gathers "information pertinent to the focused perspective…, uncertainty subsides as [*confidence* increases and] interest and involvement [deepen]."

6. **Presentation**—Once the user successfully completes the search, she has "a new understanding [and can] explain … her learning to others or … put the learning to use," so she experiences a sense of *accomplishment*. In contrast, if the user has failed to find the information she needs, she feels *disappointment*.

Kuhlthau proposes a "principle of uncertainty for information seeking"—comprehending both "affective and cognitive" uncertainty. During the early stages of the search process, an information seeker's uncertainty typically increases. The periods during which users experience increased uncertainty offer opportunities to assist users with their searches, decrease their level of uncertainty, and improve their search experience. [29]

Information foraging

Peter Pirolli and Stuart Card devised their *information-foraging theory* to understand how people adapt "strategies and technologies for information seeking, gathering, and consumption … to the flux of information in the environment. … People, when possible, … modify their strategies or the structure of the environment to maximize their rate of gaining valuable information. … The basic hypothesis of information-foraging theory is that, when feasible, natural information systems evolve [toward] stable states that maximize gains of valuable information," while minimizing the costs of obtaining that information. People who engage in information foraging exhibit adaptive tendencies in trying to gather a high volume of valuable information as quickly as possible.

Pirolli and Card's information-foraging theory comprises the following three models:

1. **Information-patch model**—This model considers how foragers allocate their time by identifying clusters of valuable information and filtering out less valuable information.

2. **Information-scent model**—This model addresses how foragers identify valuable information by recognizing *proximal cues* within an information space—such as bibliographic citations or links on the Web, which represent sources of information.

3. **Information-diet model**—This model looks at foragers' decisions regarding the information they pursue and the items they choose to consume.

To make use of information scent in ways that maximize the success of their information-foraging activities, information seekers must optimize both time allocation and item selection. [30]

Information-patch model

Within an information space, an *information patch* is a cluster of valuable information. An information space might comprehend different types of information patches, having different scopes. For example, an information patch might consist of a collection of documents or chunks of information within an individual document.

Information foragers pursue two discrete types of information-seeking activities:

1. **Between-patch activities**—The patch model assumes that information foragers must spend some time going from one patch to another after exhausting the useful information in one patch. An example of an activity that might occur between patches is a forager's searching for the next item of interest. Decreasing between-patch activities—for example, by choosing an information space with a higher prevalence of useful information patches—increases the amount of valuable information a forager can gather.

2. **Within-patch activities**—One example of a within-patch foraging activity is a forager's investing time in constructing and refining keyword queries for a search engine to improve foraging results within a patch. Increasing the profitability of within-patch activities increases the amount of valuable information a forager can gather. A forager's overall benefit from a given patch depends, in part, on the time the forager spends foraging there, which is under the forager's control. Once the forager is within a patch, the forager must decide whether to continue foraging in that patch or leave that patch for a new one. A forager should remain within a patch only as long as foraging is fruitful. Once a forager has exhausted the valuable information in a patch, the forager should move on to another patch. [31]

Information-scent model

Pirolli and Card define *information scent* as foragers' "imperfect perception of the value, cost, or access path of information sources from *proximal cues*" that represent sources of information. "[Because] information foraging often involves navigating through [physical or virtual] spaces ... to find high-yield patches..., [foragers use] imperfect information at intermediate locations ... to decide on paths through a library or [Web site]...."

Foragers use "the proximal perception of information scent ... to assess the profitability and prevalence of information sources. These scent-based assessments inform the decisions about which items to pursue ... to maximize the information diet of the forager." Pirolli and Card's information-scent model is dynamic. "As the state of the forager changes through the foraging process, the forager must make search decisions based on imperfect proximal information. ... If scent is sufficiently strong, the forager will be able to make the correct choice at each decision point. If there is no scent, the forager would perform a random walk ... in abstract search space." [32]

According to the *User Interface Engineering* report, "Designing for the Scent of Information," when information scent is strong, foragers navigate information spaces confidently and find the information they need. However, when information scent is weak, foragers lack confidence because they're just guessing what path to take, and they're unlikely to find what they need. When foraging for information, people scan page titles, section headings, hyperlinks, and search results for *trigger words*—words and phrases that convey information scent because the forager associates them with the information they need. [33]

Information-diet model

Information sources differ in their prevalence, the cost of accessing them, and thus, in their profitability. The *profitability* of an information source is the value of the information the forager gains in relation to the cost of processing the information from that source.

As part of their information-diet model, Pirolli and Card define two principles of diet selection:

1. **Principle of lost opportunity**—Foragers should ignore a class of information items if the profitability of seeking those items would be less than the expected rate of gain from seeking other classes of information items. If foragers were to seek unprofitable information classes, they would incur opportunity costs because they would lose opportunities to seek more profitable classes of information items.

2. **Independence of inclusion from encounter rate**—Foragers make decisions to seek a class of information items independent of the prevalence of that class of information items. Therefore, foragers' decisions to include a class of items in their information diet should depend only on its profitability, *not* the rate at which they would encounter that class of items within an information environment. However, a reduction in the availability of more profitable classes of information items could make it necessary to seek less nutritious classes of items. However, being selective and seeking more profitable classes of information items generally produces optimal results. Therefore, with increases in the prevalence of the most profitable information classes, foragers' information diet would likely become narrower. [34]

Integrated model of browsing and searching

In "Information Seeking on the Web: An Integrated Model of Browsing and Searching," Choo, Detlor, and Turnbull wrote: "People who use the Web as an information resource to support their daily work activities engage in a range of complementary modes of information seeking, varying from undirected viewing that does not pursue a specific information need, to formal searching that retrieves focused information for action or decision-making. Each mode of information seeking on the Web is distinguished by the nature of information needs, information-seeking tactics, and the purpose of information use."

This integrated behavioral model of browsing and searching on the Web comprises four main modes of information seeking:

1. **Undirected viewing**—The moves, or tactics, that are characteristic of this information-seeking mode include *starting* actions—identifying, selecting, and starting to read Web pages and sites—and *chaining* actions—following links on initial Web pages. "*Starting* occurs when [users] begin their Web use on preselected, default home pages, or ... visit a favorite page or site to begin their viewing—such as news, newspaper, or magazine sites. *Chaining* occurs when [users] notice items of interest, often by chance, ... then follow hypertext links to more information on those items. [Such instances of] *forward chaining* [are] most typical during undirected viewing. *Backward chaining* is also possible, [using] search engines ... to locate other Web pages that point to the site that the user is currently [viewing]."

2. **Conditioned viewing**—The moves, or tactics, for this information-seeking mode include *browsing* actions—browsing entry pages, headings, and site maps; *differentiating* actions—bookmarking, printing, copying, or going directly to a known Web site; and *monitoring* actions—revisiting favorite or bookmarked sites to obtain new information. "*Differentiating* occurs as [users] select Web sites or pages that they expect to provide relevant information..., based on prior personal visits or recommendations by others.... Differentiated sites are often bookmarked. When visiting differentiated sites, [users] browse the content by looking through tables of contents, site maps, or [lists] of items and categories. [Users] may also monitor highly differentiated sites by returning regularly to browse or by keeping abreast of new content..., for example, [by] subscribing to newsletters...."

3. **Informal search**—The moves, or tactics, for this information-seeking mode include *differentiating* actions—bookmarking, printing, copying, or going directly to a known Web site; *monitoring* actions—revisiting favorite or bookmarked sites for new information; and localized *extracting* actions—using local search engines to extract information. "*Informal search* [typically occurs] at a small number of Web sites that [users] have ... differentiated..., based on [their] knowledge about these sites' information relevance, quality, affiliation, [and] dependability.... *Extracting* is relatively informal in the sense that searching would be localized to looking for information within the selected sites. Extracting is also likely to make use of the basic, simple search features ... of the local search engine, ... to get at the most important or most recent information, without attempting to be comprehensive. *Monitoring* becomes more proactive if the [user] sets up push channels or software agents that automatically find and deliver information based on keywords or subject headings."

4. **Formal search**—The moves, or tactics, for this information-seeking mode include *monitoring* actions—revisiting favorite or bookmarked sites for new information; and systematic, thorough *extracting* actions—using search engines to extract information. "*Formal search* makes use of search engines that cover the Web relatively comprehensively and [provide powerful search] features that can focus retrieval. Because [users want to avoid missing] any important information, there is a willingness to spend more time in the search, to learn and use complex search features, and to evaluate the sources that are found in terms of quality or accuracy. Formal search may be two-staged: multisite searching that identifies significant sources..., followed by within-site searching. Within-site searching may involve fairly intensive foraging. *Extracting* may be supported by monitoring activity ... through services such as Web-site alerts, push channels [or] agents, and [email] announcements, ... to keep up with late-breaking information."

This behavioral framework "relates *motivations*—the strategies and reasons for viewing and searching—and *moves*—the tactics [people use] to find and use information...." [35]

Pearl-growing model

Sheryl Ramer defines *citation pearl growing* as follows: "The process of using the characteristics of a relevant and authoritative article, called a *pearl*, to search for other relevant and authoritative materials." In the same way, people can identify a relevant, authoritative Web site's key characteristics, then use them as keywords in searching for other similarly relevant, authoritative Web sites. [36]

According to Ralf Schlosser and his Northeastern University colleagues, "Citation pearl growing begins with a specific document or document set that is ... relevant to the topic at hand—[the] *pearl*. The searcher reviews the characteristics of that document or documents [and] adds their keywords to the search ... to retrieve additional [similar documents]—growing more pearls."

They define the process of citation pearl growing as follows:

1. "Find a relevant article.
2. "Find the terms under which the article is indexed in [a] database.
3. "Find other relevant articles in [the same] database by using the index terms in a building-block query"—that is, constructing a query by breaking up an information need into facets and concatenating the facets using AND operators, then generating multiple terms for each facet and concatenating these terms using OR operators.
4. "Repeat steps 2 and 3 in other databases.
5. "Repeat steps 1–4 for other relevant articles.
6. "[Stop] when articles retrieved provide diminishing relevance." [37]

You can see the pearl-growing approach in action on sites such as *ResearchGate* [38] that publish research papers and enable users to view their citations in other research papers. Social tagging provides another variation of this information-seeking model.

Sensemaking

The focus of all the models we've looked at so far has been on information seeking, but as Tony Russell-Rose and Tyler Tate point out in their book, *Designing the Search Experience*, "Finding information is only half of the equation; users must also make sense of what they encounter." *Sensemaking* is the cognitive process through which people derive meaning from information and assimilate new knowledge. [39]

Brenda Dervin defines *sensemaking* as the way "people construct sense of their worlds and, in particular, ... construct information needs and uses for information in the process...." A combination of cognitive and procedural behaviors enables people to make sense of and construct paths through both physical and digital information spaces. Both information seeking and the use of information are central to sensemaking.

Dervin's sensemaking model comprises the following elements:

- **Situations**—The contexts—time and space—within which people construct sense
- **Gaps**—The information gaps that people need to bridge—that is, the information needs or questions they have—as they construct sense and move through time and space
- **Uses**—The uses to which people put information once it makes sense [40]

In their paper, "The Cost Structure of Sense Making," Russell, Stefik, Pirolli, and Card describe *sensemaking* as "the active process of constructing a meaningful representation—[that is], making sense of some complex aspect of the world." [41]

Pirolli and Card, in a later paper, define an iterative process for sensemaking tasks that comprises four stages:

1. **Information**—Gathering or foraging for pertinent information
2. **Schema**—Representing the information in a way that assists with its analysis—whether in the forager's mind or in a sketch
3. **Insights**—Developing insights by analyzing and evolving the schema
4. **Product**—Creating a knowledge product on the basis of those insights

Elaborating on step 2 of the sensemaking process, the authors describe an internal process for representing the information in a schema—which is an iterative process that they refer to as a *learning-loop complex*—as follows:

1. **Generation loop**—Search for a good representation of the information.
2. **Data-coverage loop**—Attempt to encode the information in the representation, identifying information items that do *not* fit. This is the key step of this process.
3. **Representational-shift loop**—Adjust the representation to provide better coverage.

The result of this sensemaking process is a more concise, internal representation of the information that is essential to the sensemaking task. [42] However, complex sensemaking tasks often require much more detailed, external representations of information schemas such as affinity diagrams, cognitive maps, concept maps, or mind maps. [43]

Pirolli and Card conceived of a model for complex, highly iterative sensemaking tasks that require the transformation of raw information into reportable conclusions. Their model consists of an *information-foraging loop*, whose activities focus on gathering information, and a subsequent *sensemaking loop* that focuses on making sense of, or building a mental model of, the information.

The information-foraging loop comprises the following key stages:

1. Searching *external data sources* for raw evidence by rapidly skimming information and rejecting obviously irrelevant information

2. Rapidly gathering a subset of possibly relevant information into a metaphorical *shoebox* for later processing

3. Reading and extracting relevant snippets of information from the information items in the shoebox and adding them to an *evidence file*

The sensemaking loop consists of the following key stages:

1. Developing an external *schema* or mental model that optimally represents the information in the evidence file to facilitate the production of insights

2. Generating and testing multiple *hypotheses* for possible conclusions and providing supporting arguments for them

3. Presenting the final sensemaking conclusions in a *presentation*, report, or other work product [44]

While you might or might not find certain of these information-seeking models useful in your work, I hope they've given you some food for thought. Now, let's look at some practical aspects of information seeking.

Information-seeking strategies and tactics

People's information-seeking strategies and tactics derive from their current goals, tasks, motivations, and information needs, as well as the context within which they're currently seeking information. What overarching goal or task is a user trying to accomplish? What information-seeking approach would the user prefer to employ in accomplishing that information-seeking goal? How much time and effort is the user willing to invest in finding the information he needs? Depending on people's prior information-seeking experiences, they might choose to employ different information-seeking strategies and tactics in different situations.

Marcia Bates provides some definitions that are useful for this discussion:

- **Search strategy**—"A plan for [an entire] search."
- **Search tactic**—"[A] move a person makes toward the goal of finding desired information [or] to further a search."
- **Search behavior**—"What people do and … what they think when they search." [45]

Let's explore people's information-seeking strategies and tactics in greater depth.

Supporting searching, browsing, and asking strategies

Any effective information architecture supports two key information-seeking strategies: browsing and searching. Information-seeking strategies comprise combinations of tactics and behaviors, and there are many variations and combinations of these two basic approaches to looking for information. Information seeking is often a highly iterative process and, as a user's information needs evolve across iterations, the user may adopt different information-seeking strategies, tactics, and behaviors.

Now, with users' adoption of artificial-intelligence (AI) powered chatbots and search applications, digital information spaces can support a third common information-seeking strategy: asking. Users can ask basic questions directly of an ever-present helper and get an immediate response rather than asking a colleague, friend, or family member for help or calling or emailing Customer Support. As the capabilities of AI advance, chatbots and search applications will be able to handle increasingly difficult questions. [46]

Taking a holistic look at information-seeking strategies

In "Toward an Integrated Model of Information Seeking and Searching," Marcia J. Bates outlines four modes of information seeking. I've adapted her information-seeking modes as the foundation for my discussion of information-seeking strategies.

People receive information through many channels—both passively and through active information seeking. According to Bates, "People use the *principle of least effort* in their information seeking, even to the point that they will accept information they know to be of lower quality—less reliable—if it is more readily available or easier to use. … Throughout human history, most of the information a person needed came to him or her without requiring active efforts to acquire it." [47] So, let's consider users' information-seeking strategies from the perspective of the effort that users must expend in satisfying their information-seeking needs—starting with the least effortful strategies and progressing to the strategies that require the most effort.

Context awareness

Through *context awareness*, people passively absorb much of what they learn simply by *being aware* of their surroundings—the contexts in which they live and work. This requires no conscious effort on their part. We satisfy many of our information needs by talking with the people around us. We learn from the examples of the people we observe—especially from those with whom we have close relationships. "We get so much information through the natural conduct of our lives, from the flow of people and events around us," explains Bates. [48]

Monitoring information sources

Monitoring simply requires directing our attention to specific information sources. As Bates wrote, "In monitoring, we maintain a back-of-the-mind alertness for things that interest us and for answers to questions we have. We do not feel such a pressing need that we engage in an active effort to gather the information we are interested in; we are content to catch [it] as it goes by.... We also may have a question in mind and not act to find an answer, but notice when information comes along that is relevant to the question. ... We often arrange our physical and social environment ... to provide the information we need when we need it. ... We make it possible to be reminded ... of next steps or appropriate behaviors. ... The availability of these [contextual] supports cuts down on the need for active information seeking. [We learn to monitor our] environments for the infrastructural triggers for the next behavior. ... [We] come across a great deal of useful information just in the process of interacting socially ... [with] people who have a lot of common areas of knowledge [and] can suggest [useful] information or resources."

Monitoring can either be passive or semi-active—in cases where monitoring requires users to set up software tools to deliver information to them. *Passive monitoring* requires only that people be alert to incoming information and decide what information deserves their attention, so requires very little effort.

Semi-active monitoring requires deliberate action during setup, but once the user has installed and set up the necessary software, monitoring the resulting stream of information becomes passive. We receive information from our social-media feeds and notifications; email newsletters, RSS (Really Simple Syndication) feeds, and texts from our favorite content sites and apps; email messages, texts, and notifications from the ecommerce stores where we shop; and more.

"These serendipitous encounters are not truly by chance...," acknowledges Bates. "Rather, they are the product of proximity, either electronically or physically, that has come about through people organizing for common goals and needs." [49]

Seeking known items

Seeking known items is a goal-directed searching or browsing activity that lets people get the right answers to specific questions. People can either ask others for the information they need, which requires minimal effort, or attempt to discover the answers on their own through exploration, which could require significant effort that might prove unfruitful. People's willingness to invest the effort necessary to satisfy their information needs requires that they have either great interest in the information they're seeking or an urgent need for that information. [50]

The ability to *refind*, or *return to*, previously discovered information that is of value to the user is a special case of known-item seeking.

Exploratory seeking

In *browsing*, or *exploratory seeking*, which is the most effort-intensive form of information seeking, people may either lack a specific information need, but actively seek new information out of simple curiosity, *or* just be unsure what information would satisfy their need for information. [51]

Barbara Kwasnik defines *browsing* as "movement in a connected space—the strategic and adaptive technique that people use to search, scan, navigate through, skim, sample, and explore information systems." She has identified six functional browsing behaviors, as follows:

1. **Orientation**—Learning about an information space's structure and content through browsing

2. **Place marking**—Marking content the user might want to view again

3. **Identification**—Deciding whether the content is potentially interesting or definitely *not* of interest

4. **Resolution of anomalies**—Remedying any lack of clarity that exists or resolving mismatches that arise in relation to an information space's structure and content

5. **Comparison**—Making comparisons at all levels—whether between individual items of information or the structure of entire information environments—orients the user and helps her to identify or clarify her information-seeking goals

6. **Transitions**—Moving toward information—in the expectation of achieving an information-seeking goal—or away from information—either after identifying and rejecting the information or because the user has already found enough information to meet her needs or has exhausted an information space's available information resources [52]

Exhaustive research is a form of exploratory seeking that demands maximal effort from people. When pursuing this information-seeking strategy, people cannot *satisfice*—simply deciding that they already have enough information of sufficient quality to satisfy their needs—and still fulfill the goal of exhaustive research. However, when people are doing exhaustive research, they can apply their learnings from their previous explorations of information spaces in pursuing further information within other information spaces.

Contrasting information-seeking strategies

Bates summed up the utility of these four information-seeking modes, or strategies, as follows: "Monitoring and directed searching are ways we find information that we know we need to know, and browsing and being aware are ways we find information that we do not know we need to know." [53]

Information search tactics

In "Information Search Tactics," Marcia Bates distinguishes four types of search tactics, as follows:

1. **Monitoring tactics**—These are tactics that make a search persistent, ensuring its efficiency.

2. **File-structure tactics**—These tactics help people navigate an information space's file structure to find the files or information they need.

3. **Search-formulation tactics**—These tactics can help people either formulate or refine their search queries.

4. **Terminology tactics**—These tactics help people select optimal terminology in formulating or refining their search queries. [54]

Information-seeking behaviors

People can also employ different *information-seeking behaviors* in pursuit of the information they need to achieve their goals. [55] People choose to employ particular information-seeking behaviors depending on their current information needs and motivations, as well as the context within which they're seeking information. Plus, because of their unique information-seeking experiences in the past, individuals favor different ways of looking for information.

While the information-seeking behaviors I'll describe in this section are broadly applicable, I'll focus on their use within the context of a designed digital information space—whether a Web site, a mobile app, or a broader information ecosystem. There are two broad classes of information-seeking behaviors that people can employ:

1. Looking for specific, known items of information.

2. Exploring an information space either to satisfy a particular information need or to serendipitously discover interesting information of which they were previously unaware.

Known-item information seeking

People can look for specific, known items of information that provide the correct answers to basic questions. Such answers are typically facts that reside in databases. While these are the simplest information-seeking behaviors, they do usually require that people know, at least in general, where to find the information they need, how to formulate a search query or identify links that would enable them to find that information, how to use a Web site or app's functionality in seeking that information, and be able to recognize the information they're seeking when they find it.

Searching for known items

When people are looking for a specific, known item of information, or *fact-finding*, searching is usually the simplest, most appropriate information-seeking behavior to employ. Figure 2.2 depicts known-item searching. Because a user already knows there is a correct answer to his question, conducting a site search generally suffices—unless an information ecosystem comprises multiple Web sites and apps that might contain the information and the user is unsure where the information resides. In the latter case, searching for a known item could become an iterative process, requiring multiple searches across different sites and apps, as shown in Figure 2.3.

Figure 2.2—Searching to find a known item of information

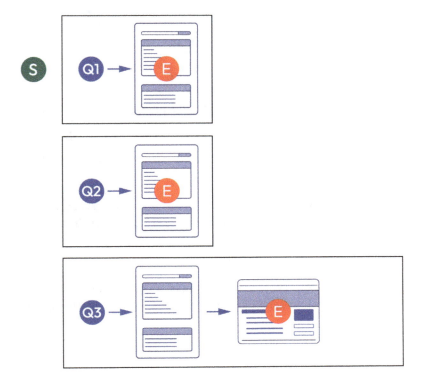

Figure 2.3—Iteratively searching multiple sources to find a known item of information

Browsing to find known items

When people are seeking a specific, known item of information and know either the precise location of that information or the location of a site directory with a link to that information, they can browse to find that known item of information, as shown in Figure 2.4. This behavior is less common than known-item searching.

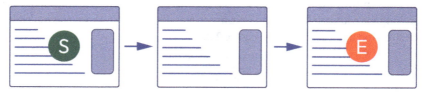

Figure 2.4—Browsing to find a known item of information

Returning to known items

When a user wants to return to a specific, known item of information and knows the location at which he previously found it, he might either search for *or* browse to find that information item again.

Seeking updated information

When people want to get updates to a known item of information, depending on a site or app's functionality, they might do the following:

- Bookmark the information to ensure they can easily return to it
- Save the information or add it to a wishlist to ensure they know where to find it
- Flag, favorite, or like the information, making it easier to identify visually
- Set up push notifications to receive updates to the information automatically via an email, text, or on-screen notification

With the exception of the last of these approaches, each of these methods of seeking updated information relies on an information-seeking behavior that involves browsing for known items.

Exploratory information seeking

By exploring a Web site, mobile app, or information ecosystem, people can satisfy their general or specific information-seeking goals or serendipitously discover valuable information of which they were previously unaware.

In contrast to known-item information seeking, people who are engaging in exploratory information seeking might or might not know what information they're seeking. They don't need to know exactly where to find specific information, how to formulate a search query, or how to identify links that would enable them to find it. However, they *do* need to know how to use a site or app's functionality to browse or search for information and to be able to recognize information that is of value to them when they find it.

Exploratory information seeking tends to be a highly iterative process. People gradually learn more about an information space and its use of terminology through the exploratory information–seeking process itself. People's information needs may evolve as they learn more about a subject. When people are exploring rather than seeking a precise piece of information, the scope of exploration is not limited at the outset. Instead, they can continue exploring until they're satisfied that they've found sufficient information to meet their needs.

However, if a single digital information space doesn't contain all of the information people need, they are likely to *satisfice*—balancing the level of effort they've already expended and their satisfaction with the results they've obtained so far in deciding whether to stop exploring that space and move on to another information space or to stop exploring altogether.

There are two basic behaviors that people can employ to explore a digital information space when they don't know exactly what information would satisfy their needs:

1. Exploratory browsing
2. Exploratory searching

Exploratory browsing

Exploratory browsing is the most common information-seeking behavior when people are simply hoping to discover interesting information serendipitously as they surf the Web. This is also a useful approach when people's information needs are only loosely defined or they have only a vague idea of what information they're seeking, as is typical when they're seeking information within an unfamiliar space or on a new subject. Exploratory browsing becomes increasingly challenging as the amount of information within an information space grows, and it is difficult to know whether one has exhausted relevant sources of information. Figure 2.5 depicts this information-seeking behavior.

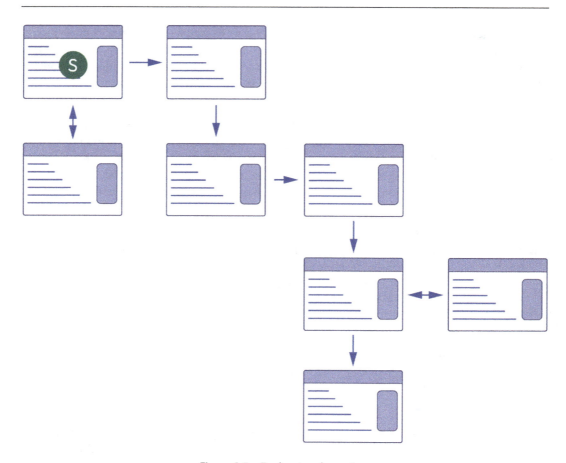

Figure 2.5—Exploratory browsing

Exploratory searching

Exploratory searching is the most common information-seeking behavior when people are looking for information on a specific topic, but don't know precisely what information they might want or where they would find it. Exploratory searching is a nonlinear process, involving complex cognitive activities. When people conducting exploratory searches don't know exactly what terminology to use, they can try various search queries. Depending on the results they get for a particular search query, they might refine their query by removing or adding search terms to broaden or narrow the results, respectively. Figure 2.6 illustrates exploratory-searching behavior.

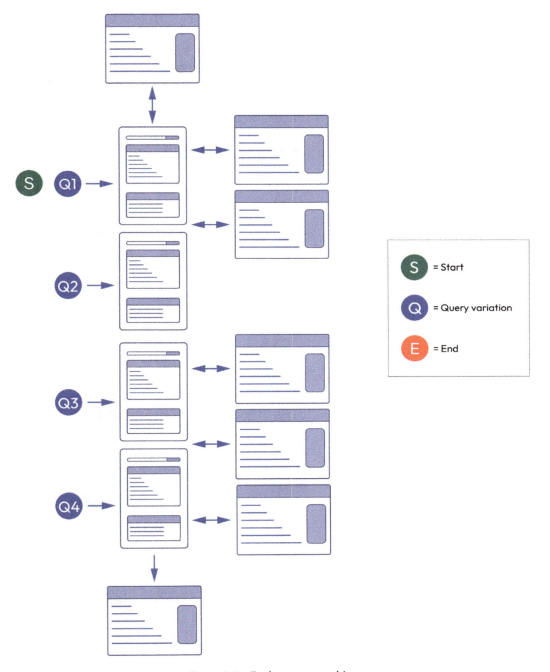

Figure 2.6—Exploratory searching

Integrating searching and browsing

When engaging in exploratory information seeking, people often alternate between searching and browsing. These two information-seeking behaviors are complementary. For example, a user might conduct a search, check out several search results, find information of particular interest on a page that includes many links to additional information on the same topic, then choose to follow some of those links. Alternatively, a user might browse a site where he thinks he would be likely to find the information he is seeking, navigate to and read a few pages that contain useful information, then search for more information by formulating queries using some of the terms from those pages. Figure 2.7 depicts integrated searching and browsing.

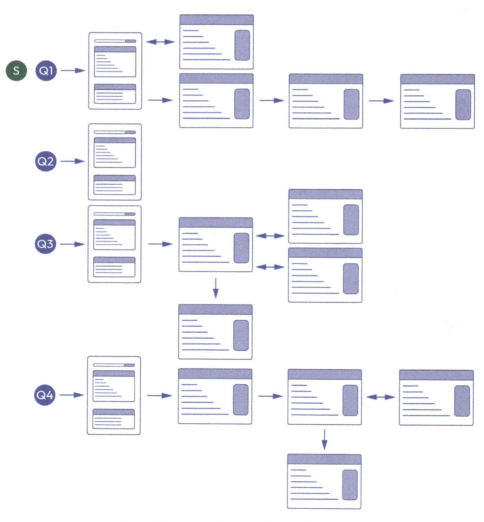

Figure 2.7—Integrated searching and browsing

Integrating searching and browsing behaviors is common among shoppers on ecommerce sites. Plus, several of the information-seeking models I described earlier in this chapter provide good examples of integrated searching and browsing, including berrypicking, information foraging, and the integrated model of browsing and searching.

Summary

In this chapter, you've learned how important it is to understand your users' information-seeking needs before designing an information architecture. We've explored a variety of useful information-seeking and sensemaking models and theories that are helpful in understanding the strategies, tactics, and behaviors that people employ when they're looking for information. We've looked at information-seeking strategies and tactics from several perspectives. Finally, you've learned about some common information-seeking behaviors that people use when they're looking for specific, known items of information versus those that they employ when exploring an information space—either to satisfy a particular information need or to serendipitously discover new information.

Next, in Chapter 3, *Design Principles*, we'll continue building the foundational knowledge you need to help you think like a designer so you can design effective information architectures. You'll learn about some user experience (UX) design principles that leverage people's cognitive and visual capabilities and, thus, should be the basis of every design decision you make, as well as about wayfinding design principles that are specific to the design of information architectures.

References

To make it easy for readers to follow links to the references for this chapter, we've made them available on the Web: `https://github.com/PacktPublishing/Designing-Information-Architecture/tree/main/Chapter02`

Further reading

Books

Andrea Resmini and Luca Rosati. *Pervasive Information Architecture: Designing Cross-Channel User Experiences*. Burlington, MA: Morgan Kaufmann, 2011.

Bella Martin and Bruce Hanington. *Universal Methods of Design*. Beverly, MA: Rockport Publishers, 2012.

Gary Marchionini. *Information Seeking in Electronic Environments*. Cambridge, UK: Cambridge University Press, 1997.

James Kalbach. *Designing Web Navigation*. Sebastopol, CA: O'Reilly Media, Inc., 2007.

Marti A. Hearst. *Search User Interfaces*. Cambridge, UK: Cambridge University Press, 2009.

Peter Morville. *Ambient Findability*. Sebastopol, CA: O'Reilly Media, Inc., 2005.

Peter Morville and Louis Rosenfeld. *Information Architecture for the World Wide Web*. 2nd ed. Sebastopol, California: O'Reilly Media, Inc., 2002.

Tony Russell-Rose and Tyler Tate. *Designing the Search Experience: The Information Architecture of Discovery*. Waltham, MA: Morgan Kauffman, 2013.

Articles and papers

Brenda Dervin. "From the Mind's Eye of the User: The Sense-making Qualitative- Quantitative Methodology." In Jack D. Glazier and Ronald R. Powell, *Qualitative Research in Information Management*. Santa Barbara, CA: Libraries Unlimited, 1992. [https://www.semanticscholar.org/paper/From-the-mind%E2%80%99s-eye-of-the-user%3A-The-sense-making-Dervin/117927016793d5d79305c0ed04f5655a0c4c9558]

Carol C. Kuhlthau. "The Role of Experience in the Information Search Process of an Early Career Information Worker: Perceptions of Uncertainty, Complexity, Construction, and Sources." (PDF) *Journal of the American Society for Information Science*, Vol. 50, No. 5, April 15, 1999. [https://asistdl.onlinelibrary.wiley.com/doi/10.1002/%28SICI%291097-4571%281999%2950%3A5%3C399%3A%3AAID-ASI3%3E3.0.CO%3B2-L]

Chun Wei Choo, Brian Detlor, and Don Turnbull. "A Behavioral Model of Information Seeking on the Web: Preliminary Results of a Study of How Managers and IT Specialists Use the Web." (PDF) Paper presented at the Annual Meeting of the American Society for Information Science (ASIS), Pittsburgh, PA, October 25–29, 1998. [https://files.eric.ed.gov/fulltext/ED438799.pdf]

David Ellis, Deborah Cox, and Katherine Hall. "A Comparison of the Information Seeking Patterns of Researchers in the Physical and Social Sciences." (PDF) *Journal of Documentation*, April 1993. [https://www.researchgate.net/publication/235802762_A_Comparison_of_the_Information_Seeking_Patterns_of_Researchers_in_the_Physical_and_Social_Sciences]

Gary Marchionini. "Exploratory Search: From Finding to Understanding." (PDF) *Communications of the ACM*, April 2006. [https://www.researchgate.net/profile/Gary-Marchionini/publication/220422328_Marchionini_G_Exploratory_search_from_finding_to_understanding_Comm_ACM_494_41-46/links/00463526ec57f5f6b2000000/Marchionini-G-Exploratory-search-from-finding-to-understanding-Comm-ACM-494-41-46.pdf]

Marcia J. Bates. "Information Behavior." In *Encyclopedia of Library and Information Sciences*, 3rd ed. Marcia J. Bates and Mary Niles Maack, eds. New York: CRC Press, Vol. 3, 2010. [https://pages.gseis.ucla.edu/faculty/bates/articles/information-behavior.html]

Peter Pirolli. "The Use of Proximal Information Scent to Forage for Distal Content on the World Wide Web." (PDF) January 2006. Retrieved January 7, 2020. [https://www.researchgate.net/publication/244515739_The_Use_of_Proximal_Information_Scent_to_Forage_for_Distal_Content_on_the_World_Wide_Web]

Peter Pirolli and Stuart Card. "Information Foraging in Information Access Environments." (PDF) *Proceedings of the ACM Conference on Human Factors in Computing Systems*. Denver: ACM, May 1995. [https://www.researchgate.net/profile/Peter-Pirolli/publication/221515012_Information_Foraging_in_Information_Access_Environments/links/0912f50d23f4080f7e000000/Information-Foraging-in-Information-Access-Environments.pdf]

T.D. Wilson. "Evolution in Information Behavior Modeling: Wilson's Model." In Karen Fisher, Sandra Erdelez, and Lynne E.F. McKechnie, eds. *Theories of Information Behavior*. Medford, NJ: Information Today, 2005. [https://www.semanticscholar.org/paper/Evolution-in-information-behavior-modeling-%3A-Wilson-Wilson/4d695d5eb58f5f269b9b0474b93a119cadfe1aab?p2df]

T.D. Wilson. "Human Information Behavior." (PDF) *Informing Science: Special Issue on Information Science Research*, Vol. 3, No. 2, 2000. [https://www.semanticscholar.org/paper/Evolution-in-information-behavior-modeling-%3A-Wilson-Wilson/4d695d5eb58f5f269b9b0474b93a119cadfe1aab?p2df]

T.D. Wilson. "The Cognitive Approach to Information-Seeking Behaviour and Information Use." (PDF) 1984. Reprint in *Social Science Information Studies*, Vol. 4, December 2018. [https://www.researchgate.net/profile/Tom_Wilson25/publication/329611261_THE_COGNITIVE_APPROACH_TO_INFORMATION-SEEKING_BEHAVIOUR_AND_INFORMATION_USE/links/5c123165299bf139c75500eb/THE-COGNITIVE-APPROACH-TO-INFORMATION-SEEKING-BEHAVIOUR-AND-INFORMATION-USE.pdf]

3
Design Principles

What are user experience (UX) design principles? *UX design principles* are basic truths, laws, or assumptions that embody the collective knowledge of the UX researchers and designers who have gone before us. They have their basis in the study of the cognitive and physical capabilities and behaviors of human beings and are the North Star for designers who place the needs of users at the center of design. User-centered design (UCD) principles set the standards and provide broadly applicable design guidance that supports more specific design guidelines. These design principles represent essential design considerations that are at the foundation of all design decisions. Effective UX designers and information architects always keep these design principles in mind—especially when solving unique or complex design problems.

Establishing what principles should apply in solving a particular design problem is a major factor in devising an effective design solution for that problem. When envisioning a new product or feature set, determine what design principles are paramount in ensuring its success. These principles define the product's key characteristics and identifying them can help your team and other stakeholders align around your design vision. They also provide a standard against which you can measure your success.

In this chapter, you'll learn about some design principles that are particularly helpful in designing information architectures. I've organized these principles into the following categories:

- **Designing for human capabilities**—These principles result in design solutions that support people's cognitive and visual capabilities.
- **Designing for wayfinding**—These wayfinding design principles include placemaking, orientation, navigation, labeling, and search principles.

Designing for human capabilities

Researchers in the disciplines of human-computer interaction and cognitive science have taught us much about the capabilities and limitations of the people who use software to consume information.

When you're designing an information architecture for a digital information space or a software user interface, you must take people's fundamental capabilities into account—especially their cognitive and visual capabilities. In this section, I'll briefly describe some generally applicable design principles that support human capabilities and, thus, can help you design information architectures and user interfaces that meet people's needs.

Accessibility

Accessible systems such as digital information spaces or software user interfaces meet the needs of people with a diverse range of physical and cognitive capabilities. In their book, *Universal Principles of Design*, William Lidwell, Kritina Holden, and Jill Butler define four characteristics of accessible designs:

1. **Perceptibility**—Everyone can perceive a user interface.

2. **Operability**—Everyone can interact with a user interface.

3. **Simplicity**—Everyone can comprehend and use a user interface.

4. **Forgiveness**—People feel safe when interacting with a user interface or exploring an information space because most interactions are reversible. [1]

When designing any information space or user interface, always strive to accommodate the full range of human capabilities.

Consistency

Consistency in the use, behavior, appearance, and layout of the elements within a digital information space or user interface improves its usability and learnability. When similar elements and interactions are consistent across a digital information space, cross-channel user experience, or user interface, people can easily transfer their knowledge and skills from one context to another.

Universal Principles of Design describes four types of consistency:

1. **Aesthetic, or visual, consistency**—Consistency in style and appearance—for example, the consistent use of color and typography—helps people to recognize similar contexts and elements.

2. **Functional, or behavioral, consistency**—Consistency in function and behavior communicates a user-interface element's functionality and behavior and can impart meaning.

3. **Internal consistency**—The aesthetic and functional consistency of similar elements within a system engenders trust.

4. **External consistency**—Consistency with similar elements outside the system or common design standards leverages people's expectations from prior experience with other user interfaces or information spaces. [2]

Consistency in the organization of content across an information space makes it easier for people to find the content they need. The consistent use of wayfinding information across an entire digital information space or a section of such a space facilitates people's navigation of a space.

Findability

Findability is the essential characteristic of any digital information space and ensures that people can find the information they need. In his book, *Ambient Findability*, Peter Morville defines *findability* as follows:

A. "The quality of being locatable or navigable.

B. The degree to which a particular object is easy to discover or locate.

C. The degree to which a system or environment supports navigation and retrieval." [3]

As Peter Morville and Lou Rosenfeld say in their book *Information Architecture for the World Wide Web*, "Findability is a critical success factor for overall usability. If users can't find what they need through some combination of browsing, searching, and asking, then the site fails. ... Users need to be able to find content before they can use it—findability precedes usability." [4]

Ensuring the findability of content within a digital information space requires the effective application of all the user-centered, wayfinding design principles that I'll cover later in this chapter.

Metaphors

Graphic user interfaces (GUIs) often employ real-world metaphors. The use of metaphors can make a software user interface feel more familiar and natural to people and thereby increase their comfort in using it. Employing clear, concrete metaphors to convey concepts and suggest the use of the features of a user interface can leverage people's existing knowledge of the world. People can take advantage of their learnings and expectations from their prior experiences in interpreting a digital environment. Animations and aural cues can sometimes be useful in communicating a metaphor clearly. Avoid extending a metaphor beyond its salient characteristics or making it too literal.

The metaphor of wayfinding is obviously applicable in information architecture. I'll explore wayfinding in depth later in this chapter.

Recognition over recall

People's ability to recall information from memory depends on how recently they've recalled that memory, how often they've recalled—or rehearsed—that memory in the past, and their current context, which may provide environmental cues that aid recall.

However, seeking information and using software need not be so cognitively demanding. People should be able to rely simply on recognizing elements of digital information spaces and user interfaces that they've experienced before rather than having to recall them from memory. When designing a digital information space, provide visual and textual cues that aid people's memory and, thus, help them to navigate the space and recognize the information they need.

Reducing cognitive load

Cognitive load refers to the amount of short-term memory, or *working memory*, a person is currently using for reasoning and decision-making in regard to accomplishing a task. Users' reliance on recognition rather than recall when using direct-manipulation interfaces—such as digital information spaces' navigation and search user interfaces—reduces cognitive load, improves performance, and reduces errors. To minimize cognitive load, do the following:

- Make available navigation options highly visible.
- Eliminate unnecessary elements and information from a page or user interface.
- Chunk content logically to make it easier to consume.

Simplicity

An essential characteristic of a usable software user interface or digital information space is simplicity. Both must be easy to learn *and* easy to use. People should easily be able to discover and learn all of an information space's functionality and find the information they need. Simple, coherent design solutions include only essential elements and features that serve the needs of users.

Achieving simplicity requires focusing on meeting users' needs and resisting pressures from business colleagues to either add features that don't serve their needs *or* differentiate their brand by creating flashy designs that users would find distracting. As Sarah Horton and Whitney Quesenbery say in their book, *A Web for Everyone*, "Simplicity can be a great differentiator where important elements stand out…." [5]

One way of managing complexity—and thus, gaining greater simplicity—is by supporting progressive disclosure. Logically chunk functionality or information, hide whatever functionality or information users do not initially or currently need, and allow users to reveal more when they need it.

User control

In most situations, users should have the ability to initiate and control interactions and software systems should respond to them. Giving users control keeps them engaged and helps them to learn. Provide users with the functionality they need to accomplish their tasks, then give them broad freedom of action so they can decide what they want to do and when.

Software systems should *not* be overprotective of users, take control, and thereby, restrict what users can do. While there are circumstances in which it can be helpful to provide a wizard to guide users through a difficult process, also provide analogous functionality that supports users' independent action. This approach supports users with different levels of proficiency in using the software, offering users more control as they gain expertise.

Nevertheless, there are a few circumstances in which the system should initiate action, by doing the following:

- Provide notifications to users when they've requested updated information or urgently need to take action.

- Caution users about and, thus, prevent user errors that would destroy data, but still allow users to proceed by confirming that destroying data is their intent.

In both cases, users still remain in control of what they ultimately choose to do.

Now that you're familiar with some basic principles for designing information-architecture solutions to support human capabilities, let's consider the wayfinding design principles that are foundational to information architecture.

Designing digital spaces for wayfinding

Information architects have adapted many design principles from the mature practice of wayfinding design for built environments and have adopted the term *wayfinding* to describe the parallel concerns of information seeking, navigation, and user orientation within digital information spaces. [6]

What is *wayfinding*? People find their way within digital information spaces and seek the information they need by making a series of interrelated decisions on the basis of personal preferences, environmental factors, and the availability of information. The first decision they make is to begin an information-seeking activity—often to find a particular item or category of information. Second, they decide what strategy to employ in seeking that information, starting with where to look for the information and choosing tactics and behaviors that would enable them to find it. Since all of the elements of an information-seeking strategy can evolve over time, this usually involves making a series of decisions. Third, they decide what path to take in seeking that information, which also involves making a series of decisions that depend on their personal preferences, what they already know, and both the wayfinding aids and the usefulness of the information they encounter within an information environment. [7]

Digital wayfinding stages

Let's briefly consider the stages for wayfinding in digital information spaces, the information users need to make wayfinding decisions at each stage, and the interactions that enable users to advance to the next stage. There are three types of wayfinding information, which correspond to the three stages of a wayfinding journey within a digital information space, as shown in Table 3.1.

Wayfinding Stage	Wayfinding Information	Interaction
Accessing a digital information space	A URL, generally gleaned from other media, or a hyperlink	The user might type a URL into a **Search** box; click or tap a bookmark; or click or tap a link either in a search result in a Web search engine, on another Web site, or in an email message, text, chat, or social-media post.
Navigating a digital information space	A navigational link or a series of navigational links acting as signposts at decision points on a wayfinding journey Alternatively, a link in a site map or directory	Depending on the complexity of an information space and the user's information-seeking strategy, tactics, and behaviors, the user might take a very direct path to a particular destination or a meandering path to multiple destinations.
Identifying useful information	Easily distinguishable information such as an organization's logo, page titles, and section headings, as well as visual and textual content	Once a user arrives at a page, the user can scan or skim the page's content to find information of interest. [8]

Table 3.1—Wayfinding information and stages

A retrospective on wayfinding in physical spaces

In his book, *The Image of the City*, renowned urban planner Kevin Lynch coined the term *wayfinding* to describe people's "consistent use and organization of definite sensory cues from the external environment" in navigating physical spaces. [9]

Romedi Passini and his colleagues provide a detailed definition:

"*Wayfinding* is the ability to reach desired destinations in the natural and built environment. Defined in terms of spatial problem solving, wayfinding is composed of three interrelated processes:

1.   Decision-making and the development of a plan of action,
2.   Decision [execution]—transforming the plan into action and behaviors at the right place and time, and Information gathering and treatment [that] sustains the two decision-related processes." [10]

In 1948, psychologist Edward C. Tolman hypothesized that people find their way by building "a mental representation of the spatial layout of the environment," or "a mental spatial model," for which he coined the term a *cognitive map*. [11] However, subsequent research has disagreed about the area of the brain in which such a map might exist, questioned the ability of people to generate or use such a map, and posited simpler explanations for how people navigate spatially complex environments.

For example, behavioral psychologists have studied the effect of *stimulus control*—the way in which the reinforcement of a prior stimulus within a complex spatial environment affects the probability that a person would repeat the same behavior in response to the same stimulus. People respond differently to different types of visual-spatial stimuli—for example, looking at a map of a physical space or using landmarks to navigate through an actual physical space or a virtual depiction of that space. While the stimulus control of visual-spatial stimuli predominates, other sensory inputs also assist in people's successful movement through complex spaces.

Some researchers find considering the *process of cognitive mapping* a much more useful approach than hypothesizing about an unobservable, cognitive map. [12] According to Roger M. Downs and David Stea:

"*Cognitive mapping* is a process composed of a series of psychological transformations by which an individual acquires, codes, stores, recalls, and decodes information about the relative locations and attributes of phenomena in his everyday spatial environment." [13]

"The typical wayfinding individual is using the environment itself more for recognition than recall and attending to only the minimum information necessary," says Andrew Hinton, in his book, *Understanding Context*. "The environment itself serves as an external map of physical and semantic information cues, most of [them] beyond our conscious awareness. … The environment that makes up [people's] context is inseparable from their ability to understand, learn, and navigate that environment." [14] The environment is the map, so there is no need to build a map in one's head.

Now that you have a general understanding of digital wayfinding and have learned a little about wayfinding in physical spaces—the study of which has provided the foundation for the design of digital wayfinding—let's consider some design principles that contribute to successful wayfinding in digital information spaces.

Design principles for wayfinding

Note—In adapting wayfinding design principles for physical spaces to the design of digital information spaces, I've taken great inspiration from Paul Arthur and Romedi Passini—who are the authors of the seminal work, *Wayfinding: People, Signs, and Architecture*, [15, 16] which is regrettably out of print—and Mark Foltz's Bachelor's thesis, [17], which provides an excellent collection of design principles for wayfinding in physical spaces.

Now, let's explore wayfinding design principles that support users' successful wayfinding within digital information spaces such as Web sites, intranets, extranets, and information-rich applications, including the following types of principles:

- Placemaking principles
- Orientation principles
- Navigation principles
- Labeling principles
- Search principles

Placemaking principles

What makes people return to the same digital information spaces again and again? People build long-lasting relationships with information spaces that clearly convey a sense of place, communicate their trustworthiness, and provide the reliable information people need.

As Jorge Arango says in his book, *Living in Information*, "When [people] inhabit … an environment, use it for its intended purpose, and interact with other people there, it becomes part of [their] mental model of the world. … When … there, [they] feel, think, and act in ways that are particular to that environment. We call such environments places…. Our effectiveness as individuals and societies greatly depends on how well these places serve the roles we intend for them… Approaching software design as a placemaking activity—with a focus on intended outcomes and behavior…—results in systems that can serve our needs better in the long term." [18]

Placemaking defines the contexts within which people seek information and, thus, reduces information seeker's feelings of disorientation and encourages them to build relationships with digital information spaces. Now, let's look at some design principles that support placemaking.

Give each digital information space a unique visual identity that distinguishes it from all other digital information spaces.

To clearly convey a sense of place and ensure that people know where they are:

- Prominently display an organization's logo, which functions as a landmark for an entire information space.
- Use common page elements such as headers, organizational logos, navigation bars, and footers across all pages.
- Make the use of color within a digital information space consistent with an organization's branding.

Clearly establishing a sense of place for a digital information space ensures that people know where they are and can quickly recognize that information space when they return to it, which is especially important to ensure coherence across the multiple channels of a cross-channel information ecosystem.

Visually distinguish the various sections of a large-scale digital information space.

To convey a sense of place for each individual section of a digital information space, enable people to distinguish the different sections of an information space, and ensure that people know what section they're currently in, do the following:

- Ensure that the layout of a section's pages prioritizes features or information that emphasize the section's unique purpose or content.

- Consider using unique color coding to identify each section. This both unifies the pages in a section and separates each section from the others. However, be aware that people with color-deficient vision won't perceive these colors as you intend.

- Prominently display the title of each section on all the pages within that section to ensure that users know they're currently in that section.

Clearly distinguishing each of the sections within a digital information space ensures that users know they're in a section and can quickly recognize a section when they return to it. Doing so also communicates the organizational structure of an information space.

Orientation principles

For people to be able to successfully navigate a digital information space, they must first orient themselves to that space and their current location within it. Thus, providing users with effective means of orienting themselves is key to their wayfinding success. Now, let's consider some design principles whose implementation facilitates users' ability to orient themselves so they can confidently and successfully move through a digital information space.

Make the purpose of an information space clear so people can easily ascertain whether they're in the right place to find the information they need.

When people visit an information space for the first time, they need to orient themselves to that information space and determine whether it might offer the information they need. If an information space adequately supports these goals, visitors can learn about its purpose by looking over the links on its primary and any secondary navigation bar. Its navigation system should comprehend the full scope of the information that is available on that space. Visitors should also be able to learn more about the information that is available on a space by skimming the content on its home page.

Once visitors have determined that an information space might be able to meet their information needs, they must next decide what path to follow in seeking that information. Visitors' ability to make wayfinding decisions successfully depends both on how well an information space is organized and the strength of the information scent that its content and navigation system provide.

To ensure that people visiting an information space for the first time can readily orient themselves to that space and find the information they need, use familiar organization schemes, navigation design patterns, and labels. Designing the wayfinding information for an information space with the needs of first-time visitors in mind results in a navigation system that works well for everyone.

Provide previews of some of the content within a digital information space.

To communicate the purpose of a digital information space—or a section within it—and thus, orient people to that information space, give them an overview of the larger space by showing representative examples of its content on the home page. Providing previews of content that resides deep within an information space, but might be of interest to people encourages them to explore a space or section more deeply.

While previews can take a variety of forms, they usually attract people's attention through images and very meaningful text, comprising concise headlines and perhaps brief descriptive text or excerpts. Unfortunately, many such images are what I refer to as *decorative images* because they convey *no* useful information.

Previews provide entry points into an information space or section that let people go directly to content that might be of interest to them and determine whether it actually is. By enabling information seekers to sample content, previews can draw them more deeply into an information space. Previews are especially helpful to first-time visitors who are unfamiliar with an information space, infrequent visitors who don't know a space well, and people seeking new or updated content—in the latter case, especially if those previews are personalized to satisfy the needs of individuals.

Create context to ensure a navigable digital information space.

As Andrew Hinton says in his book, "Whenever we're trying to figure out *what one thing means in relation to something else,* … we're trying to understand its context. … We need context to *be clear and to make sense.*" [19]

When people are seeking information within a *navigable* digital information space, their surrounding context provides such wayfinding aids as a well-designed navigation bar and descriptive page titles, enabling people to easily determine their current location, orient themselves to the broader information space, and successfully perform wayfinding tasks. The use of imagery and color—at least for people who do *not* have severe forms of color-deficient vision—can make different contexts readily distinguishable from one another. Once people know exactly where they are, understand their current context, and know where they can go from there, they can successfully engage in what Paul Arthur and Romedi Passini refer to as *spatial problem-solving* in their book, *Wayfinding: People, Signs, and Architecture*—that is, people can make the right navigation decisions to take them by the most direct route to the information they need. [20]

Designing an effective information architecture is essential to creating a navigable digital information space that facilitates wayfinding.

Establish visually prominent, memorable landmarks.

Landmarks help people to orient themselves within the context of a digital information space and often correlate to decision points at which people must choose which path to follow. In his Bachelor's thesis on wayfinding, Mark Foltz—now a software engineer and technical lead at Google—wrote, "A system of landmarks helps to organize and define an information space." [21] Do the following to orient people using landmarks:

- Convey important location information using visual elements on pages.

- Use *global landmarks* such as page headers, organizational logos, navigation bars, and footers to ensure that people know they're on a particular information space.

- Use *local landmarks* such as highlighted links on navigation bars and page title bars to ensure that people know where they are within an information space.

- Prominently display the title of each page on a digital information space to ensure that people know exactly where they are within that space.

- On the Web, use Accessible Rich Internet Application (ARIA) landmarks, which specify role attributes that enable people to navigate the key areas of a Web page—including the header, navigation bar, main content, and footer—using a screen reader. [22]

Within a digital information space, most landmarks manifest the organizational structure of the space, and they demarcate the paths that people follow when seeking information.

Navigation principles

Navigability refers to people's ability to make their way successfully through a digital information space and depends in large part on the usefulness and usability of its navigation system and other wayfinding aids—the features that enable people to move from place to place within that space. The design of these features provides essential support to people's wayfinding process. We'll now look at some principles that support the design of effective navigation systems.

Provide the appropriate navigation options at each wayfinding decision point.

Within a digital information space, the labels of navigation links are analogous to signposts in the physical world. At wayfinding decision points, people determine how to proceed on their wayfinding journey by choosing navigation options—whether they decide to explore further along their current path, choose a different path, or reverse direction and backtrack to an earlier decision point.

The labels of navigation links should clearly communicate what people would find at their destination. Depending on a person's current level within a navigation hierarchy, a link destination might be either a section of an information space, a general category of information, *or* specific content at the link's destination.

Providing the appropriate navigation options for a given context requires including all necessary links, but no more. Be sure to include both links that meet people's key needs, as well as links that make it easy for people to find information that supports an organization's key objectives. To avoid overloading the primary, or global, navigation bar, determine what links would be most useful within a particular context and, thus, belong at a lower level in the navigation system—for example, in a secondary-navigation sidebar or embedded within the content on a page.

To prevent people from choosing a less-than-optimal path at a decision point, ensure that the appropriate navigation options are available where people need them, that the links are clearly labeled, and that navigation options are easily distinguishable from one another—and, thus, unambiguous.

An information space's wayfinding signposts—coupled with other sources of information scent that people pick up from their environment along the way—ensure that people can find the information they need. When a digital information space provides effective wayfinding, people's exploration of that space becomes safe.

Avoid presenting too many navigation options to people at once.

To ease the wayfinding decisions that people make, limit the number of navigation links at any wayfinding juncture to only those that are essential at that point. The links on a primary or any secondary navigation bar should provide access to *all* of the available information—either within an entire digital information space or a section of that space—without overwhelming people with too many choices. Accomplishing both of these goals at the same time requires balancing the breadth and depth of the information architecture that underlies an information space's navigation system.

When there are more navigation options than people could easily manage at once, limit the number of navigation options that people see at the same time by employing progressive disclosure. Logically group navigation links—giving the groups clear labels that provide good information scent—and initially hide the links within those groups. Users can then reveal the links within groups when they need them. For example, provide a navigation system that comprises both primary and secondary navigation, then when the user clicks or taps a link on the primary navigation bar, display subordinate links on a secondary navigation bar or drop-down menu.

Navigation options should be quite limited in cases where the goal is either to guide people down a particular path, as in a tour, *or* even to constrain them to a path, as when following a procedure using a step-by-step wizard. For a tour, while it might be desirable to allow some flexibility in what people view next or to provide some branches off the main path, it is important to ensure that they see the information in an order that makes sense—and that makes its meaning clear. It is also important to encourage people to view *all* of the available information by clearly indicating what they have and have not yet seen. In contrast, wizards typically restrict people to either choosing the next option or retracing their previous steps. Both tours and wizards provide an overview of a predetermined path, show a person's current location on that path, and show how much distance remains to the ultimate destination.

Simplify navigation systems by grouping related links.

To help people more quickly make sense of an information space's navigation system, group related links together. Groupings define the relationships that exist between certain links and distinguish related links from unrelated links.

The logical grouping of links serves the needs of people who have different purposes for visiting an information space and are looking for different types of information. Providing logical groups of links simplifies information seeking for people by making it easier for them to either recognize information scent or reject entire groups of links that are *not* of interest to them.

Make sure that each of the links in a group of links is clearly distinguishable from the others.

It is important to choose words or phrases for link labels that are clearly distinguishable from one another. Using distinctly different labels for all of the links in a group of links ensures that people can easily distinguish the links from one another. It is particularly important to avoid repeating the same words at the beginning of the labels for each of the links in a series of links. The text of each link label should be unique to avoid confusing people about which link they should click or tap.

Ensure that wayfinding information is clear and consistent across an entire digital information space.

The clarity and consistency of wayfinding information are essential to engendering people's confidence and trust in a digital information space. Nobody wants to experience becoming lost in hyperspace. Clear, consistent wayfinding aids can prevent people from becoming confused.

Therefore, the use of visual and textual wayfinding information should be consistent across an entire digital information space or section, as well as at each stage of a user's particular wayfinding journey. As Andrew Hinton says in his book, *Understanding Context*, "Consistency isn't about just the details, but really about the coherence of meaning from one context to the next." [23]

Wayfinding information should also be consistent with familiar design standards, so people seeking information within a particular information space can apply their learnings from their experiences with other information spaces.

Within an information space, it is usually possible to navigate to a single destination via multiple paths. These paths might take people through disparate sections or parts of an information space. The user experience should be consistent regardless of the path someone takes.

Consider providing an overview of a digital information space.

An overview of a digital information space can be a valuable navigation aid, helping people to understand the scope and organization of the content within that space, the relationships between collections of content and individual content elements, the available destinations, what's nearby, and possible routes to those destinations. [24] Such an overview might take the form of a site map, a directory, an alphabetical index, or a list of topics, *and* provide a fairly comprehensive collection of links—making it an exception to the previous principle.

Of course, within any digital information space, the primary navigation bar and any secondary navigation bar already provide a high-level overview of the entire space, as well as information about the user's current location within that space.

Location breadcrumbs can serve a similar function, by showing the position of the current page within an information hierarchy. *Path breadcrumbs* show users how they got where they currently are and provide an easy way to get back to where they've been. *Attribute breadcrumbs* comprise product or other attributes and are common on ecommerce Web sites and apps. [25]

Back in 2008, UIE's Jared Spool advised that product teams would do better to invest in improving their information architecture and redesigning their navigation system to provide stronger information scent rather than in the implementation of either a site map or breadcrumbs. [26, 27] While Jared certainly had his priorities right, that doesn't mean implementing site maps and breadcrumbs offers no value.

At about the same time, Jakob Nielsen of the Nielsen Norman Group (NN/g) wrote about the usefulness of site maps as a supplement to primary navigation because they "offer a visual representation of the information space, [helping] users understand where they can go." Site maps can be especially helpful in navigating large, complex information spaces. Jakob recommends implementing site maps even though people rarely use them. [28] Still, for some digital information spaces, especially those of limited scope, a site map is unnecessary.

Jakob and NN/g have consistently advocated the use of location breadcrumbs as a supplementary navigation aid—but *not* path breadcrumbs. Location breadcrumbs are especially helpful when users click or tap an external link that takes them to a page that is deep within a destination information space because they provide a means for users to "visualize the current page's location" within its information hierarchy. [29, 30]

Labeling principles

In Chapter 2, *How People Seek Information*, I described how people follow information scent when they're foraging for information. Creating clear, meaningful labels for navigation links, site-section titles, page titles, and section subheadings within pages is essential to providing information scent. In this chapter, I've already discussed the importance of labeling in helping people orient themselves to digital information spaces and how labels function as signposts in navigation systems.

Now, in this section, I'll provide some design principles for creating effective labels—especially labels for the most tangible elements of an information architecture: the wayfinding elements. These elements include labels for navigation links, site-section titles, page titles, and section subheadings within pages.

Create clear, easy-to-differentiate link labels.

To make people's navigation choices clear, the labels of navigation links should be easy to differentiate from one another. Therefore, in creating link labels, you should use words or phrases that are distinct from one another. It is especially important to avoid repeating the same words at the beginning of a series of link labels, which would make perusing links less efficient. The text of each link label should be unique. Otherwise, people might accidentally click or tap the wrong link.

Create link labels that clearly convey what information is at their destination.

The labels of navigation links must clearly and concisely communicate what people should expect to find at their destination. Labels that provide strong information scent encourage people to click or tap links because they feel confident they'll find the information they need at their destination.

Clearly identify destination pages.

Depending on whether the destination of a navigation link is a directory page for a site section *or* a content page, the site-section title or page title, respectively, for that destination page must clearly identify that destination, using similar terminology to that of the label for the navigation link that led to it. People should be able to immediately discern whether they're in the right place: a destination page that provides the information they were seeking.

Use familiar terminology in all labels.

In creating labels for navigation systems, site-section titles, page titles, section subheadings within pages, and embedded links, use simple words and phrases that most people who belong to your target audience understand. Using the same language as your users ensures that they can avail themselves of the content your digital information space provides.

Organizations working in the same business domain typically use similar terminology, so most people belonging to their audience would probably be familiar with that terminology. However, in designing digital information spaces for use by the general public, never use esoteric business jargon in labels for wayfinding elements.

Create consistent labeling systems.

One key to the success of a labeling system for a digital information space is consistency in its use of terminology across *all* wayfinding elements. As I mentioned earlier, consistency makes an information space more usable and easier to learn. Consistency in word choice, syntax, and the use of style, typography, and color all give users the impression that labels are part of a coherent system.

In labeling systems, the specificity of links should be similar. Labeling systems should also be comprehensive and cover the full scope of a digital information space.

Search principles

When people are seeking information within a digital information space of significant scope, they appreciate having alternative means of wayfinding and may choose to either browse or search for the information they need. Ensuring findability requires that such an information space should support both of these modes of information seeking effectively. Earlier in this chapter, I covered navigation-design principles that support successful browsing. Now, let's look at some design principles for effective search systems.

Provide search to support wayfinding for people with disabilities.

For many people who have disabilities, navigation is too slow and laborious, so searching is their preferred means of wayfinding. [31] Create a search system that adequately meets their needs.

Place the search box in a consistent location on all pages of an information space.

People shouldn't have to go looking for this essential wayfinding tool.

Adapt the design of search systems to their context.

There are many permutations of search systems. Depending on the type of information space, anything from a simple search box and results page to a complex, faceted search system might be appropriate.

Check the spelling of words in search queries.

Users should never receive no results at all because of a typo in a query.

Ignore irrelevant information in search queries.

When interpreting a search query, a search engine should ignore letter case and, in some cases, predefined stop words—which usually include words such as articles and prepositions—and punctuation characters.

Support users' ability to revise their search queries.

If users don't initially get useful search results, they might want to revise their current query rather than typing a completely new one. To ensure users can do so without navigating to another page, display the current query in a search box on the search results page.

Display the best results first on the search results page.

People often satisfice when seeking information, so they may consider only the first few search results. Therefore, it is imperative that those first few results be the best results an information space can deliver—that is, they must be the most relevant results.

Display the number of search results for a query.

This information helps users to know whether they should broaden or narrow their search query. It also communicates the scope of the information space.

Provide a solution for situations where there are no results.

If a search query produces no results at all, display a search results page that clearly indicates there are no results, suggests that the user might broaden the search query, and lets the user immediately revise the query.

Ensure that search results provide sufficient information scent.

Each search result should include sufficient information to provide good information scent so people can determine whether a search result is relevant. The most salient information in a search result is the link, whose link text should be the title of the link's destination page. Summary information or a brief excerpt also provide a strong information scent. Highlight terms in the search results that match the user's query.

Depending on the context, including other information in the search results might be helpful. For example, for a Web magazine, an article's publication date and author are also important information. The information that you include in search results for a particular type of information should be consistent.

Quick reference: Wayfinding design principles

Now that we've explored these wayfinding design principles in depth, here is a summary of the principles for your reference.

- **Placemaking principles**:
  - Give each digital information space a unique visual identity that distinguishes it from all other digital information spaces.
  - Visually distinguish the various sections of a large-scale digital information space.

- **Orientation principles**:
  - Make the purpose of an information space clear so people can easily ascertain whether they're in the right place to find the information they need.
  - Provide previews of some of the content within a digital information space.
  - Create context to ensure a navigable digital information space.
  - Establish visually prominent, memorable landmarks.

- **Navigation principles**:

  - Provide the appropriate navigation options at each wayfinding decision point.

  - Avoid presenting too many navigation options to users at once.

  - Ensure that wayfinding information is clear and consistent across an entire digital information space.

  - Simplify navigation systems by grouping related links.

  - Make sure that each of the links in a group of links is clearly distinguishable from the others.

  - Consider providing an overview of a digital information space.

- **Labeling principles**:

  - Create clear, easy-to-differentiate link labels.

  - Create link labels that clearly convey what information is at their destination.

  - Clearly identify destination pages.

  - Use familiar terminology in all labels.

  - Create consistent labeling systems.

- **Search principles**:

  - Provide search to support wayfinding for people with disabilities.

  - Place the search box in a consistent location on all pages of an information space.

  - Adapt the design of search systems to their context.

  - Check the spelling of words in search queries.

  - Ignore irrelevant information in search queries.

  - Support users' ability to revise their search queries.

  - Display the best results first on the search results page.

  - Display the number of search results for a query.

  - Provide a solution for situations where there are no results.

  - Ensure that search results provide sufficient information scent.

Summary

Every information architect should have a solid grounding in UX design principles, which have their basis in our understanding of human capabilities. Learning design principles teaches you to think like a designer. Once you learn them, design principles provide a sound foundation for every design decision you make.

In this chapter, you first learned some design principles that support human capabilities and, thus, user-centered design (UCD). Then, we explored design principles for wayfinding in depth—including placemaking, orientation, navigation, labeling, and search principles.

Next, in Chapter 4, *Structural Patterns and Organization Schemes*, you'll learn about some common structural patterns and organization schemes that are useful in designing information architectures.

References

To make it easy for readers to follow links to the references for this chapter, we've made them available on the Web: `https://github.com/PacktPublishing/Designing-Information-Architecture/tree/main/Chapter03`

Further reading

Books

Apple. *iOS Human Interface Guidelines*. (PDF) Cupertino, CA: Apple, Inc., March 23, 2011. Retrieved January 17, 2020. [`https://tableless.github.io/exemplos/pdf/guidelines-interface-mobiles/MobileHIG.pdf`]

Apple Computer. *Apple Human Interface Guidelines*. (PDF) Cupertino, CA: Apple Computer, September 8, 2005. Retrieved January 17, 2020. [`https://tableless.github.io/exemplos/pdf/guidelines-interface-mobiles/MobileHIG.pdf`]

Apple Computer. *Macintosh Human Interface Guidelines*. (PDF) Cupertino, CA: Apple Computer, 1995. Retrieved January 17, 2020. [`http://interface.free.fr/Archives/Apple_HIGuidelines.pdf`]

Apple Computer. *Apple Human Interface Guidelines: The Apple Desktop Interface*. Menlo Park, CA: Addison-Wesley Publishing Company, Inc., 1987.

Apple Computer. *Macintosh Human Interface Guidelines*. Menlo Park, CA: Addison-Wesley Publishing Company, 1992.

Kevin Lynch. *The Image of the City*. Cambridge, MA: The M.I.T. Press, 1960.

Microsoft. *Microsoft Windows User Experience*. Redmond, WA: Microsoft Press, 1999.

NeXT Publications. *NeXTSTEP User Interface Guidelines*, Release 3. Menlo Park, CA: Addison-Wesley Publishing Company, 1993.

Romedi Passini. *Wayfinding in Architecture*. New York: John Wiley & Sons, Inc., 1984.

Romedi Passini. *Wayfinding: People, Signs and Architecture*. New York: McGraw-Hill, 1992.

Sarah Horton and Whitney Quesenbery. *A Web for Everyone: Designing Accessible User Experiences*. Brooklyn, NY: Rosenfeld Media, 2013.

Sun Microsystems. *Java Look and Feel Design Guidelines*, Second ed. Boston: Addison-Wesley Professional, 2001.

Sun Microsystems. *Java Look and Feel Design Guidelines: Advanced Topics*. Boston: Addison-Wesley Professional, 2001.

William Lidwell, Kritina Holden, and Jill Butler. *Universal Principles of Design*. Beverly, MA: Rockport Publishers, Inc., 2003.

Articles and papers

Apple. "iOS Design Themes" *Apple, Inc.*, undated. [https://apple.co/35A13Aa]

Apple. "macOS Design Themes" *Apple, Inc.*, undated. [https://apple.co/2OAMQtx]

Apple. "tvOS Design Themes" *Apple, Inc.*, undated. [https://apple.co/33rt8qM]

Apple. "watchOS Design Themes" *Apple, Inc.*, undated. [https://apple.co/2H6KjD1]

Design Principles FTW, undated. [https://www.designprinciplesftw.com/]

Google. "People + AI Guidebook." *Google*, undated. [https://pair.withgoogle.com/guidebook/]

Google. "*Principles of Mobile App Design: Engage Users and Drive Conversions*." (PDF) *Google*, March 2016. [https://www.thinkwithgoogle.com/_qs/documents/23/principles-of-mobile-app-design-engage-users-and-drive-conversions.pdf]

Google. "Principles of Mobile Site Design: Delight Users and Drive Conversions." (PDF) *Google*, 2014. [https://www.thinkwithgoogle.com/_qs/documents/538/multi-screen-moblie-whitepaper_research-studies.pdf]

Interaction Design Foundation. "Design Principles." *Interaction Design Foundation*, undated. [https://www.interaction-design.org/literature/topics/design-principles]

Microsoft. "Our Inclusive Design Principles." *Microsoft*, undated. [https://www.microsoft.com/design/inclusive/]

Pabini Gabriel-Petit. "Color Theory for Digital Displays: A Quick Reference: Part I." *UXmatters*, January 23, 2006. [https://www.uxmatters.com/mt/archives/2006/01/color-theory-for-digital-displays-a-quick-reference-part-i.php]

Pabini Gabriel-Petit. "Color Theory for Digital Displays: A Quick Reference: Part II." *UXmatters*, January 23, 2006. [https://www.uxmatters.com/mt/archives/2006/01/color-theory-for-digital-displays-a-quick-reference-part-ii.php]

Pabini Gabriel-Petit. "Applying Color Theory to Digital Displays." *UXmatters*, January 23, 2006. [https://www.uxmatters.com/mt/archives/2007/01/applying-color-theory-to-digital-displays.php]

Pabini Gabriel-Petit. "Ensuring Accessibility for People With Color-Deficient Vision." *UXmatters*, January 23, 2006. [https://www.uxmatters.com/mt/archives/2007/02/ensuring-accessibility-for-people-with-color-deficient-vision.php]

Usability.gov. "Accessibility Basics." *Usability.gov*, undated. [https://www.uxmatters.com/mt/archives/2007/02/ensuring-accessibility-for-people-with-color-deficient-vision.php]

W3C. "Accessibility Principles." *W3C*, May 1, 2019. [https://www.w3.org/WAI/fundamentals/accessibility-principles/]

W3C. "Web Content Accessibility Guidelines (WCAG) 2.0." *W3C*, June 2010. [https://www.w3.org/WAI/GL/WCAG20/]

4

Structural Patterns and Organization Schemes

This chapter provides some additional foundational knowledge that is essential to designing effective information architectures for digital information spaces, focusing on some of the common ways in which you can structure and organize such spaces.

The electronic *page* is the basic unit of content on any digital information space. Because today's software lets organizations create digital content independently of its presentation, organizations can publish the same content through a variety of media such as Web pages, mobile-app pages, ebook pages, or PDF documents. Such pages, or content objects, are the basic building blocks of any digital information space.

To make it easy for people to find the pages they need, information architects employ some common structural patterns that give information spaces a coherent, navigable structure, as well as familiar schemes for organizing information in a logical, easily comprehensible way.

The *organizational model* for an information space consists of its structural pattern and organization scheme—plus any classification scheme that the structural pattern employs, as we'll see in Chapter 7, *Classifying Information*. This model determines how people perceive and understand that information space, establishes the context in which its information exists, and sets people's expectations regarding what information they're likely to find there. [1]

In this chapter, we'll explore the following key topics:

- **Structural patterns**—These patterns define the types of structural relationships that can exist between groups of related content objects. You can employ common structural patterns, or *models*—alone or in combination—to provide an intelligible framework for the pages of a digital information space. We'll look at each of these patterns in some depth, and you'll learn how best to structure particular types of digital information spaces.

- **Organization schemes**—These schemes employ some abstract, shared characteristics of content objects in creating logical groupings of objects. You can use exact, ambiguous, or hybrid schemes in organizing the content of a digital information space. [2] You'll learn about these different types of organization schemes, the schemes that belong to each type, their characteristics, and some typical applications of each organization scheme.

- **LATCH**—Otherwise known as the *five hat racks*, LATCH is Richard Saul Wurman's model for organizing information and stands for *Location, Alphabet, Time, Category,* and *Hierarchy.* I'll briefly define and consider the usefulness of this model. [3]

Let's first look at some common structural patterns for digital information spaces.

Structural patterns

What is a *structural pattern*? A solution that information architects can apply to a particular design problem they frequently encounter in endeavoring to define a coherent, navigable information architecture for the pages of a digital information space. Structural patterns derive from common information-architecture solutions.

The structural patterns you employ provide the foundation for a digital information space's information architecture. When you're designing an information architecture, you should choose the structural pattern or a combination of patterns that is most suitable for an information space's content and best meets the needs of users. Particular sections of an information space might benefit from the use of different structural patterns. The primary impact of the structural patterns you use is that they determine how users can navigate.

We'll now consider the following structural patterns, which you can apply to your information–architecture projects:

- Hierarchy, or taxonomy
- Relational database:
  - Metadata
  - Faceted metadata
- Hypertext
- Linear sequence
- Hub and spoke
- Matrix
- Hybrid structures

As we look at these structural patterns, you'll learn about their characteristics, strengths, and weaknesses, as well as their usefulness for particular types of content on digital information spaces.

Hierarchy, or taxonomy

A *hierarchy*, or *taxonomy*, is a top-down information architecture—with an information space's home page at the top—and comprises content objects belonging to categories that have logical parent–child relationships. Within a hierarchy, higher-level categories are the parents of related, lower-level, child categories. These categories progress from the general to the specific.

The hierarchical pattern is the most common structural pattern for digital information spaces, as well as for a variety of real-world organizational systems, so hierarchies feel familiar to people and help them find useful information. When hierarchies are well designed, they clearly convey a person's current location within an information space and, thus, are easy to navigate and learn—even when people don't know what information they're looking for. However, creating a complex hierarchy can be challenging.

Because the hierarchy is by far the most common structural pattern for information spaces, considering whether you should create a hierarchy is a logical starting point when you're creating a new information architecture. By determining the highest-level categories of content in a hierarchy, you can quickly understand the overall scope of an information space.

Using a hierarchy as the overall structure for an entire digital information space can work well for everything from small information spaces that comprise just a few levels to large, content-rich information spaces that comprise many levels and a broad variety of content. However, hierarchies are not infinitely scalable. Try to limit the number of categories at one level to fewer than fifty. If a category is too large, divide it into subcategories. If there are too many categories, combine similar categories.

As the information within a large hierarchy becomes increasingly heterogeneous, the hierarchy can become more difficult to maintain. Inconsistencies in the categorization of information can cause a hierarchy to lose its coherence and impair its usability.

A hierarchy is also an effective structure for information spaces that provide content at various levels of detail or complexity—for example, a landing page that previews several topics and detail pages that cover each topic.

While some information spaces comprise content objects that let an information architect create a strict hierarchy, those comprising content objects with diverse interrelationships require greater flexibility. That's where polyhierarchies come in. We'll look at both of these types of hierarchies next.

Strict hierarchies

In designing a strictly hierarchical information architecture, your goal is to create mutually exclusive categories of content that do *not* overlap with one another. In a *strict hierarchy*, each category of information exists in just one place in the hierarchy, and each page of content belongs to just one category, as shown in Figure 4.1. However, it's not always possible to adhere to a strict hierarchy. Maintaining a strict hierarchy is especially difficult for ambiguous organization schemes, which I'll discuss later in this chapter.

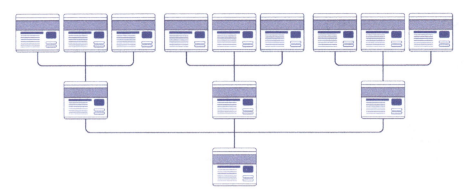

Figure 4.1—A strict hierarchy

Polyhierarchies

The inherent ambiguity of language and, thus, the classification schemes that we use in categorizing content sometimes necessitates placing content in more than one category—perhaps even several categories. In a *polyhierarchical taxonomy*, a page of content can belong to multiple categories, as shown in Figure 4.2, and thus have multiple parent pages. Polyhierarchy makes it easier for people to find information wherever they expect to find it. For large digital information spaces, some degree of polyhierarchy is inevitable. However, avoid placing the same page of content in so many categories that you make it difficult for people to build a mental model of the hierarchy's structure.

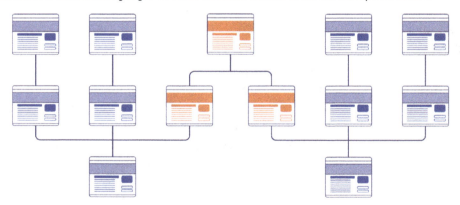

Figure 4.2—A polyhierarchy

One advantage of digital versus physical information spaces is our ability to create the illusion that a page of content exists in more than one place at a time within a digital information space. The classification scheme I use for the topics of articles on my Web magazine, *UXmatters*, is a polyhierarchy, which lets me assign multiple categories to an article if appropriate and, thus, allows readers to find articles under multiple topics. Figure 4.3 shows the Topics page on *UXmatters*.

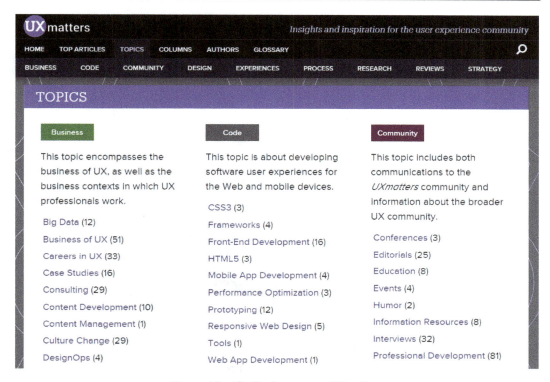

Figure 4.3—The Topics page on *UXmatters*

Balancing breadth and depth

When you're designing a hierarchical structure, it is important to balance its *breadth*—the number of items at each level—and its *depth*—the number of levels in the overall hierarchy. Creating a hierarchy that is too broad and presents too many choices to take in visually at once can make people feel overwhelmed. Figure 4.4 shows a hierarchy that is overly broad and lacks depth. In contrast, a hierarchy that is too deep requires users to click through too many levels to find the content they need. However, if there is good information scent at each decision point, people generally persist in seeking the information they need rather than giving up. Figure 4.5 shows a hierarchy that is very narrow and deep.

Figure 4.4—A hierarchy that is broad and shallow

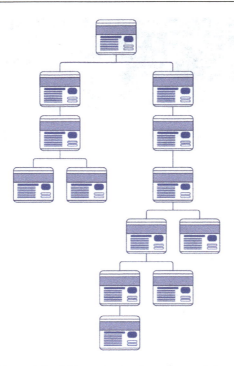

Figure 4.5—A hierarchy that is narrow and deep

For optimal usability and user outcomes, balance the breadth and depth of your information space's hierarchy while ensuring that you create coherent groups of categories at each level.

If you're designing a new digital information space that you expect to grow significantly, err on the side of breadth to ensure that the hierarchy you create is sustainable over time and won't require restructuring.

If you need to provide different points of entry for specific audiences, you'll likely create a fairly narrow hierarchy. However, each audience branch in such a hierarchy could broaden significantly to address all the needs of that specific audience.

Relational database

Relational database–management systems (RDMSs) provide the underlying technology for most content-management systems (CMSs) that support large-scale, homogeneous digital information spaces. Relational databases function as their content repositories. A *relational database* comprises a collection of records, each of which consists of the same fields. Each record in a relational database for a CMS is a content object whose fields contain content components or metadata for the content object. The records in a CMS's database define the structure of the information space it supports. The consistent structure of the records in a database facilitates fast, efficient navigation, search, and retrieval, as well as the reuse of content components on different pages.

A relational database's content model, or *schema*, defines the collection of entities that reside in the database and the relationships between them. These entities take the form of tables, in which each row represents a *record*—that is, an instance of that type of entity, with a unique key identifying that row—and each column represents a *field*, or attribute, and has a predefined data type. A row in one table can be linked to a row in another table by adding a column that contains a *foreign key*—the unique key of the linked row. [4, 5]

The database is the most effective structural pattern for creating scalable digital information spaces that comprise somewhat homogeneous content, for which it is desirable to display the same content components in different ways on different pages. For example, similar information might appear on both the search results and product pages of an ecommerce Web site or mobile app.

The database structural pattern also makes it easier for people to find the information they need in different ways. Figure 4.6 depicts the database structural pattern, which can also be useful for smaller information spaces such as Web magazines and blogs. My Web magazine, *UXmatters*, employs the database structural pattern.

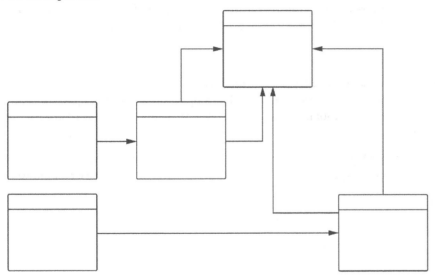

Figure 4.6—Example of the database structural pattern

While the database structural pattern offers many benefits, it also has some limitations. In a database, there must be at least one field for which each of the records contains a clearly distinguishable value—for example, the title of an article. A database is *not* an appropriate structure for heterogeneous data that lacks uniformity. For people to be able to navigate or search successfully, a database must comprise known items of information.

A CMS's database may contain either conventional or faceted metadata. Let's briefly look at these two types of metadata.

Metadata

Metadata is information about other data that enables you to understand and use that data. For each record, or content object, in a CMS, there are three types of metadata:

1. **Structural metadata**—This type of metadata determines the discrete content components, or chunks of information, that a content object comprises and, thus, what appears on a content page. The CMS database for *UXmatters*, for example, includes fields for the title and the three main content components of each article: the excerpt field, whose content appears both on the home page and at the top of an article; then the body and extended fields, which contain the remainder of the article.

2. **Descriptive metadata**—This type of metadata describes each content object; enables the generation of lists of related content objects; and facilitates navigation, search and retrieval, and filtering. For example, on *UXmatters*, the settings for a categories field determine one or more topics to which an article belongs.

3. **Administrative metadata**—This type of metadata provides the business context for a content object. On *UXmatters*, there are status, publication date and time, author, and coauthor fields. [6]

Faceted metadata

Facets are independent sets of flat or hierarchical categories that you can use to describe the same content object from different perspectives. *Faceted metadata* is descriptive metadata that employs facets. It supports much greater flexibility in navigation, filtering, and search systems and, thus, offers many ways of finding the same content.

A common example of a type of Web site that takes advantage of the power of faceted metadata is a recipe Web site. Such a site might include the following facets, each of which would have multiple values:

- Meal
- Course
- Type of dish
- Cooking method
- Main ingredient
- Dietary restriction
- Cuisine
- Holiday

Hypertext

A *hypertext* information space has no overarching structure. Associative links between content objects, or pages, define the *only* relationships between those pages. Therefore, defining *meaningful* associative links between each new page and other pages is essential in creating a hypertext system. Together, the links connecting pages form a web, as shown in Figure 4.7.

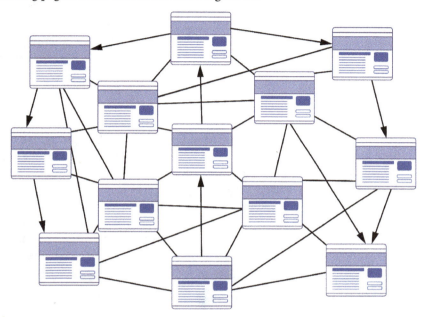

Figure 4.7—A hypertext

The central characteristic of a hypertext is its flexibility and scalability, which can be both a strength and a weakness. A hypertext structure might be a logical choice if an organization doesn't know exactly what content they'll create over time and consequently doesn't know what structure to create. However, the lack of a clear structure can be confusing to people—especially in a large, complex hypertext system—making it difficult for them to find the information they need and potentially resulting in their becoming lost in hyperspace.

People's ability to navigate a hypertext depends on the content creators' embedding useful associative links within their content and, thus, relating the content objects they create to other content objects. A hypertext link can take the form of link text, a button, or an image such as an icon. Link text can comprise individual words, phrases, or the titles of articles—depending on the space available—as can alt text. Whether a link consists of text, a button, or an image, it should clearly convey the link's meaning, appear clickable, and attract sufficient attention.

A *hypermedia* system comprises content objects that may include text, image, video, and audio objects, as well as other types of data.

While it is possible to use associative hypertext links to form a hierarchy, relying primarily on the hypertext structural pattern is unlikely to result in a successful overall structure for an information space. Because hypertexts do not convey a clear sense of the user's current location within an information space, it is difficult for users to find their way back to the same content. Therefore, in most cases, the use of embedded hypertext links merely supplements the primary structural pattern of an information space.

Wikis provide excellent examples of the hypertext structural pattern. They do not have a predefined structure, but instead grow organically. Organizational knowledge bases offer other great examples of hypertexts. It is difficult to predict in advance how the content in a knowledge base might evolve over time.

Linear sequence

A *linear sequence* is a path that comprises an ordered series of procedural steps, content objects, or pages that users *must* view in sequence and usually has a fairly limited scope. Figure 4.8 depicts the linear-sequence structural pattern. Create a linear sequence when you need to guide users through a series of steps—whether a user must complete a step in a task before progressing to the next step or must understand certain information before reading the next information. However, beware of using this structural pattern unless users must generally adhere to a sequence. Otherwise, the constraints of this pattern would frustrate users unnecessarily. As a result of its constraints, this is not a common pattern.

Figure 4.8—A linear sequence

Employing the linear-sequence structural pattern is analogous to defining a well-structured *path* in wayfinding. According to Mark Foltz, who wrote about wayfinding in physical spaces, a well-structured path is continuous and has directionality, with a clear beginning, middle, and end. The beginning of the path should provide an introduction; the end, a conclusion. A well-structured path enables the people traversing it to maintain their orientation in relation to the next signpost or landmark along the path, as well as toward the ultimate destination. The path's wayfinding stages should always be apparent—no matter where users' are on the path—enabling them to assess both their progress along the path and the distance to their destination. [7] All of these requirements for a well-defined physical path apply equally well to designing any well-defined linear sequence of pages.

Some examples of the linear-sequence structural pattern include online surveys, *wizards* that guide users through step-by-step procedures, online tests, instructional Help, tours, and timelines. Site search is also a linear sequence: the user must enter a search query to view one or more pages of search results.

Some design guidelines for designing a well-structured linear sequence of pages include the following:

- Create a page title that clearly describes the overall sequence and use the same title at the top of all pages in the sequence. This page title should reflect the text of the link or the alt text for an image link that the user clicked to view the sequence.

- On each page in the sequence, provide a separate subtitle immediately above the content object on that page. This subtitle should clearly describe the current step in the sequence.

- Consider providing an overview page at the beginning of the sequence to introduce the purpose of a complex sequence.

- Display a list of *all* of the steps in a sequence.

- Allow the user to navigate the steps of the sequence by clicking the next step in the list of steps *or* clicking a **Next** button. If the user is on the final page of the sequence, the **Next** button should appear dimmed.

- Show the user's progress by clearly indicating the steps the user has already completed.

- Clearly indicate the user's current step in the list of steps.

- If allowing the user to navigate back to previously completed steps would be helpful in the context, consider letting the user go back by clicking a previous step in the list of steps *or* by clicking a **Previous** or **Back** button. If the user is on the first page of the sequence, this button should appear dimmed.

- Provide a **Cancel** button to allow the user to end the sequence without completing all the steps, discard any user input, and return to the page from which the user began the sequence.

- For a sequence that allows the user to input data, provide a **Finish** button so the user can complete the sequence, submit the data, and return to the page from which the user began the sequence.

- For a sequence that does *not* allow the user to input data, provide a **Close** button to let the user end the sequence at any time, with or without completing it, and return to the page from which the user began the sequence.

- Clearly indicate the final step in the sequence.

- Consider providing a summary page at the end of the sequence. [8]

There are several variants of the linear-sequence structural pattern, as follows:

- **A linear sequence with alternative paths**—This pattern, shown in Figure 4.9, provides some flexibility by offering alternative paths, or branches—one of which the user must follow—then continues along the main path of the sequence.

- **A linear sequence with options**—This pattern, shown in Figure 4.10, provides access to optional information, then returns the user to the main path of the sequence.

- **A linear sequence with backtracking**—This pattern, shown in Figure 4.11, lets the user repeat part of the sequence.

- **A linear sequence with shortcuts**—This pattern, shown in Figure 4.12, lets the user skip certain steps in the sequence.

- **A circular linear sequence**—This pattern, shown in Figure 4.13, lets the user view the same sequence of information repeatedly, following the same path. It is also useful for sequences that have no clear beginning or end. [9]

Figure 4.9—A linear sequence with alternative paths

Figure 4.10—A linear sequence with options

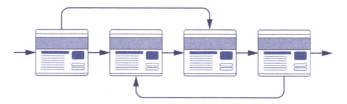

Figure 4.11—A linear sequence with backtracking

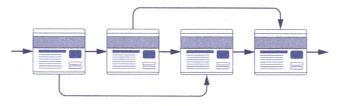

Figure 4.12—A linear sequence with shortcuts

Figure 4.13—A circular linear sequence

Hub and spoke

In the *hub-and-spoke* structural pattern, a central *hub*, or landing page, is the single entry point from which the user initiates navigation to individual pages that are just one step away—the *spokes*—from which the user then returns to the hub. This pattern reflects a common Web-browsing behavior in which users click a link on a page, view the content on the link's destination page, then return to the original page to click one or more additional links. [10] Figure 4.14 shows a simple hub-and-spoke structure.

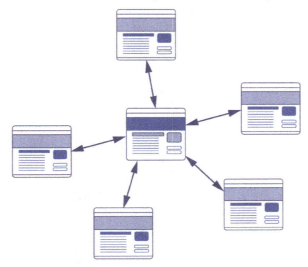

Figure 4.14—A simple hub-and-spoke structure

While hub and spoke is a very common structural pattern for mobile apps and is sometimes useful for Web and desktop applications, [11] its use for digital information spaces is less common. However, this pattern can be useful for information spaces that comprise discrete content objects. You can use a hub-and-spoke structure for an entire information space—perhaps one that has a fairly limited scope or consists of different sections—or for just one or more individual sections of a larger space. Some hub-and-spoke structures have a large number of spokes, while others have just a few. In more complex hub-and-spoke structures, each spoke radiating from a hub might itself have a hub-and-spoke structure, as shown in Figure 4.15. The overall structure of an information space could consist of a series of hubs, as shown in Figure 4.16.

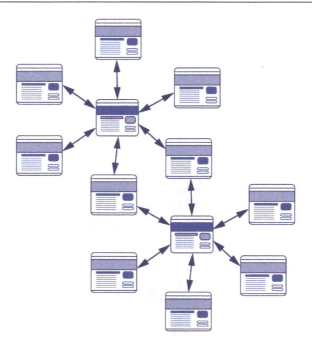

Figure 4.15—A complex hub-and-spoke structure comprising multiple hubs

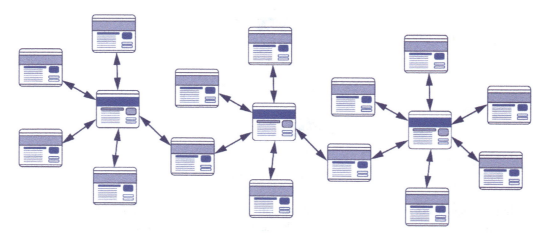

Figure 4.16—An information space that comprises a series of hubs

Within an information space with a hub-and-spoke structure, a hub commonly provides a list of content objects, while each spoke is a page that displays an individual content object. For example, a hub might comprise a list of articles, while each of the content pages displays an article. Links do *not* generally connect these content pages to one another. Instead, the user returns to the hub to choose another article.

In a hub-and-spoke structure comprising several sections or topics, each spoke might be a hub for an individual section or topic, and each of the spokes for a section or topic, a content page or article belonging to that section or topic.

Matrix

In *The Elements of User Experience*, Jesse James Garrett wrote about the *matrix* structural pattern shown in Figure 4.17, which lets users move from page to page along two or more dimensions.

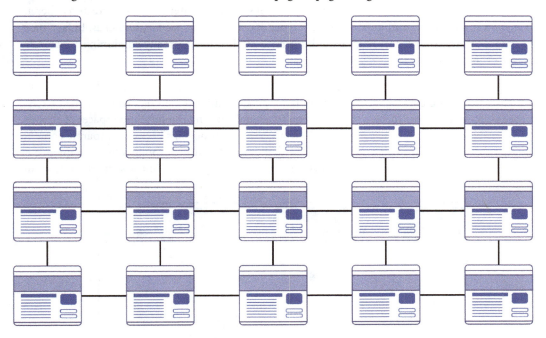

Figure 4.17—A 3D-matrix structural pattern

In the 3D matrix shown in Figure 4.17, the user can move in three dimensions, as follows:

1. **Horizontally**—from side to side
2. **Vertically**—up and down
3. **Laterally**—backward and forward

You can provide ways for users with different needs to navigate an information space's content by creating a matrix structure and associating each axis of the matrix with a particular user need. A matrix lets users look at the same content from different perspectives. Thus, the matrix structure fulfills a very similar purpose to a database structure that uses faceted metadata.

Garrett cautioned that using a matrix of more than three dimensions for an information space's primary navigation system would present problems for users. [12] However, if each dimension of a matrix addresses the needs of a different audience, you can use a matrix to create audience-focused points of entry to an information space.

Hybrid structures

Hybrid structures integrate or combine multiple, complementary structural patterns to create more complex information architectures. By using a variety of structural patterns, you can create a coherent information space that accommodates different types of content and specific user needs. In fact, most large digital information spaces employ multiple structural patterns—many combining a hierarchy with one or more other structural patterns.

For example, while the overarching structure of most information spaces is hierarchical, a space might also include sections comprising collections of homogeneous information that would benefit from a database structural pattern. This is a very common structural pattern for information spaces of all sizes. Most information spaces also use hypertext links to create relationships between disparate content. Plus, an information space might include a tour that guides the user through a linear sequence of pages. Figure 4.18 shows an example of a hybrid structure, comprising a linear sequence and a hub and spokes.

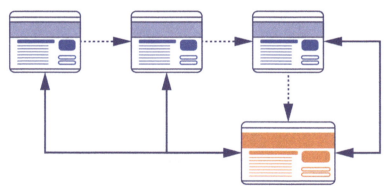

Figure 4.18—A hybrid structure

When designing an information architecture, you must consider an information space's structure and its organization scheme in tandem. Now that you've learned about the most common structural patterns for digital information spaces, let's look at another key component of an organizational model: the organization scheme.

Organization schemes

The organization schemes that we'll consider in this section let you create logical groupings of content objects according to their shared characteristics.

In their landmark book *Information Architecture for the World Wide Web*, whose first edition came out in 1998, Peter Morville and Lou Rosenfeld devised a very useful way of classifying different types of organization schemes based on their studies in library information science (LIS), as follows:

- Exact schemes

- Ambiguous schemes

- Hybrid schemes [13]

Let's explore these types of organization schemes, what schemes belong to each type, their characteristics, and their typical applications. I'll provide some examples from my Web magazine *UXmatters*.

Exact schemes

Exact, or objective, organization schemes result in information spaces that comprise discrete, mutually exclusive categories of information. These organization schemes are useful primarily for information spaces that comprise fairly homogeneous content. Both exact organization schemes and their categories are generally well known and in common use, and the boundaries between these categories are generally clear. Therefore, exact organization schemes are easy to design and maintain. Applying such categories to content objects is also relatively easy because each decision about what category to choose is an objective decision.

Even in the few cases where the definitions of the boundaries between exact categories might be subjective, once these categories are clearly defined, they become easy to apply. For example, an alphabetical organization scheme might require finer divisions of the alphabet than whole letters to prevent certain pages from becoming excessively long *or* one page might comprehend entries for several letters of the alphabet. Once the people who use an information space understand the categories that you've defined, it will be easy for them to find the information they need.

Key to deciding whether to use an exact organization scheme and determining which organization scheme to use is conducting user research to learn what organization scheme would best help people find the information they need.

One limitation of exact organization schemes is that they require users to know what information they're looking for—that is, they support only *known-item* information seeking. Exact organization schemes include the following:

- Alphabetical

- Numerical

- Chronological

- Geographical

- Spatial

- Organizational
- By media

Alphabetical

An *alphabetical*, or A-to-Z, organization scheme provides a familiar, well-defined structure. However, it does not convey any meaningful information about the relationships between the content objects within an information space. It just breaks down a large collection of information into smaller groups. To use an alphabetical organization scheme, people must know exactly what they're looking for, as well as the proper terms for items, including how to spell them. Therefore, alphabetization is useful only when people know the names of items. But alphabetical lists enable people to quickly find the items they need. Another limitation of alphabetical schemes is that they don't scale well. Long alphabetical lists are hard to use. Whenever possible, keep an entire alphabetical list on one page.

Dictionaries, thesauri, glossaries, and encyclopedias; phone, student, customer, and employee directories; and contact-management apps use alphabetical organization as their primary organization scheme. However, in most cases, alphabetical organization is more useful as a secondary organization scheme. For example, as Figure 4.19 shows, the *Authors* page on *UXmatters* lists authors alphabetically by last name. An alphabetical index at the top of the page lets users click a letter to scroll automatically to the items that begin with that letter.

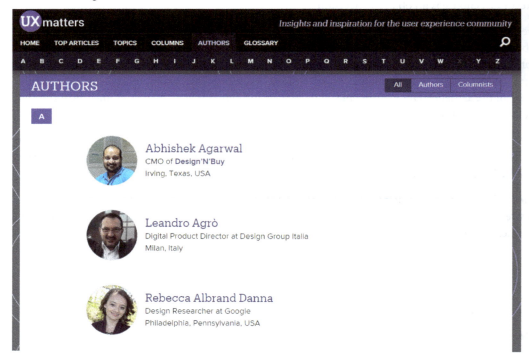

Figure 4.19—An alphabetical organization scheme

Numerical

In a *numerical* organization scheme, categories are based on meaningful ordinal or cardinal numbers. These numbers might represent ratings or popularity—as on the *Top Articles on UXmatters* page, which is shown in Figure 4.20—individual values, or ranges of values. In most cases, numerical organization schemes provide a secondary means of structuring a digital information space.

Figure 4.20—A numerical organization scheme

Even though a numerical organization scheme is an exact scheme, choosing the values to use as categories and defining the boundaries between them is entirely subjective and can be challenging. The boundaries that you define for categories have a significant impact on the meaning of the categories. But once you've defined the categories and their boundaries, it is easy to apply the categories to content objects. It is also easy to find information using these categories.

Chronological

Chronological organization schemes are appropriate for digital information spaces whose primary organizing principle is time—whether in the past, present, or future. There are several useful ways of displaying information chronologically.

The content objects in online publications such as Web magazines, blogs, and news sites appear either in chronological order, by date and time, or in reverse chronological order, with the most recent content object first. The recency of an article's publication is often important. On my Web magazine *UXmatters*, articles appear in reverse chronological order on the home page, as well as on the pages that display recent editions, as shown in Figure 4.21.

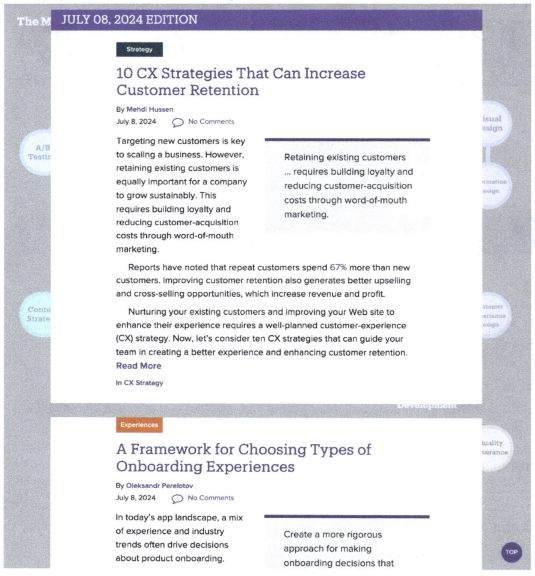

Figure 4.21—A chronological organization scheme

Chronological schemes would also be useful for organizing historical events on a history Web site, corporate histories of events or press releases, TV guides, listings of future events, diaries, or any other information for which date and time are key elements of the content.

Events appear on calendars or in chronological lists. However, very long lists of events are hard to use, so try to limit them to fifty or fewer items. Alternatively, you can break up such lists into labeled spans of time—for example, by eras, centuries, decades, months, weeks, or milestones.

Another way of organizing information chronologically is to display events on a timeline—for which you must determine the appropriate spans of time. These timespans constitute the categories on a timeline and defining them can be challenging, as can depicting the passage of time consistently. You can also show the durations of events on a timeline.

Chronological organization schemes often work well as the primary way of organizing a digital information space, but they can also be useful as a secondary organization scheme.

Geographical

The basis of a *geographical* organization scheme is *locale*—for example, a place, town, city, county, state, province, country, administrative or geographic region, or continent—or *location*—that is, a position or site on a map. Because information spaces with geographical organization schemes reflect the actual world in which we live, they are generally easy to design and use. However, designing them becomes more difficult as their geographical scope and complexity increase. Trying to achieve consistency across all the nations of the world can present challenges. Plus, defining regions can be a rather subjective exercise.

Geographical organization schemes are broadly applicable. Some common examples of information spaces that use a geographical organization scheme include many that provide tourism, travel, weather, event, and real-estate information, as well as geographical survey sites such as the USGS. Location-aware apps on our mobile phones now push location-based information to us. Using geographical organization schemes can provide effective primary or secondary ways of organizing a digital information space.

Many apps that employ a geographical organization scheme display data on a map—for example, directories of local stores, restaurants, hotels, and services such as Yelp. While map user interfaces are very useful to people who are familiar with particular geographical areas, using a map is difficult for someone who lacks geographical knowledge. As Figure 4.22 shows, the Zillow iPad app provides an excellent example of the use of a geographical organization scheme and displays a map as part of its primary user interface, which lets users display *and* navigate real-estate data.

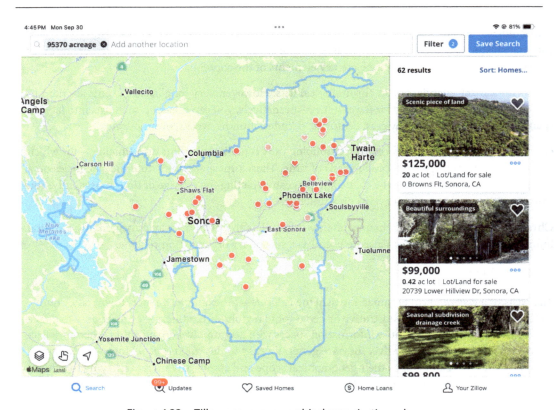

Figure 4.22—Zillow uses a geographical organization scheme

Spatial

The categories for *spatial* organization schemes are analogous either to the spaces within an actual built environment *or* the spaces that particular types of built environments share in common. Similar to geographical organization schemes, spatial organization schemes are modeled on real-world environments, so they are easy to design and use. However, a spatial organization scheme is rarely the primary organizational scheme for a digital information space. Examples of spatial organization schemes include a Web site whose structure is based on a museum's galleries or an ecommerce furniture store such as IKEA, which is organized by the rooms of a house, as shown in Figure 4.23.

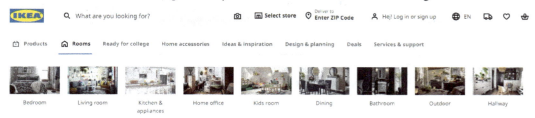

Figure 4.23—IKEA uses a spatial organization scheme

Organizational

An *organizational* scheme replicates the actual structure of a company, governmental body, university, or other organization. Modeling such schemes on an existing organization makes them easy to design. For people who are very familiar with an organization—for example, the users of its intranet—such an organizational scheme might also be easy to use. However, these organizational structures tend to be siloed, which can isolate information from many of the people who need to use it because they don't know who is responsible for creating it. On the other hand, managers and their teams know just where to look for the information they've authored.

However, if an organization structured its public-facing Web site primarily according to the organization's internal structure, customers would be confused because they wouldn't know enough about the organization to be able to find the information they need. Instead, the organization should conduct generative user research to determine what structure would make sense to customers. Still, an organizational scheme can occasionally be useful as a secondary means of organization, as on the *Colleges and schools* page on the University of Washington Web site, shown in Figure 4.24.

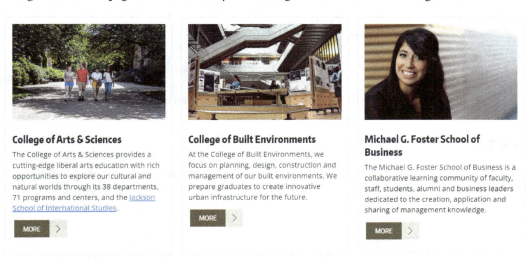

Figure 4.24—An organization scheme that reflects an organization's structure

By media

A *media-based* organization scheme structures information according to its file format. Digital-asset management (DAM) systems such as Adobe Experience Manager Assets, which is shown in Figure 4.25, epitomize the media-based organizational scheme that provides their primary structure. Many training services and publishers use a media-based organization scheme as a useful secondary means of structuring information—often using a filtering user interface to let users view only content in a particular format such as video. The primary organization scheme for such services is likely to be topical because users first decide what they want to learn about, then choose the type of media they want.

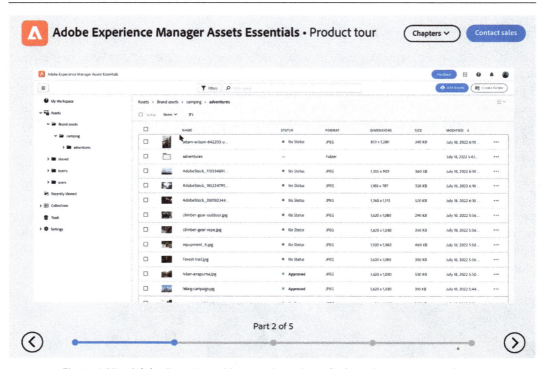

Figure 4.25—Adobe Experience Manager Assets's media-based organization scheme

Ambiguous schemes

Ambiguous, or subjective, organization schemes usually address unique requirements for digital information spaces that demand custom design solutions. Information architects tailor an ambiguous organization scheme to the content of a specific information space and the needs of its audience. As a result, ambiguous organization schemes are usually more useful to people than exact organization schemes. However, information architects must define the categories that make up ambiguous organization schemes, which requires them to make challenging, subjective decisions. Because ambiguous organization schemes are difficult to design *and* maintain, creating sustainable organization schemes requires the skills of a professional information architect.

The ambiguity of language and the subjectivity of people's differing perspectives on both the meaning of language and what constitutes a logical organization scheme can result in some ill-defined categories with unclear boundaries. Therefore, applying the most appropriate categories to content objects also requires challenging, subjective decisions. Some content could fit into two or more different categories, while other content would not fit well in any category, making it difficult for users to find that content. Plus, as time goes by, creating new content would likely necessitate creating some new categories.

Ambiguous organization schemes are especially useful for heterogeneous information spaces that comprise widely varied content. They are in prevalent use across diverse information spaces. One important reason for their broad use is their familiarity. People frequently use them in the real world and sometimes even create their own subjective schemes. Another reason is that, for people to use ambiguous organization schemes successfully, they need *not* know exactly what information they're looking for or how to describe it.

As Morville and Rosenfeld said in *Information Architecture for the World Wide Web*: "Information seeking is often iterative and interactive. What you find at the beginning of your search may influence what you look for and find later in your search. This information-seeking process can involve a wonderful element of associative learning. … Ambiguous organization supports this serendipitous mode of information seeking by grouping items in intellectually meaningful ways."

They make a strong case for the general superiority of ambiguous organization schemes over exact organization schemes. Common ambiguous organization schemes include the following:

- Topical
- Categorical
- Audience-specific
- Task-oriented
- Sequential
- Metaphorical

Topical

Most large digital information spaces use a *topical* organization scheme to organize similar content objects by subject matter. While this is the most common and broadly useful of the organization schemes, it is often just one of several organization schemes an information space employs. Topical organization schemes provide optimal support for exploratory information seeking, or browsing, because they do not require that people know precisely what they're looking for.

Devising an optimal topical organization scheme for an information space that can address both current and future content needs can be very challenging and involves making highly subjective decisions. The greater the breadth of an information space, the more challenging the task of breaking down a collection of content into logical groups, or topics, becomes. You must ensure comprehensive coverage of the content, determine the right granularity at which to define topics, and assign meaningful labels to topics that clearly distinguish them from all other topics.

Once you've defined the topics, you must assign the appropriate topic or topics to each content object. This can be difficult when content covers multiple topics. While you can employ polyhierarchy to solve this problem, you should avoid its overuse. Otherwise, the use of topics on an information space can lack clarity and consistency.

Topical organization schemes are in widespread use, across disparate types of digital information spaces. A few examples of digital information spaces that employ a topical organization scheme include corporate Web sites, training Web sites, business directories that are organized by types of businesses, newspapers, and Web magazines. The *Topics* page on *UXmatters*, shown in Figure 4.26, lists the topics of the magazine's articles and, thus, conveys the scope of the magazine's content.

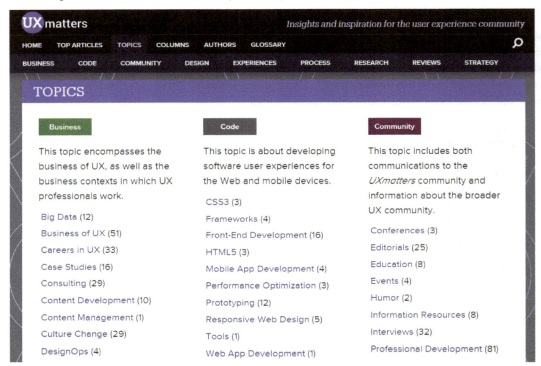

Figure 4.26—The topical organization scheme on *UXmatters*

Categorical

A *categorical* organization scheme is very similar to a topical scheme. But while topics can be numerous and quite fine grained, categories are usually few in number and tend to be domain specific. For example, on an ecommerce store such as Wayfair, people can shop by department, as shown in Figure 4.27.

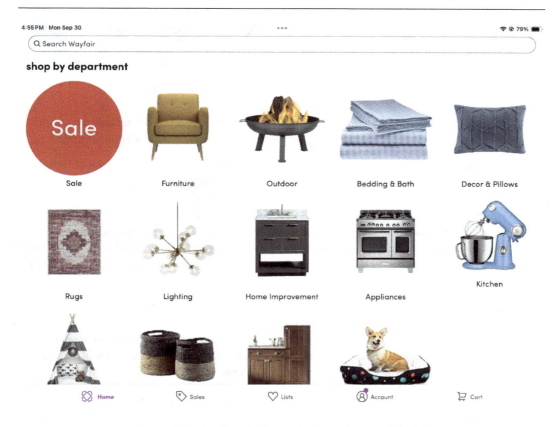

Figure 4.27—A categorical organization scheme on Wayfair

Other common examples of categorical organization schemes include product categories and genres of music, films, books, and games.

Categorical organization schemes are usually fairly easy to design and use because of their limited scope and the fact that they often employ existing, domain-specific categories or are modeled on real-world structures.

Audience-specific

Digital information spaces—typically Web sites, intranets, and extranets—that serve several clearly defined audiences can employ an *audience-specific* organization scheme to establish their primary structure. For example, the Web site of the healthcare-insurance provider Anthem uses an audience-specific organization scheme and comprises information that targets the following audiences: **For employers**, **For producers**, and **For suppliers**, as shown in Figure 4.28. As is typical, this site's secondary organization scheme is topical.

Figure 4.28—Anthem uses an audience-specific organization scheme

Avoid mixing audience segments with other navigation links, as Anthem's site unfortunately does, which could cause confusion regarding whether, for example, a section is *for* providers or *about* providers.

Audience-specific organization schemes break up an information space into several distinct sections according to audience segments—targeting as few as two or as many as several different audiences, each of which must be clearly labeled. Audience-specific organization schemes let you customize each section's navigation system and collection of content for a specific audience. Decomposing an information space into audience-specific sections can be quite challenging because it's necessary to include *all* of the content an audience would need and *none* of the content that wouldn't be relevant to them. Creating an audience-specific organization scheme is not appropriate if much of an information space's content would be of interest to multiple audience segments.

If people can easily identify the audience to which they belong, using an audience-specific organization scheme enables people to easily distinguish the content that addresses their needs. Once they choose an audience link on the primary navigation bar, they'll see only content that might be relevant to them. Thus, information spaces that employ audience-specific organization schemes are usually easy to use—as long as you include *all* of the information a particular audience needs.

Audience-specific sections of many information spaces are *open* to everyone, so people belonging to one audience can view the information for all other audiences. However, if you need to limit access to audience-specific sections of an information space—either to protect sensitive information or because viewing the information requires a subscription—you can create a *closed*, audience-specific organization scheme. To access the information in a closed, audience-specific section of an information space, people must first become a member of information space, then as a member of that audience, sign in to access the section for that audience. Becoming a member of an information space might or might not require paying for a subscription.

Task-oriented

Task-oriented organization schemes have the tasks that people perform as their basis. Thus, common examples of task-oriented organization schemes include digital information spaces that comprise instructional materials of various kinds—such as tutorials, Help systems for software products, and product-support sites. Sometimes, the same information is useful within the context of different tasks. Creating a polyhierarchy lets you connect supporting information to all of the tasks for which it is relevant.

Companies that produce a single product might use a task-oriented scheme as the primary organization scheme for Help—for example, Figure 4.29 shows Outreach's Help topics on their *Support* page. However, companies offering multiple products would probably use a topical scheme that organizes content objects by products or services as the primary organization scheme for Help and a task-oriented scheme as its secondary organization scheme, as on Apple's *Support* pages, shown in Figures 4.30 and 4.31

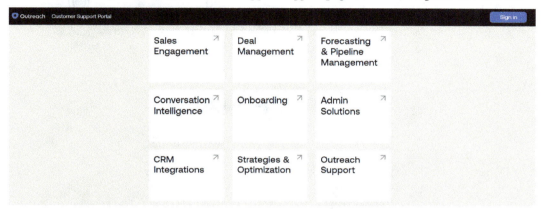

Figure 4.29—Categories of Help topics on Outreach's task-oriented Support page

Figure 4.30—Apple's topical Support page, which is organized by product

iPhone Support

[Forgot passcode ›](#)

Forgot Apple Account
password ›

Subscriptions and
refunds ›

Learn more about
Find My ›

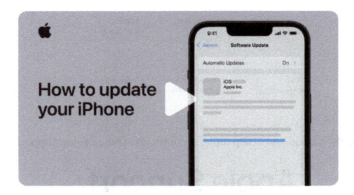

Figure 4.31—Apple's task-oriented *iPhone Support* page

The broader the scope of a digital information space and the more varied the tasks that its information supports, the more challenging it becomes to create a task-oriented structure. You must create logical task groups that collectively comprehend the full range of tasks.

Metaphorical

In *metaphorical* organization schemes, a real-world metaphor provides the organizing principle for a digital information space. The use of metaphorical organization schemes is rare and is *not* generally recommended. Their most common use is for simple digital information spaces of limited scope, for young children who cannot read.

For a metaphorical organization scheme to be successful, its metaphor must be broadly familiar to people. It is very difficult to identify and implement a robust metaphor that everyone would interpret in the same way. The larger the information space, the greater the likelihood that a metaphorical organization scheme would not be sustainable. Some information simply would not fit the metaphor—particularly that regarding abstract ideas—so would introduce inconsistencies. Therefore, it's best to avoid using them.

Hybrid schemes

Hybrid organization schemes employ multiple complementary organization schemes that allow users to find various types of information in different ways. In a digital information space that comprises a variety of information, it's not always possible to make a single organization scheme serve the requirements of presenting disparate content or meet the needs of a diverse audience.

When creating a hybrid organization scheme, consider using different organization schemes at the different levels of a hierarchy—for example, as an information space's primary and secondary organization schemes. This is a very common approach. However, limit the use of hybrid schemes to shallow information spaces.

In some cases, information spaces employ different information schemes for different types of information, as I've demonstrated through several examples from *UXmatters*. Similarly, many Web and mobile applications provide a combination of functionality and information, so their primary navigation systems must successfully integrate topical and task-oriented organization schemes. This enables users to easily access both an application's functionality and its information, as in the LinkedIn iPad app shown in Figure 4.32. Such hybrid schemes can work well if the primary navigation bar includes only a limited number of options.

Figure 4.32—Combining topical and task-oriented organization schemes

In other cases, information spaces use hybrid schemes to organize the same information. *The United States House of Representatives* Web site provides an interesting example. **By State and District** is the primary organization scheme of the *Directory of Representatives* page, shown in Figure 4.33, which is a geographical organization scheme—although district is also a numerical organization scheme. The alternative organization scheme, **By Last Name**, is a topical organization scheme.

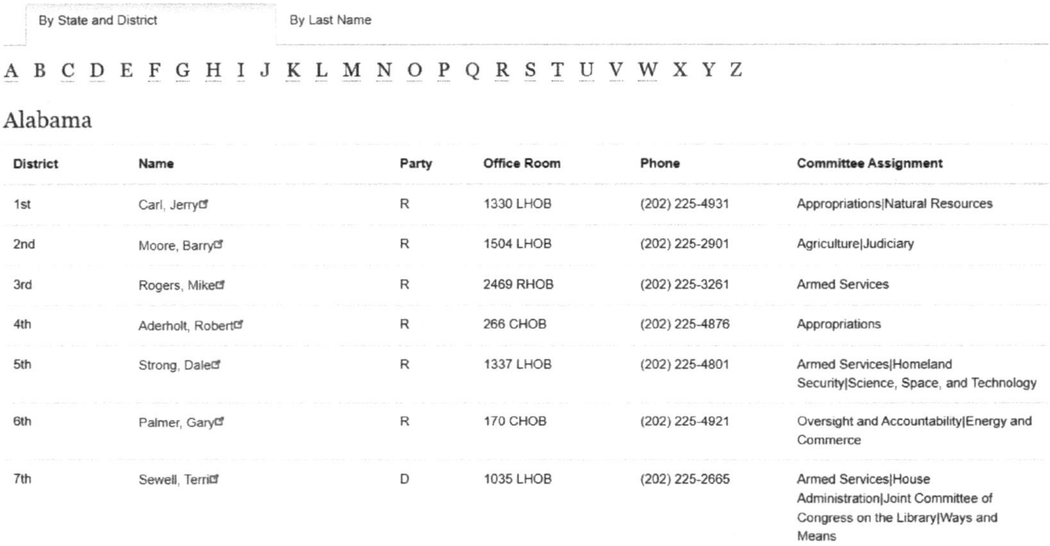

By State and District	By Last Name

A B C D E F G H I J K L M N O P Q R S T U V W X Y Z

Alabama

District	Name	Party	Office Room	Phone	Committee Assignment
1st	Carl, Jerry	R	1330 LHOB	(202) 225-4931	Appropriations|Natural Resources
2nd	Moore, Barry	R	1504 LHOB	(202) 225-2901	Agriculture|Judiciary
3rd	Rogers, Mike	R	2469 RHOB	(202) 225-3261	Armed Services
4th	Aderholt, Robert	R	266 CHOB	(202) 225-4876	Appropriations
5th	Strong, Dale	R	1337 LHOB	(202) 225-4801	Armed Services|Homeland Security|Science, Space, and Technology
6th	Palmer, Gary	R	170 CHOB	(202) 225-4921	Oversight and Accountability|Energy and Commerce
7th	Sewell, Terri	D	1035 LHOB	(202) 225-2665	Armed Services|House Administration|Joint Committee of Congress on the Library|Ways and Means

Figure 4.33—*The United States House of Representatives* Web site's *Directory of Representatives* page

While the hybrid scheme in the example in Figure 4.33 works well, be careful in creating hybrid organization schemes. Otherwise, you might create an unusable Frankenstein-monster mashup.

Despite the potential usefulness of hybrid organization schemes, you should always bear in mind that using a single organization scheme across an entire information space—at least in organizing its primary content—makes it easier for people visiting the information space to understand its content.

LATCH

In his book *Information Anxiety 2*, Richard Saul Wurman defined a model for organizing information, describing its modes of organization as the *five hat racks*. Wurman coined the acronym *LATCH* for the following five modes of organization, each of which is most appropriate for particular types of information:

1. **Location**—Organizing information by location lets users find information that is either from or relates to different geographical locales. Examples include atlases, tourism and travel Web sites, real estate apps, and location-aware mobile apps.

2. **Alphabet**—This mode is useful for organizing very large collections of information for which no other mode of organization is appropriate—for example, reference information such as dictionaries, encyclopedias, contact information, and telephone directories. Organizing information alphabetically is most useful when an audience comprises people who might *not* have the necessary knowledge to find information based on location or time or share the same understanding of different categories.

3. **Time**—This mode of organization helps users to find information that has its basis in a chronology of historical events, a schedule of current or future events, or information that is published on a schedule. It is especially useful for information that gets frequent updates. Examples include historical timelines, newspapers, TV guides, calendars, train timetables, email apps, and step-by-step instructions.

4. **Category**—This mode is useful for organizing items of information that are equally important and similar or related to one another. However, users must share a common understanding of the meanings of category labels and the scope of the information that different categories comprehend. Categorization is useful in organizing items such as products on ecommerce sites, genres of music or movies, and topics of books or articles.

5. **Hierarchy**—This mode of organization is useful in assigning value or importance to information—for example, in organizing items by order of importance, from most to least popular or expensive, or from largest to smallest. Examples include the most highly rated products on an ecommerce site, the cheapest flights on a travel site, and the most relevant search results in a search engine. [14]

In my view, LATCH is excessively prescriptive. Despite Wurman's statement to the contrary, there are more ways of organizing information than the five modes that he defined. While the organization modes location, alphabet, time, and category are clear, his description of a hierarchy actually sounds like an ordinal mode of organization. Wurman had originally used the term *continuum* for his fifth mode of organization, which makes much more sense, but changed it to *hierarchy* to enable him to use the acronym LATCH. [15] The term *hierarchy* is much more usefully applied in describing the most common structural pattern for digital information spaces. Using the term differently in the context of LATCH is confusing.

Summary

In this chapter, you learned about two key components of an organizational model for a digital information space. First, we looked at common *structural patterns* that information architects use in defining the structural relationships between groups of related content objects. You also learned how to create coherent, navigable structures for particular types of information spaces. Then, we considered familiar *organization schemes* that information architects use in organizing the content of digital information spaces—including exact, ambiguous, and hybrid schemes—as well as their typical applications. Finally, I briefly described LATCH, Richard Saul Wurman's model for organizing information.

Now that you have the basic knowledge you need to design effective information architectures for digital information spaces, let's shift gears and look at the broader, strategic context in which information architects work.

In Chapter 5, *UX Research Methods for Information Architecture*, I'll describe some UX research and content-analysis methods that you'll find particularly helpful in designing an information architecture. Plus, I'll describe some data analytics that are useful in improving an information architecture.

References

To make it easy for readers to follow links to the references for this chapter, we've made them available on the Web: `https://github.com/PacktPublishing/Designing-Information-Architecture/tree/main/Chapter04`

Further reading: Books

Avi Parush. *Conceptual Design for Interactive Systems: Designing for Performance and User Experience.* Waltham, MA: Morgan Kaufmann, 2015.

Bob Baxley. *Making the Web Work: Designing Effective Web Applications.* San Francisco: New Riders Publishing, 2002.

James Kalbach. *Designing Web Navigation.* Sebastopol, CA: O'Reilly Media, Inc., 2007.

Peter Morville and Louis Rosenfeld. *Information Architecture for the World Wide Web.* 3rd ed. Sebastopol, California: O'Reilly Media, Inc., 2007.

Douglas K. Van Duyne, James A. Landay, and Jason. I. Hong. *The Design of Sites: Patterns for Creating Winning Web Sites.* 2nd ed. Upper Saddle River, NJ: Prentice Hall, 2007.

Part II:
Foundations of
Information-Architecture
Strategy

This part explores the foundations of IA strategy, which include UX research methods that focus on information architecture, how to understand and structure an information space's content, and the classification of information, which inform information-architecture (IA) strategy and design. It then discusses how to define an IA strategy in depth. Part II comprises the following chapters:

- *Chapter 5, UX Research Methods for Information Architecture*
- *Chapter 6, Understanding and Structuring Content*
- *Chapter 7, Classifying Information*
- *Chapter 8, Defining an Information-Architecture Strategy*

5

UX Research Methods for Information Architecture

While many broadly applicable UX research methods can be helpful in the design of information architectures for digital information spaces and products, UX researchers have devised a number of useful methods of both generative and evaluative UX research for the specific purpose of improving the quality of information architectures. During the *Discovery* phase of a project, before defining requirements, conduct generative user research to better understand how actual or potential target users would organize and label particular classes of information and gain other insights that inform the design of an information space or product's information architecture. Throughout your *Design* process, you can iteratively evaluate and refine an information architecture's structure and labeling, as well as improve its resulting findability.

In this chapter, I'll discuss some UX research methods that apply specifically to the design of information architectures for information spaces or products. I'll provide an overview of each of the following broad categories of methods, then describe some important methods in each category:

- Card-sorting methods for understanding and evaluating information hierarchies
- Other UX research methods for discovering optimal structure and labeling
- UX research methods for evaluating findability
- Data analytics that can help you understand and improve information architectures

Card-sorting methods

I'll first provide an overview of the various aspects of card sorting that apply broadly to all methods of card sorting, then cover the specifics of three common card-sorting methods that you can use in designing and evaluating information architectures, as follows:

1.   Open card sorting

2.   Modified-Delphi card sorting

3.   Closed card sorting

Purpose—Card-sorting activities can help you advance your primary goal in designing an information architecture: understanding and employing the mental models of the target users of a digital information space or product to ensure that they can find the information they need. Card sorts enable you to learn how people use information and organize it into categories. You can discover what categories of information would be meaningful to your target users—and, thus, should make up an information space or product's organizational structures, taxonomies, and navigation schemes.

For a new information space or product, you can explore what categories of content you should create. You can also identify optimal terminology to use in labeling specific content objects, categories, and navigation links—especially the key links in an information space or product's primary and secondary navigation systems. At any stage of design, card-sorting activities can help you understand the relationships between different content objects or concepts and learn what content objects would be difficult to categorize. For an existing information space or product, you can evaluate the categories of information that provide its structure and the relationships between them. Thus, conducting card sorts provides insights into participants' mental model of an information architecture, as well as their expectations regarding what content or functionality an information space or product should include.

Unfortunately, card-sorting activities lack the context of participants' pursuing their own wayfinding goals or performing their actual tasks using an existing information space or product.

Description—In preparing for any card-sorting activity, create a set of about 30 to 100 cards that represent the content objects for an existing or proposed information space or product. Be sure that your cards represent *all* of your most important content. Creating the cards requires complete familiarity with an information space or product's content. Each card should contain a word or phrase that describes—or even a photo that represents—a content object *or* a category, topic, concept, or task. For example, depending on their level of granularity, the cards might contain words or phrases that describe the sections or pages of an information space or product—or proposed titles for those sections or pages—or even the content elements within pages.

Choose words and phrases that would be meaningful to your target audience. The terms you choose could derive from the findings of prior generative or evaluative UX research— for example, user interviews or usability testing—or from project stakeholders, customer-support feedback, competitive analyses, or the labels for content objects on any existing information space. The terminology you use should avoid introducing any bias by suggesting certain group labels—for example, through the repetition of similar terms across multiple cards or the use of brand names rather than generic descriptions of products—or even by describing content objects at different levels of granularity.

However, for an early-stage card sort whose goal is gathering rich, qualitative data, you may want to create cards that represent content objects at various levels of granularity. Then, at a later stage of your project, for a card-sorting activity whose goal is collecting quantitative data such as the percentages of participants who grouped certain cards, you should create cards at the same level of granularity—for example, cards that represent pages at the same level in a hierarchy. Be sure that all the words and phrases on your cards consistently describe content objects that reside at the same level, one level beneath any group labels.

If the scope of an information space or product is very large, consider creating multiple sets of cards to prevent the number of cards from becoming overwhelming to participants. This is especially important if different sections of an information space or product have distinct target audiences for whom different terms would be most meaningful.

Do a pilot card sort to ensure that your cards have enough common attributes to allow your participants to group them logically. Ask the person who is piloting your card-sort activity to think aloud so you can identify any problems that would introduce confusion or bias.

Set up the necessary equipment to record video—or at least audio—for all of your card-sorting activities. Plus, you should take notes that capture participants' key comments, as well as any insights that occur to you during the sessions.

When conducting a card-sorting activity, give each participant a set of cards that represent the content objects in your information space or product. Ask participants to organize the cards into groups that are meaningful to them. To learn more about *why* participants are making each of their sorting decisions, ask them to think aloud as they work. Provide some blank cards and marker pens, so participants can create additional cards if they notice any content gaps. Once your participants have finished sorting all of their cards, take photos of their work. Then give them some Post-it notes and small binder clips or rubber bands so they can bundle the groups of cards together and label them.

If the purpose of a card sort is to consider alternative navigation paths, you might want to allow your participants to create copies of certain cards, then add them to multiple groups or categories. Ask them to describe their reasons for placing a card in multiple locations.

If, after about ten card-sort sessions, consistent groupings have failed to materialize, you could try creating some new cards, using your participants' terminology, then conduct a sufficient number of additional card sorts to discover consistent groupings. Also, consider whether the inconsistencies might indicate that there are differences in your participants' profiles. Perhaps the inconsistencies represent the groupings and terminology of different personas that represent specific segments of your audience.

Once each cart-sorting session is complete, you should ask your participants questions about *why* they've grouped certain cards together, which groups seemed most and least obvious to them, which of the cards in each group are most and least representative of the group, and what they might have found confusing or frustrating.

To strengthen your findings from moderated card-sorting activities with statistical evidence, consider conducting some unmoderated card-sorting activities as well.

When—Card-sorting activities ideally take place during the latter part of the *Discovery* phase of the development process—once you've determined what types of information a digital information space or product should comprise, but prior to your *Design* phase. Plus, conducting card sorts can be helpful whenever you need to make changes to an information architecture—for example, if you're adding new sections or more categories of information to an existing information space or product.

Where—You should preferably conduct moderated card-sorting activities in a lab setting, face to face with your participants, so you can ask them to think aloud as they sort the cards and follow up by asking them questions about their specific choices. Participants can either sort physical cards manually *or* digital cards on a computer. Face-to-face card sorts typically generate rich, qualitative data, but also some useful quantitative data.

However, you could alternatively conduct *unmoderated, remote card sorts*, using card-sorting software that automatically captures, consolidates, and analyzes the card-sorting data from *all* of your card-sorting sessions, then analyzes and presents your study's overall findings. You can conduct remote card sorts with larger numbers of participants, who can be anywhere in the world. However, because you can neither observe participants, request that they think aloud and listen to their comments as they sort the cards, nor ask them questions about *why* they're making certain sorting decisions, remote card sorts generate predominantly quantitative data.

Or you could use a combination of both approaches, supplementing the data from a large number of remote sessions by moderating a few face-to-face card-sorting sessions, through which you could obtain more qualitative data regarding participants' grouping decisions.

Participants—Your participants should be potential or actual target users who would be interested in the content of the information space or product for which you're conducting a card sort. You can conduct card-sort sessions with either individual participants or small groups of participants. When you're conducting card sorts with *groups* of participants, each group should ideally comprise three or fewer participants and no more than five.

Most UX researchers typically conduct card-sorting sessions with between six and twelve *individual* participants. However, participants might sort the cards differently—especially when there are various possible organizational structures—and this inconsistency can reduce the clarity of the relationships between categories and content cards. Therefore, if you're getting inconsistent results, you should consider conducting card sorts with 15 or more individual participants to obtain reliable results. As you increase the number of participants, take note of any diminution of value from your sessions with these additional participants. If you're not gaining any new insights, you should consider ending your card-sorting activity. However, deriving findings that you could reliably generalize to an entire user population would require at least 30 to 50 participants. If your target users include multiple distinct roles, you should conduct card sorts with participants in each role—typically, between six and twelve participants per role. Bear in mind that using larger numbers of participants significantly increases the complexity of your analysis and, thus, the overall duration and cost of your card-sorting activity.

You could conduct card sorts with multiple, individual participants simultaneously to gather data from many participants more quickly, but doing so would prevent their thinking aloud as they work. When you do card sorts with small groups of participants, you can listen to the discussions they have as they make sorting decisions and obtain more data from fewer card-sorting sessions—however, the card sort ultimately represents the consensus of the group.

To benefit from getting both individual participants' unique perspectives and learnings from group discussions, you could do simultaneous card-sort sessions with individual participants, then gather all of those participants together for a group discussion about their reasons for grouping the cards as they did.

In contrast to *face-to-face* card sorts—which require participants to come to your location and, thus, limit their geographic diversity—participants in *remote* card sorts can be anywhere in the world. You can easily include a large number of participants in a remote card-sorting activity, making this approach particularly useful if you need to extrapolate your data across a large, homogeneous user population.

Type of data—By conducting card-sorting activities, you can gather both qualitative and quantitative, self-reported data—*and*, if you conduct your card sorts in a lab, observed data as well. In general, face-to-face, card-sorting activities generate more qualitative data, while *remote* card sorts generate more quantitative data, but conducting card-sorting activities in either of these settings can generate both types of data. One type of quantitative data you can typically gather by conducting card sorts is the percentages of participants who have grouped certain cards.

Data analysis—You can analyze the data you've gathered from your card-sorting activities either qualitatively or quantitatively, depending on the amounts and types of data you've collected. Particular card-sorting methods generate different volumes and types of data, suggesting the optimal approach to take in analyzing patterns in your data. For example, there are more unknowns in regard to open card sorts, so their data is more difficult to analyze than that from closed card sorts. (You'll learn more about these two methods of card sorting later in this section.)

To make it easier to refer to specific cards during analysis, add an unobtrusive number or code to the same corner on each card. When you're numbering content cards, an Arabic numeral is usually sufficient. However, if you've created multiple card sets, number the content cards using a letter for each set—for example, *A1, B1*.

When you're moderating face-to-face, manual card sorts, you'll absorb a significant amount of *qualitative* data as you observe and listen to participants as they sort cards into groups, and you can begin to derive key insights from your observations. Once you've moderated *all* of your card-sorting sessions, you've probably gathered a large amount of data. To make it easier to look at *qualitative* data in different ways, add each participant's groupings to a spreadsheet, then conduct an informal analysis by following these steps:

1. Examine individuals' or teams' groupings of cards to identify patterns and themes across different card sorts and understand the relationships between the various cards. High-frequency associations between pairs of cards accurately represent users' mental models. Note any cards that participants were unable to place in groups, as well as cards whose placement was inconsistent across participants. Consider the cause of participants' lack of clarity about a concept you're trying to convey or the meaning of a card's label. Are there some cards that have no clear relationships to other cards?

2. Assess individuals' or teams' placements of specific cards in groups. Did participants consistently place certain cards in the same groups?

3. For an *open* card sort, evaluate participants' labels for groups of cards. Have they consistently used the same words or phrases to describe similar groups? Are there any similar groupings that participants have labeled differently? Consider whether the reasons for these differences have arisen from participants' attributing different meanings to content cards or group labels.

Alternatively, if the amount of *qualitative* data that you've gathered through a card-sorting activity is not too large, you could use Post-it notes to create a simple, easy-to-understand representation of the activity's overall results in the form of an affinity diagram, focusing on an information space or product's key pathways or tasks. You could follow the same three steps that I've outlined for spreadsheet analysis. Concentrate on your most salient findings.

Now, let's look at some *quantitative* approaches to data analysis. If you've conducted card-sorting sessions with a large enough number of participants, consider using a spreadsheet to measure the similarities and differences between various participants' card sorts. You could automatically calculate, as a percentage, how often certain cards appear in specific groups; tally the frequency with which participants have placed two cards in the same group; or assess what groupings participants agree about most strongly. By doing such analyses, you can determine the strengths of the relationships between specific cards and categories or specific pairs of cards.

If you've conducted an unmoderated, *remote* card sort, you can use the same card-sorting software to automatically capture, analyze, and display the *quantitative* data from all of the card-sorting sessions.

If you've compiled digital card-sort data and have sufficient *quantitative* data to warrant statistical analysis, you could use cluster-analysis software to calculate the frequency with which participants have placed cards in the same groups and visually represent relationships between cards and groups, or clusters, that would otherwise be difficult to see. Figure 5.1 shows an example of a cluster-analysis diagram.

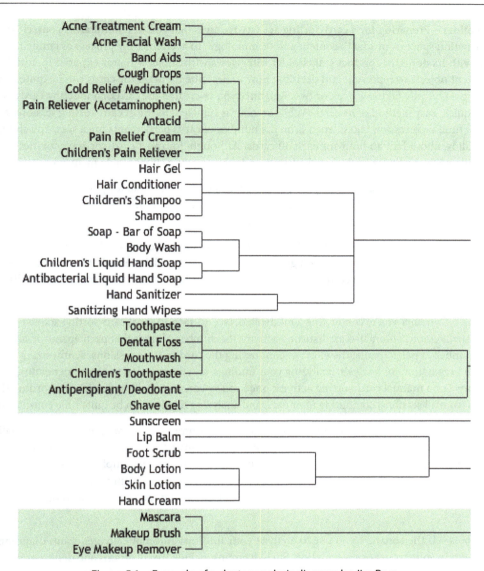

Figure 5.1—Example of a cluster-analysis diagram by Jim Ross

Image source: "Comparing User Research Methods for Information Architecture." *UXmatters*, June 7, 2011.

https://www.uxmatters.com/mt/archives/2011/06/comparing-user-research-methods-for-information-architecture.php

The results of a card-sorting activity should serve only as a general guide in determining an information space or product's structure and labeling.

Level of effort—Preparing for a card-sorting activity requires that you first familiarize yourself with the information space or product's content and terminology. To accelerate the process of familiarizing yourself with its domain, conduct stakeholder interviews with subject-matter experts. Identifying what content objects to represent and deciding how to describe them on the cards can be quite time consuming—typically taking a week or two. But, once you've created the cards, conducting card sorts is a fairly quick, easy method of research. While an hour is sufficient for most card-sorting sessions, the amount of time each session takes depends on the number of cards participants must sort. You should typically allow about half an hour for each 30 cards. Although, for most card-sorting activities, you would moderate between five and 15 card-sorting sessions, you could choose to do many more sessions.

Card sorting is an iterative process for your participants—and even more so for participants in *remote* card-sorting activities, whose limited screen real estate restricts the number of cards they can see at once, making sorting the cards into categories more difficult.

The overall level of effort for face-to-face, individual card sorts is moderate, while that for group card sorts is low. It usually takes less time for a group to do a card sort collaboratively than for individual participants to complete their card sorts. However, you could conduct card sorts with multiple individuals simultaneously to save time.

Depending on whether you're conducting a moderated, face-to-face, manual card-sorting activity or an unmoderated, remote, automated card-sorting activity, the number of cards that participants must sort, *and* the number of participants, the effort of conducting the card-sorting sessions, synthesizing your findings in preparation for analysis, analyzing your findings, and making design recommendations can vary greatly. For a manual card-sorting activity, once you've conducted all the sessions, recording the groupings from all sessions, combining their data, and analyzing the data can be quite time consuming.

The card-sorting software you would use when conducting unmoderated, *remote* card sorts usually generates predominantly quantitative data—automatically capturing, analyzing, and representing the data and, thus, minimizing both your own effort and the time it takes to complete all of the sessions and enabling you to complete your analysis quickly. Ideally, you should pair unmoderated card–sorting activities with moderated card–sorting activities to avoid sacrificing the richly detailed data that the latter provides.

Deliverables—If the software you use to analyze your findings automatically generates diagrams, include them in your report or presentation. Such diagrams include tree diagrams and dendrograms—branching diagrams that represent a hierarchy of similar categories. Also include the findings from each individual card-sort session in a table, representing the cards within each category, as well as any changes that participants made to cards and cards they added or were unable to group.

Investment—Card-sorting activities are relatively inexpensive and, because they require only a modest level of investment, it is often easier to obtain approval and funding for them. However, doing card-sorting sessions with larger numbers of participants—or increasing your *sample size*—can significantly increase the cost of card-sorting activities. Conducting card-sorting activities remotely reduces their cost—in part, because participant stipends for remote research are typically lower—even factoring in the cost of subscribing to an online card-sorting tool.

[1, 2, 3, 4, 5, 6, 7, 8, 9, 10, 11]

Recommended book—Donna Spencer. *Card Sorting: Designing Usable Categories.* Brooklyn, NY: Rosenfeld Media, 2009.

Open card sorting

Purpose—Open card sorting is a form of *generative* user research—whose purpose is discovery—and is the most common method of card sorting. During an open card–sorting activity, participants group cards that represent existing content objects into categories that they create and label rather than into predefined categories. Thus, you can learn both what cards participants associate and group with one another, as well as how they label the groups that they create. By conducting an open card sort, you can better understand participants' mental models of an information space or product, then leverage their mental models in creating meaningful categories and subcategories for its content. Because your participants assign labels to their groups of cards, you can gain insights into the optimal terminology for labeling specific groups of content objects or categories of information, as well as various levels of navigation links.

Another reason for conducting an open card–sorting activity is to explore various potential design solutions for an information space or product's organizational structure, navigation system, or labeling system. For example, you might be considering alternative groupings of content objects and different categories and labels for them. Plus, by better understanding the relationships between the content objects in an information space or product, you can identify more useful links to related information.

Description—When conducting an open card sort for either an existing or a proposed information space or product, give participants a set of cards that represent its most important content objects, ensuring that the cards adequately represent the full breadth of the information space's content objects. Then, ask your participants to group the cards in whatever ways seem natural to them. To prevent your participants from thinking too much about possible labels as they group the cards, you should initially avoid telling them that you'll be asking them to label the groups they create. Once participants have finished grouping the cards, ask them to try to label or describe the groups that they've created, using Post-it notes, then bundle each group of cards together. Figure 5.2 shows an example of an open card sort, with Post-it notes providing labels for groups. Finally, you could optionally ask participants to gather any related groups of cards into larger groups, labeling the higher-level groups and, thus, creating a hierarchy of possible group labels.

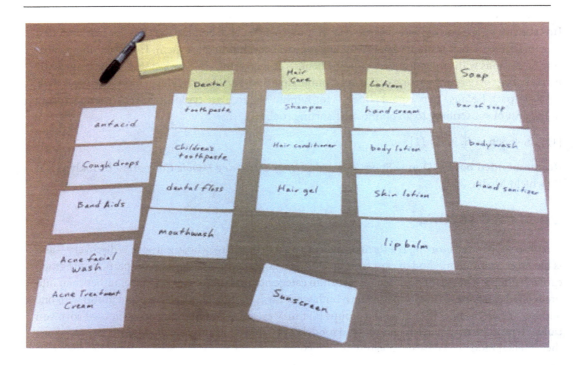

Figure 5.2—Example of an open card sort by Jim Ross

Image source: "Comparing User Research Methods for Information Architecture." *UXmatters*, June 7, 2011.

```
https://www.uxmatters.com/mt/archives/2011/06/comparing-user-research-
methods-for-information-architecture.php
```

Once participants have completed their open cart sort, consider asking them questions about *why* they've chosen certain labels, whether the process of choosing those labels made them question any of their initial groupings, and whether there are any missing categories.

Data analysis—Because each participant's card sort and group labels represent that person's own mental model of an information space or product, the primary challenge of data analysis for an open card–sorting activity is reconciling the differences among and combining those disparate mental models to create a single model for that information space or product. Unless you've conducted card-sorting sessions with only a small number of participants, this typically requires using statistical-analysis techniques.

During your analysis, you need to be able to refer to specific cards easily. For a manual, open card–sorting activity, in addition to the numbers or codes you've assigned to each content card, you should add an unobtrusive code to the same corner on each Post-it note that indicates a group label—for example, *L1*, *L2*, and so on. Before analyzing the data from a manual open card sort, list the numbers of all the cards a participant has placed in each group on the Post-it note that indicates its label.

Because participants have created their own labels for their groupings of cards, assessing the frequency with which participants have collectively placed particular cards in specific categories as a percentage becomes more difficult. Prior to calculating these percentages and conducting a *quantitative* analysis, you must *normalize*, or bring consistency to, any similar labels for similar groups. For example, *UX Design*, *User Experience Design*, and *User-Interface Design* might be equivalent categories that you could combine to create a new category, giving the new category card the label *UX Design*. Once you've combined and normalized the labels of similar groups, it is easier to calculate percentages accurately and, thus, to see meaningful patterns in your data. On the resulting new category cards, be sure to note *all* the labels that participants assigned to their original groups.

If any participants have created groups with labels such as *Miscellaneous* or *Other*, you should either determine why the cards in these groups are difficult to categorize and place them in other groups *or*, if there really isn't a category into which a particular card fits, you might need to exclude the corresponding content from the information space or product. [12, 13, 14, 15]

Modified-Delphi card sorting

Purpose—Researcher Celeste Lyn Paul devised modified-Delphi card sorting as a more reliable variation of open card sorting. The purpose of this method of generative user research is to understand participants' mental models of an information space or product and inform the design of an information architecture.

Description—Paul based this method of card sorting on the *Delphi method*—a structured communication technique in which the individuals on a panel of experts successively work toward answering a question. This method assumes that group decisions are more reliable than individuals' independent decisions.

When conducting a modified-Delphi card-sorting activity—rather than asking individual participants to perform their own card sort independently, as in an open card sort—you'll request a succession of individuals in a *group* of participants to sort a single set of content cards. As with any card sort, these participants sort the cards to conform to their mental model of an appropriate structure for the content of an information space or product.

A modified-Delphi card-sorting activity comprises the following steps:

1. Select a *seed participant*, who could be either an individual participant working alone *or* with the help of an information architect, a pair or a group of participants working collaboratively, *or* an information architect. For a set of cards that would likely be difficult to classify in categories or an information space or product whose content is unfamiliar to participants, it can be helpful to have a pair or a group of participants act as the seed participant.

2. Ask the seed participant to conduct an initial card sort, proposing a structural model for an information space or product for which you're conducting a card sort. If your seed participant is *not* an information architect and is having difficulty creating a coherent structural model for an unfamiliar information space or product, you could optionally ask an information architect to assist the seed participant in sorting the cards, but only if this is absolutely necessary. Because

this initial card sort could limit the potential structural models that the group might explore, causing participants to overlook other possible structural models, it might be beneficial to conduct multiple modified-Delphi card-sorting activities with different seed participants.

3.   One by one, request each subsequent participant in your group of participants to provide feedback on the group's current card sort—which has resulted from the combined efforts of *all* previous participants—then either modify that structural model *or* propose a new model while thinking aloud about their changes. The structural model evolves throughout the card-sorting activity, as each participant refines the card sort.

4.   Once the group achieves consensus on the structural model and agrees that no further significant changes are necessary, the card-sorting activity ends. However, if your participants are struggling to resolve their disagreements and reach a consensus on a complex structural model, you could consider asking an information architect to help them. But you should avoid asking for an information architect's assistance unless it is the only way to reach a consensus.

Figure 5.3 shows an example of a modified-Delphi card, with Post-it notes providing labels for groups.

Figure 5.3—Example of a modified-Delphi card sort by Jim Ross

Image source: "Comparing User Research Methods for Information Architecture." *UXmatters*, June 7, 2011.

```
https://www.uxmatters.com/mt/archives/2011/06/comparing-user-research-
methods-for-information-architecture.php
```

Data analysis—Your group of participants creates only a single structural model for an information space or product, which lets you move forward with analysis immediately after the card-sorting activity ends and greatly simplifies analysis.

[16, 17, 18]

Closed card sorting

Purpose—Closed card sorting is a form of *evaluative* user research. Thus, by conducting a closed card–sorting activity, you can evaluate and, ultimately, validate the usability of either a proposed *or* an existing information architecture. Because your participants group content cards in predefined or existing categories, you can learn what content objects they associate with specific categories, whether they would group them as you have, and, if not, where they would place them instead. Closed card sorting lets you easily assess whether your organizational structure conforms to participants' mental model, as well as whether your labeling for categories makes sense to them.

Description—When you're preparing for a closed card–sorting activity, you must create both a set of *content cards* that represent the content objects in an information space or product *and* a set of *category cards* that have predefined or existing category labels, comprising individual words *or* phrases, and represent the categories into which you'll ask participants to organize the content cards.

At the beginning of a closed cart sort, give your participants two sets of cards: the category cards and the content cards. You could optionally ask participants to describe the meaning of each of the category labels. Next, ask participants to group the content cards under the categories—and, in some cases, in subcategories as well.

If, after you've conducted ten or so card sorts, your participants are *not* creating consistent groupings of cards, you should consider creating some new category cards or using different terminology to label the categories.

Figure 5.4 shows the digital equivalent of a closed card sort.

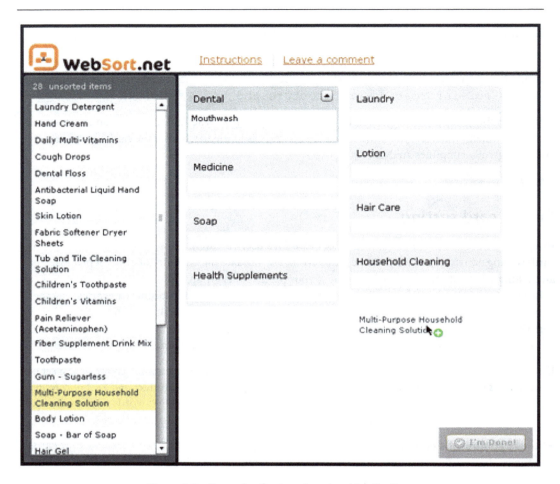

Figure 5.4—Example of a closed card sort by Jim Ross

Image source: "Comparing User Research Methods for Information Architecture." *UXmatters*, June 7, 2011.

https://www.uxmatters.com/mt/archives/2011/06/comparing-user-research-methods-for-information-architecture.php

When—Conduct closed card–sorting activities later in the design process, once you have become familiar with the terminology of your target users and have designed at least a preliminary version of an information architecture for an information space or product.

A closed card–sorting activity is most helpful when you're adding new content to an existing information space or product whose information architecture has good findability. Although you wouldn't want to make significant changes to the existing information architecture, you might need to expand or make minor modifications to it. Conducting a closed card–sorting activity lets you test these changes.

Data analysis—When analyzing the data from a closed card–sorting activity, *if* you ask participants to describe the meanings of the category labels, you can assess the similarity of participants' descriptions *and* compare them to your own. The more similar participants' descriptions of a label, the better that label is. Your data analysis should consider both the frequency with which participants have placed certain cards under specific categories, as well as the language participants used in describing the category labels.

[19, 20, 21]

In the next section, I'll briefly discuss what UX research methods are helpful in discovering the optimal categories in which to place content about specific topics or types of content, as well as the best labeling for those topics, types of content, or links in navigation systems for digital information spaces and products.

UX research methods that inform structure and labeling

You learned about *open card sorting* earlier in this chapter, which is an excellent method of determining where your target users would place certain content and how they would refer to particular categories of content. Later in this chapter, we'll look at *site search–log analytics*, which is a type of data analytics that enables you to learn what terminology your actual users employ when searching for information on your information space or product.

In the era of Web 2.0, some popular Web sites and intranets enabled users to tag articles or posts to categorize and share their content—thus, creating user-generated folksonomies that defined their categories of content and dictated their structure. Information architects could conduct a tag analysis to ascertain the vocabulary of actual users, then use their terminology in labeling a site's content objects and navigation links. However, because the use of algorithms that push content to users has largely superseded the use of tagging, tag analysis is no longer a common method of UX research for information architecture. [22]

When you're conducting UX research to inform the design of a labeling system, notice the terms and any jargon or acronyms that your participants commonly use in describing the content within the domain of a digital information space or product. If you're developing a controlled vocabulary, it can also be helpful to note common misspellings. [23]

Now, I'll discuss *free-listing*, a method of UX research that lets you explore participants' common understanding of an information space or product's domain to inform structure *and* labeling.

Free-listing

Purpose—In her *Boxes and Arrows* article *"Beyond Cardsorting: Free-listing Methods to Explore User Categorizations,"* Rashmi Sinha describes free-listing, a method of *generative* user research that lets you ascertain participants' common understanding of a *semantic domain*, including the domain's content, scope, and boundaries, as well as its structure. Free-listing can reveal how participants perceive the domain of your information space or product. By conducting a free-listing activity, you can also become familiar with the target users' vocabulary for the domain, which is essential to designing a useful labeling system.

Description—Prior to a free-listing activity, you should list all of the categories of content on your digital information space or product, then decide what subset of those categories you want to study. For a new information space or product, endeavor to focus on the most important categories. For an existing information space or product to which you're adding new content, focus on the new categories.

During a free-listing activity, you can ask your participants to either tell you or write down a list of all the things that belong to specific categories for example, "Name all of the types of products you can think of." Let participants take their time doing this. If they stop listing items after listing just a few, prompt them to continue by saying, "Can you list any more?"

The items that are most highly salient to users are those that are of primary importance for a domain. Two key factors determine items' salience, as follows:

1. **Average rank distance**—During their free-listing session, participants list certain items *earlier* than others. The closer items are to one another in a list, the more similar a participant perceives them to be to one another. The order in which *all* participants have listed specific items determines their *average rank distance*. For an overview of all items' average rank distance, calculate the totals for the numbers of times that participants have listed them at each level of ordinal ranking, as shown in Figure 5.5. You can also note the average rank distance for specific pairs of items.

	cat	dog	horse	rabbit	rat
cat	7	6	2	4	5
cat	6	7	4	3	2
cat	2	4	7	3	1
cat	4	3	3	7	4
cat	5	2	1	4	7

Figure 5.5—Example of average rank distance, a type of similarity matrix, by Rashmi Sinha

Image source: "Beyond cardsorting: Free-listing methods to explore user categorizations." *Boxes and Arrows*, February 24, 2003.

`https://boxesandarrows.com/beyond-cardsorting-free-listing-methods-to-explore-user-categorizations/`

2. **Co-occurrence**—Overall, participants list some items *more frequently* than others, giving them a higher level of *co-occurrence* across all participants. Items that only a single participant has listed are outliers so are not pertinent to a domain. Calculate co-occurrence metrics for all pairs of items, as shown in Figure 5.6.

	cat	dog	horse	rabbit	rat
cat	5				
cat	4	5			
cat	2	2	5		
cat	2	2	2	5	
cat	1	1	1	1	5

Figure 5.6—Example of co-occurrence, a type of similarity matrix, by Rashmi Sinha

Image source: "Beyond cardsorting: Free-listing methods to explore user categorizations." *Boxes and Arrows*, February 24, 2003.

```
https://boxesandarrows.com/beyond-cardsorting-free-listing-methods-
to-explore-user-categorizations/
```

As you conduct free-listing sessions with your participants, you could record the data for each of your domain's categories in a category matrix, similar to that shown in Figure 5.7. First, create a similarity matrix that lists all participants by number in the first column and the category's items across the top. In this matrix, a *1* indicates that a participant has included an item in their list; a *0* shows that he has not. Once you've completed all of your sessions, tally the totals for each of the category's items in the last row of the matrix.

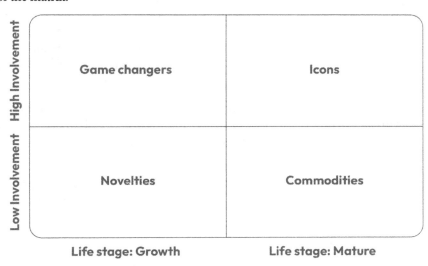

Figure 5.7—Example of a category matrix by Marius Dønnestad

Image source: "Category Matrix: A Holistic Approach to Innovation & Channel Strategy." *LinkedIn*, July 9, 2015.

```
https://www.linkedin.com/pulse/category-matrix-holistic-approach-
innovation-channel-marius-donnestad-6024809878755762176
```

Alternatively, you could capture simple lists of individual participant's items for each of the domain's categories, listing the items in the order in which participants named them.

When—Free-listing activities should ideally take place early during the *Discovery* phase, either in lieu of or prior to any card-sorting activities. For example, you could conduct a free-listing activity to inform the labeling of your category cards for a closed card–sorting activity. You could alternatively incorporate free-listing into your user interviews.

Where—You can conduct a free-listing activity either face-to-face or remotely, using an online survey tool.

Participants—Because most people find free-listing natural, you can conduct free-listing activities with many different user populations. If you're conducting free-listing activities verbally, you can even do them with young children or people who are illiterate.

Similar to card-sorting activities, you'll typically need between six and twelve target users for a free-listing activity. However, the number of participants you actually need depends on users' general level of agreement on a domain's categories and what items belong within them. Therefore, with a small number of participants, a single participant with a unique perspective could make it difficult to recognize patterns in your data.

To determine whether agreement exists among your participants, list all the items they have placed within each category, ordering them by the frequency with which participants have included them in a category. Then, as you conduct sessions with more participants and tally the data from their sessions, notice whether the order of items in categories changes. If their order remains fairly consistent across your additional sessions, you can end your free-listing activity. But, if this agreement doesn't exist after your six initial free-listing sessions, it might be necessary to conduct free-listing sessions with as many as 15 to 30 participants to obtain reliable results.

Type of data—This method provides both self-reported, qualitative data—a domain's salient categories and the items within them—*and* quantitative data—the priority with which participants mention particular items and the rankings of items within categories according to the frequency with which individual participants mention them.

Data analysis—As with card-sorting data, you can conduct your data analysis using a spreadsheet to identify patterns in your data or examine similarity matrices using cluster-analysis software to show the relationships between items in each category.

The larger the number of participants in your free-listing activity, the greater the complexity of your data analysis. You could include *all* items that multiple participants have mentioned in their lists in your analysis. Alternatively, you might be able to identify points in your data below which items lack salience, then include only the most salient items in your analysis.

In preparing to analyze your data from a free-listing activity, list all salient items for each category—that is, those items that more than one participant has placed in a category. Order the items by the frequency with which participants have included them in the category.

As you conduct sessions with additional participants, tally the items from those sessions under your categories. Notice whether the order of the items in each of the categories changes, according to the frequency of their occurrence.

Level of effort—You can quickly and easily conduct free-listing activities. However, doing such activities with larger numbers of participants significantly increases both their overall duration and cost, as well as the complexity of your analysis. You could conduct free-listing activities iteratively to improve your understanding of the structure of your content domain. Use each item from your first activity as the starting point for your next free-listing activity.

Deliverables—Include your visualizations of your data—such as your similarity matrices—in your report or presentation.

Investment—Conducting a free-listing activity face-to-face is a very low-cost method of UX research. However, the expense of subscribing to an online-survey tool makes using this method remotely considerably more costly.

[24, 25, 26]

UX research methods for evaluating findability

Findability is the degree to which users can easily find the specific information they're seeking when using a digital information space or product. Findability is an essential characteristic of a successful information space or product and a precursor to its usability. The findability of the content objects on an information space or product depends on how effectively each page's structure and metadata describe and uniquely identify it. In this section, I'll describe some methods of UX research that are useful in evaluating findability, as follows:

- Reverse card sorting
- Card-based classification evaluation
- Tree testing

Reverse card sorting

Purpose—*Reverse card sorting*, also known as *inverse card sorting*, is a variant of closed card sorting. This evaluative method of research lets you quantitatively rate *or* validate either a proposed information hierarchy *or* an existing information space or product's findability. Reverse card sorting provides insights that are useful in refining or redesigning the top levels of an information architecture or navigation system.

Description—Ideally, before conducting a reverse card sort, you should conduct a closed card sort to validate the existing or proposed information architecture that you'll evaluate during a reverse card–sorting activity. To prepare for your reverse card–sorting activity, create a set of category cards that represent the top level of the existing or proposed information architecture, then lay them out on a table. Alternatively, you could create a structural diagram, labeling just the top-level categories, as shown in Figure 5.8. Or, if you've already designed and validated the structure of the top two levels of a complex information space or product, you could label both levels. Be aware that conducting a reverse card–sorting activity for a complex information space or product can be difficult.

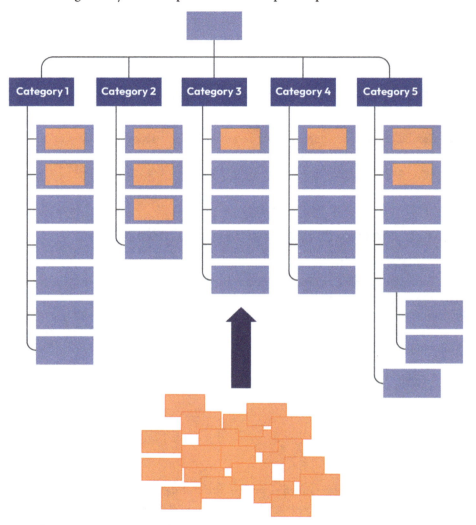

Figure 5.8—Example of a structural diagram for a reverse card sort by Chauncey Wilson

Image source: "Method 9 of 100: Reverse Card Sorting." *100 User Experience (UX) Design and Evaluation Methods for Your Toolkit*, March 17, 2011.

https://dux.typepad.com/dux/2011/03/method-9-of-100-reverse-card-sorting.html

In either case, you could create a set of content cards that represent your information space or product's subcategories, content topics, tasks, or navigation links. When creating content cards, avoid labeling them using the same terminology as is on the category cards. Ask your participants to sort the cards into the appropriate top-level categories or subcategories of your predefined information architecture. Be sure to shuffle the content cards thoroughly before each of your card-sorting sessions. Alternatively, you could simply ask participants a question about a topic or describe a task scenario to them, then have them tell you the top-level category or subcategory in which they would expect to find related information. Let them know that there are no wrong answers. Place the corresponding content card in that location.

When—Ideally, you should conduct a reverse card–sorting activity early when defining your design strategy, once you've designed an initial version of the information hierarchy for a *new* information space or product. Doing so gives you the opportunity to iteratively test and refine the information hierarchy throughout *Design*. You could also conduct a reverse card–sorting activity either just prior to or immediately after launching a new information space or product.

To evaluate the findability of an *existing* information space or product, conduct a reverse card–sorting activity during *Discovery*, before expanding the scope of its content and, thus, its information hierarchy or making any other structural changes. This lets you establish a baseline for the findability of the information space or product prior to your initial redesign.

Where—Preferably, you should conduct a moderated reverse card–sorting activity in a lab, face-to-face with participants, so you can ask them to think aloud as they sort cards either manually or digitally and ask them follow-up questions. But you could alternatively conduct *unmoderated, remote, reverse card sorts* or combine both approaches.

Participants—Refer to the section "Card-sorting methods," earlier in this chapter, for detailed information about working with participants when conducting card-sorting activities.

Type of data—By conducting a reverse card–sorting activity, you can gather quantitative, self-reported data that indicates the frequency with which your participants place particular cards in specific categories and subcategories—*and*, if you conduct a reverse card sort in a lab, you can obtain observed data as well. You can also gather self-reported, qualitative data about the terminology participants use in referring to specific content.

Data analysis—During analysis, assess how many participants have placed cards in the categories and subcategories you intended. Calculate the mean percentage of cards that participants have sorted correctly.

Level of effort—For details about the level of effort that a card-sorting activity requires, refer to the section "Card-sorting methods," earlier in this chapter.

Deliverables—Include any diagrams that your analysis software generates in your report or presentation, as well as tables summarizing the findings from each individual card-sorting session.

Investment—While reverse card–sorting activities generally require only a modest level of investment, doing them with large numbers of participants can significantly increase their cost. Conducting a reverse card–sorting activity remotely reduces its overall cost—despite the additional expense of subscribing to an online card-sorting tool.

[27, 28, 29, 30]

Card-based classification evaluation

Purpose—Donna Spencer devised this method of evaluative research so she could assess an information space or product's *taxonomy*, or information hierarchy, in isolation from its user interface. By conducting a card-based classification evaluation, you can focus on the findability of an information hierarchy—whether you're creating a new digital information space or product or proposing changes to an existing information hierarchy.

During a card-based classification–evaluation activity, your participants complete realistic, information-seeking task scenarios—rather than an information-organizing activity, as for card-sorting activities. Because this activity more closely resembles what users actually do when navigating an information space or product, you can gain a deep understanding of participants' perceptions about the classifications of information and their labels. A card-based classification evaluation also provides insights into where in an information hierarchy participants expected to find specific items of information. Therefore, you can focus your usability testing on evaluating the *user interface* of an information space or product—a purpose for which it is much better suited—rather than its information hierarchy.

Description—Prior to conducting a card-based classification evaluation, you should first propose an information hierarchy for an entire information space or product. Document the information hierarchy, using decimal numbers to indicate hierarchy—for example, *1.0, 2.0, 3.0, 4.0, 5.0*, and so on, for top-level categories; *1.1, 1.2, 1.3, 1.4*, and *1.5*, for second-level categories; *1.5.1, 1.5.2, 1.5.3, 1.5.4*, and *1.5.5*, for third-level categories; and so on, for as many levels as the information hierarchy requires.

Represent each of the numbered categories and subcategories in the hierarchy on an index card, writing legibly and large enough that participants can read the cards from a distance. If you're unsure about the placement of some subcategories, you can create duplicate cards and place them in more than one location. Bundle the cards for the levels of the hierarchy under each category and subcategory, using binder clips or rubber bands.

Also, prepare a set of realistic information-seeking task scenarios that is large enough to evaluate the full scope of the information hierarchy. Document a question representing each task scenario on a scenario card. For example, you might ask a question such as *What US state lost the most acreage to wildfires in 2020?* Avoid using the same terms in your questions that you used on your category cards. Assign a scenario ID to each card, using capital letters.

As with a reverse card–sorting activity, you'll show the highest level of your information hierarchy to participants when you're moderating card-based classification–evaluation sessions. However, card-based classification evaluation differs from reverse card sorting in that your participants must navigate *all* the levels of an information hierarchy, drilling down to the subcategory in which they would expect to find the information you've requested.

At the beginning of each card-based classification–evaluation session, lay out the top-level cards representing your information hierarchy on a table. Then, ask your participants the questions you've represented on the scenario cards, prompting them to complete these realistic information-seeking tasks. Ask participants to navigate the information hierarchy to find the information they're seeking by successively tapping category and subcategory cards to indicate where they would first look for the content that would enable them to answer the questions—the categories and subcategories that have the strongest information scent. Each time participants tap a category or subcategory, lay out the cards for that part of the information hierarchy, then ask them where they would look next. Repeat this process until your participants reach the lowest level of the hierarchy. Managing all of these cards and laying out the right cards for each category and subcategory can be quite challenging and requires great familiarity with an information hierarchy.

If participants don't see a subcategory that looks promising, allow them to retrace their last step, then choose a different subcategory. But, if they still don't see a likely subcategory, conclude that scenario and move on to the next one.

If it's not clear to you why participants have made particular choices, ask them about the reasons for those choices. Ask participants what any problematic labels mean to them. If some part of a hierarchy clearly is not working, you can try some alternative structures or different labeling during your sessions. Keep some blank index cards and markers handy in case you need to create cards for these new subcategories.

At the end of each scenario, clear away all but the top-level categories, bundling the cards for each subcategory. Then, ask the participant a question that initiates the next scenario. To ensure that your participants collectively complete *all* of your task scenarios, always pick up wherever you left off as you progress from one participant's session to the next.

As participants complete your task scenarios, your notetaker should record the scenario ID and the numbers of the category and subcategories that participants chose when navigating the information hierarchy for each scenario, clearly indicating any choices that did not yield satisfactory results. Your notetaker should also capture all comments that participants make during their sessions.

Card-based classification evaluation focuses on testing an information space or product's findability and is an effective complement to usability testing, which is best for evaluating the user interface.

When—A card-based classification–evaluation activity should usually take place early in your *Design* phase. For a *new* information space or product, conduct this activity after you've completed your information hierarchy's first design iteration. However, if you're evaluating the hierarchy of an *existing* information space or product, you could conduct a card-based classification evaluation during

Discovery, before making any revisions to its information hierarchy. You could do several rounds of iterative design and evaluation prior to turning over your design to Development for implementation. You could alternatively conduct a card-based classification–evaluation activity either just prior to or immediately after launching a new information space or product to ascertain where participants look for the information they're seeking.

Where—This method requires that you conduct face-to-face sessions with participants—for example, in a lab or a conference room at a customer's site.

Participants—Recruit 20 or more potential or actual target users to participate in your card-based classification–evaluation activity.

Type of data—This method elicits observed, primarily quantitative data for each of your task scenarios—such as the numbers of participants who chose particular category and subcategory cards when completing each scenario. However, you might also be able to gather some qualitative data regarding structure and labeling.

Data analysis—Create a spreadsheet in which the first column comprises a numbered list of categories and subcategories and the top row contains letters representing scenario IDs. Then, record the data from all participants' sessions in the spreadsheet, as shown in Figure 5.9. In the spreadsheet, for each participant who chose a category or subcategory for a particular scenario, insert an uppercase X in the cell at that intersection for a first choice; insert a lowercase x for a second choice. Place the X for any choice with an unsatisfactory result within parentheses.

	Scenario A	Scenario B	Scenario C	Scenario D
1				
1.1				
1.1.1	XXOO			
1.1.2				
1.1.3			X	
1.2			X	
1.2.1			O	
1.2.2	XX			
1.2.3		XXOO		
1.3			OO	
1.3.1				XO
1.3.2				
1.3.3				

Figure 5.9—Example data for a card-based classification evaluation inspired by Donna Maurer Spencer

Image source: "Card-Based Classification Evaluation." *Boxes and Arrows*, April 7, 2003.

https://boxesandarrows.com/card-based-classification-evaluation/)

Note any clusters among participants' choices, which indicate strong information scent. If many participants have chosen the same category or subcategories that you intended for a scenario, your information hierarchy is working well for that scenario. If participants have chosen several different categories or subcategories for a scenario, their choices are worth considering, but the optimal information hierarchy for that scenario is unclear. You can ask participants why they chose or did *not* choose certain categories or subcategories. Similarly, if no participants have chosen your intended category or subcategories for a scenario, you should ask participants why, then redesign the hierarchy or the labeling for that scenario.

Level of effort—This quick, easy method requires only 10 to 15 minutes per session. Participants can complete 10 to 15 task scenarios during each session. Thus, if you're conducting sessions with 20 participants, this activity requires approximately three and a half to five hours overall. You can use this method iteratively during *Design*.

Deliverables—Write a brief report about your learnings from your evaluation.

Investment—Conducting a card-based classification–evaluation activity requires only a small investment and provides a high return on that investment. [31, 32, 33, 34]

Tree testing

Tree testing, also known as *tree jacking*, is an online variant of card-based classification evaluation and resembles that method in most respects. Therefore, in this section, I'll cover only the differences between card-based classification evaluation and tree testing. Refer to the previous section, "Card-based classification evaluation," for more information.

Purpose—Tree testing is both an evaluative and a generative method of research. You can conduct a tree-testing activity to evaluate the effectiveness of an information hierarchy and its labeling in enabling participants to complete realistic, information-seeking task scenarios. Once tree testing has identified any weaknesses in an information hierarchy, your team can iterate on its design, then conduct another round of tree testing to determine whether the changes have improved it. Plus, because tree testing lets you discover what categories and subcategories your participants have chosen when completing each task scenario, you can gain insights into where they expected to find specific items of information.

Tree testing can also show what first-level categories participants have chosen. Participants' first clicks have great impact on their task-completion rates. If, at the very beginning of a task scenario, participants selected the first-level category you intended, about 87% would complete the task successfully. However, if participants selected an incorrect first-level category, their task-completion success rate would drop to just 46%.

Tree testing provides specific, actionable findings that inform and can help you improve the design of an information hierarchy through iterative design and testing.

Description—Before conducting a tree-testing activity, propose an information hierarchy for a new digital information space or product *or* use that of an existing information space or product to mock up realistic, information-seeking task scenarios in an online tree-testing tool. These task scenarios should ideally let you evaluate the full scope of the information hierarchy—especially key paths through the information space or product *and* any parts of the hierarchy that you know pose challenges to users.

Ask your participants to complete several of these task scenarios using the tree-testing tool, which automates the facilitation of participants' sessions—presenting a random subset of the task scenarios to each participant—and tracks participants' clicks as they navigate the information hierarchy. At the beginning of each task scenario, the tree-testing tool displays the information hierarchy's top-level categories to a participant, who clicks a category to display its subcategories. Whenever the participant clicks a subcategory, the next level of subcategories appears. The participant continues clicking subcategories until he navigates *all* the levels of the information hierarchy, then indicates a subcategory in which he expects to find the information he is seeking.

You could optionally compare the findability of two alternative information hierarchies for a new information space or product by conducting tree testing on both design solutions in parallel, using the same task scenarios.

Although tree testing is generally a form of unmoderated research, you could optionally conduct moderated tree-testing sessions. When you're conducting unmoderated tree-testing sessions, you cannot observe participants as they work. Nor can you directly ask participants questions about how they've made specific choices. However, you could include an automated, post-test questionnaire at the end of each participant's session. Alternatively, once participants complete each individual task scenario, you could ask them to rate their experience on a Likert scale—that is, a *multipoint rating scale* comprising values that range, for example, from *Strongly disagree* to *Strongly agree* or *Very important* to *Not at all important*—by posing questions such as the following:

- How easy or difficult was this task to complete?
- What is your level of confidence that you've completed this task successfully?

Where—While tree testing is usually a method of remote research, it could alternatively take place in a lab. In either case, participants would use an online tree-testing tool to complete information-seeking task scenarios.

Participants—As when using any online UX research tool, you can more easily recruit large numbers of participants, who can be anywhere in the world. Larger sample sizes increase the validity of your findings—for example, 500 participants would let you achieve better than 95% confidence in your results—with only minimal additional expense. Most UX researchers typically recruit 20 or more potential or actual target users to participate in a tree-testing activity.

Type of data—Tree-testing activities elicit self-reported, quantitative data for each of your task scenarios. Online tree-testing tools automatically collect data about the categories and subcategories that participants choose when completing task scenarios. Depending on the tool you're using, the data you can view and the visualizations of that data differ somewhat. For each task scenario, you can view a tally of the categories and subcategories that all of your participants have chosen and probably the exact path that each participant has taken as well.

You can also see participants' overall rates of success or failure in navigating the hierarchy to complete each scenario and perhaps each participant's degree of success in completing each scenario as well—for example, *direct success*, taking the optimal path through an information hierarchy to find the desired item; *indirect success*, ultimately finding the desired item, but with some incorrect choices along the way; *direct failure*, taking a path that does *not* lead to the desired information; or *indirect failure*, ultimately failing to find the desired item, but with some correct choices.

Tree-testing tools usually indicate the median time all participants took to complete a task scenario, as well as the range between the lowest and highest amounts of time. Some tools also track the categories that participants clicked first when beginning a task scenario, as well as which categories participants clicked throughout an entire task.

Data analysis—Tree-testing tools automate the preliminary analysis of your findings. In other respects, your approach to data analysis should be similar to that for a card-based classification–evaluation activity.

Level of effort—Tree testing is a quick, easy method of research that you can employ iteratively throughout *Design*. Allocate just 15 to 20 minutes to each participant's tree-testing session. While preparing for tree testing requires a similar level of effort as card-based classification evaluation, using an online tree-testing tool greatly reduces the effort of recording data and preparing it for analysis.

Investment—Like card-based classification evaluation, tree testing requires little expense. However, the additional cost of subscribing to an online tree-testing tool does increase the investment that is required. Nevertheless, tree testing delivers a high return on investment.

[35, 36, 37, 38]

Data analytics

You can learn about the actual behaviors of the people who use your existing information space or product from data that your organization already collects, especially the following types of data analytics:

- Usage-data analytics
- Path analysis
- Search-log analytics

These analytics systems automatically log all the actions your users take, so leveraging such analytics data is particularly important for information spaces and products that comprise large collections of content.

You could also gather data about an information space or product from other sources, such as the following:

- **User-sentiment data**—Automating the process of monitoring the continuous stream of self-reported, qualitative data from relevant social-media platforms enables you to generate hypotheses about users' attitudes and perspectives, the reasons for their behaviors, and the painpoints they experience. *Social-sharing data* results from actual users' sharing an information space or product's content on various social-media platforms and indicates what content they find most valuable. The consistent use of hashtags on Twitter can facilitate the process of gathering feedback on an information space or product or its competitors. Social-sharing data can convey both user sentiment *and* engagement. *Comments* and *reviews* that users post about an information space or product or its content provide positive and negative feedback that might be helpful in improving them. Users might have posted comments on an article page, discussion forum, or social-media posts about an article.

- **Support data**—Gathering internal data is another means of obtaining self-reported, qualitative data in the form of verbatim feedback from actual users, but also quantitative incident metrics. The data from an organization's Customer Support department might include various types of customer feedback, including specific requests for content or features or questions about specific topics. The data from a Technical Support department or Help Desk might include information about specific issues users have encountered or questions about how to accomplish their goals. Users can contact Support via phone, chat, or email or look for solutions to common problems on an organization's Support Web site—which your team should instrument to capture Web analytics data, enabling you to analyze the site's usage. Categorize the reasons for which users have needed support, then tally the number of incidents regarding each category.

[39, 40, 41, 42]

Usage-data analytics

Purpose—Although broadly applicable in identifying strategically important pages and costly painpoints on information spaces or products and other user interfaces, usage-data analytics is particularly informative to the design of information architectures. By tracking, measuring, and analyzing the usage data for an existing information space or product, you can learn how real *users*, or visitors, are actually engaging with it and its content across time—rather than just observe what UX research participants are doing at one moment during a research session.

You can discover and analyze users' behavior patterns. Synthesizing and analyzing usage patterns lets you learn both what an information space or product's users *are* doing and what they're *not* doing. Thus, you can gain a deeper understanding of users' needs that enables you to recognize opportunities for optimizing an information architecture and informs product and design strategies.

An evaluative form of research, usage-data analytics lets you assess how well an existing information space or product's information architecture and individual pages or groups of pages are performing with users. Analytics data provide reliable, convincing evidence that supports decision-making throughout requirements definition and iterative design and helps you to prioritize necessary changes. The immediacy of working with analytics data facilitates continuous improvement.

Analytics data such as Web analytics can help you to identify categories of content that users have had difficulty finding—perhaps because some specific content categories, their labeling, or the overall structure of the information hierarchy is confusing to users. Or, the content-management team might have assigned some content to less than optimal categories, impairing its findability and causing that content to underperform with users.

By analyzing an existing information space or product's analytics data, you can learn what *is* or is *not* currently working well. What pages are getting the most visitors? What content do users spend the most time consuming? What content are users downloading? In contrast, where are users having difficulty completing their information-seeking tasks? At what points are users encountering insurmountable obstacles that are preventing them from finding the information they need? From what pages are users choosing to exit the information space or product? Analytics data can help you to quickly identify such painpoints and other opportunities for improvement, which you should then investigate further by conducting qualitative UX research. Leveraging usage-data analytics enables project teams to quickly hypothesize and test design solutions to address the problems they identify, minimizing their negative impacts on users and reducing support costs.

Description—In preparing to implement usage-data analytics for an information space or product, you must first define the Key Performance Indicators (KPIs) you should measure to help ensure that you meet the organization's business goals. Then, you can configure the data-analytics system to provide the data you need to inform your business strategy and design decisions. Determine what metrics would be most helpful in identifying issues and redesigning the information architecture. Once you've identified issues with the information space or product's information hierarchy, content categories, and labeling, you might decide that you need to conduct additional, complementary methods of UX research to understand users' behaviors and the *whys* behind them. This and the following subsections—four subsections in total should be moved

Defining Key Performance Indicators

Before implementing usage-data analytics for the information space or product, you must set clear business goals, decide what user-experience outcomes to measure, and define actionable KPIs that clearly indicate success or failure. Determine specific numbers that you intend to achieve for each measurable outcome, or *metric*—for example, the total *number* of times a specific user behavior has occurred, the *average* time users spent completing a particular action, or the *percentage* by which a metric has increased or declined. Educate your entire project team about the effective use of these KPIs. However, bear in mind that, as your team gains a better understanding of the usefulness of particular metrics, defining your KPIs could become an iterative process.

Neil Bhapkar, a digital-content marketing professional, has defined several categories of KPIs that leverage usage-data analytics and are useful in measuring the effectiveness and impact of an information space or product's content, including those that measure the following:

- **Reach**—Data-analytics systems can measure reach—particularly data about unique visitors who have visited an information space or product or a specific page during a specified period of time, the geographic locations of visitors by country, and mobile readership on both mobile phones and tablets.

- **Engagement**—Data analytics can measure visitors' interest in and engagement with specific pages—by counting page views and providing a rough calculation of the average time users spend on a page. Heatmaps can show how visitors scan or read the content on a page, and click patterns show how they engage with specific elements on a page such as navigation and other links. Data analytics can indicate all visitors' engagement with an entire information space or product—by providing a rough calculation of the average duration of user *sessions*, or visits; the average number of pages users have visited during a session; and individual pages' *bounce rate*—the percentage of visitors who left the information space after viewing *only* that one page—or *exit rate*—the percentage of visitors who left the information space after viewing that page. Because of the typical inaccuracy of session-duration and time-on-page numbers—resulting from analytics systems' inability to perceive users' attention or reliably recognize when users have left a page or information space—you should ideally analyze trends for specific metrics rather than the actual numbers. Key engagement KPIs should relate a narrative about the ways in which target audiences interact with an information space or product's content to achieve their goals over time.

- **Sentiment**—As I described earlier in this chapter, this KPI relies on data analytics for social sharing and users' comments.

In addition to these KPIs, if an information space or product's users must subscribe or create an account and sign in to use their account, you might also want to collect usage data that supports the following KPIs:

- **Acquisition**—You can collect user-acquisition data, which captures the number of new users who have subscribed or created new accounts.

- **Loyalty**—You can gather loyalty data such as how frequently particular users sign in and how recently they've signed in, as well as *user-retention data* such as the overall percentage of users you've retained over a specified period.

Configuring a data-analytics systems

The data-analytics systems that organizations can implement to automatically log usage data include Web-analytics systems, proprietary instrumentation code, and server logs. When setting up a data-analytics system, you need to embed its analytics tracking code in the correct location on *every* page of the information space or product. To achieve optimal results, you must first determine what data you need to gather to learn what you want to learn, then customize the data-analytics system's setup to ensure that it provides all the relevant data. Configure the system to exclude, or *filter*, traffic from spiders and bots that could distort your analytics data by artificially driving up the numbers.

You can identify and follow the behaviors of individual visitors or users, *if* your information space or product requires users to sign in, uses a unique identifier such as a mobile number, *or* less reliably, uses cookies to identify users—or at least users who are using the same device *and* Web browser. However, analytics systems often cannot identify the same users across different devices or browsers. You can also compare the behaviors of new versus returning users. However, be sure to implement the instrumentation of your information space or product to capture usage data in a way that anonymizes individual users and, thus, protects their privacy and presents their usage data *only* in aggregate. No information that identifies individual users should appear in the analytics data.

To improve the efficiency with which you can analyze the performance of pages—especially pages that employ the same template—you can customize *content grouping* to segment your data by one or more page types or content types, depending on your organization's needs.

You can set up demographics reports that enable you to understand how different target audiences are engaging with the information space or product. You can *segment* the data by *user groups*, or audiences, which comprise users whose behaviors or attributes are similar, then view the data for a single user group or compare the data for different user groups. You can segment the data by individual or multiple attributes and behaviors of users. For example, you could segment users by their personal *attributes* such as their language, interests, gender, or age; their location by country or city; or the type of device and Web browser they used to visit the information space. Alternatively, you could segment the data according to users' *behaviors* such as what pages they've visited; how often they visit the information space; or their traffic source, or channel, which could be organic or paid search—plus the keywords the user typed—a link on a referring site, a social-media link, or direct traffic from users who typed a Web address into their browser. Alternatively, you could base your segments on user sessions or specific user journeys.

To determine what users are actually doing on individual pages, you must set up custom *event tracking* by adding an event-tracking code to specific page elements, which tracks users' clicks on those elements. For example, you could track users' clicks to scroll, download a PDF document, traverse an outbound link, play video or audio, view a presentation, or interact with a form field, menu, button, or icon. You could capture each event's category, action, label, and value.

You could also set up click-mapping tools to track users' behaviors on specific pages that are of strategic importance to an organization. These tools generate heatmaps that show the page elements on which users are clicking, as well as scrollmaps that visualize the time users are spending on each section of a page. Some of these tools can also record individual users' entire sessions, capturing videos of their on-screen behaviors—for example, what controls they click or tap, what links they click or tap to navigate to other pages, pointer movements, and typing.

You can configure your analytics system to track custom goals for specific interactions that correspond to particular KPIs. The system can calculate *page value*, the monetary value of individual pages, using the following formula: **Total Goal Value** divided by **Unique Page Views** = **Page Value**. A page's *total goal value* represents the combined value of the goals for that page and derives from users' visiting that page during user journeys that have resulted in conversions. For an ecommerce page, the *total goal value* includes the page's transaction revenue.

You can also set up an analytics system to track a visitor's usage of the information space or a product's internal search system.

Add annotations to the analytics data to identify any points at which significant events or changes have occurred that might have caused spikes or dips in the information space or product's traffic—for example, launches of redesigns or new features, reconfigurations of the data-analytics system, or major industry events. Such annotations are helpful during analysis.

Metrics that are particularly useful to information architects

Let's explore some of the metrics that are available in an example data-analytics system: Google Analytics, which generates Web analytics, including usage data, or *metrics*, that can help you identify issues with a site's information hierarchy, as well as its content categories and their labeling. According to Nielsen Norman Group (NN/g), some metrics that are particularly useful in evaluating an information space or product's information architecture include the following:

- **Page views**—This is the number of times within a specified period that users have visited the information space or product's pages *or* visited specific pages, enabling you to ascertain pages' popularity, identify pages that receive low traffic, monitor trends in pages' popularity over time, and perhaps determine whether specific events have prompted users to view a page—for example, a navigation-system redesign, an advertising campaign, or a news event. *Unique page views* is the number of times users have viewed specific pages, excluding repeat visits to the same page during the same user session. With the appropriate JavaScript code, you can also measure *virtual page views*, which occur whenever new content loads onto a page without reloading the entire page. Page views, in combination with the time users have spent on a page, clearly indicate users' engagement with specific pages and enable you to assess the usefulness of their content. You can evaluate users' engagement with specific categories of content by either tallying the page views for *all* pages belonging to a category *or* looking at the page views for a specific category page. Filtering out all repeat page views during the same user session prevents users' pogo-sticking from a category page to content pages and back from inflating page views for that category page. When evaluating a category page that has low traffic or is experiencing attrition in page views, consider the category's strategic importance to both the organization and users and determine whether it's getting the expected number of page views. If not, remedying the problem might require you to redesign the category page's layout or visual design, change the labeling for the category, or even restructure the information hierarchy for the information space or product. To determine whether the page views for a category page are comparable to those of other category pages, total the page views for *all* category pages, then divide that total by the overall number of category pages to calculate category pages' average number of page views. You can then compare a category page's page views to that average. If any category page receives very few page views, you could consider eliminating that category—unless it is of interest to particularly valuable users.

- **Conversions**—A *conversion rate* is the percentage of user sessions that involve a conversion. Determine what *conversions*—user actions that deliver business value to the organization and help it achieve its goals—are strategically important to the organization. To evaluate conversions—such as signups, document downloads, or purchases—your team would likely need to configure customized goals in your analytics system. A popular content category that generates few conversions might lack strategic importance. However, you should consider whether its category page or other pages that belong to the category might be driving traffic to category pages that *are* strategically important and generate valuable conversions.

- **Bounce rates**—The purpose of a category page is to provide links to the content belonging to that category and, thus, drive traffic to that content. If, for any reason, users decide that they are not interested in the links on the category page and do *not* click them, that page would have a high *bounce rate* or rate of abandonment. This often happens because a category is poorly labeled, so users do not see links to the content they had expected to find within that category. Some of the links on the content page could themselves be poorly labeled. Or, a category page that has a less-than-optimal visual design or page layout might prevent users from recognizing the content they're seeking.

- **Entrance rates**—A *landing page*, or entrance, is the page on which a user enters the information space or product at the beginning of a user session. Landing pages can provide valuable opportunities to increase the information space or product's audience. To identify what landing pages are attracting new users, filter entrances to view *only new visitors*, who have not previously visited the information space or product. A category page is one type of landing page. Category pages that have high entrance rates in comparison to other category pages could be some of the most important entrances. Category pages that have unexpectedly low entrance rates might be poorly labeled or have ineffective search-engine optimization (SEO). To prevent your eliminating categories whose category pages are valuable entrances, always check a category page's entrance rate first. Also, avoid eliminating a category page that has a low entrance rate, but a high conversion rate among the audience it attracts.

- **Search-query volume**—Generate a list of an information space or product's most common search queries, as well as lists of the queries of specific user groups. A high volume of site-search queries that employ keywords relating to strategically important content could indicate that users are having difficulty finding that content by browsing the information space or product. Search queries also indicate content that is of interest to your users so could suggest some categories of content that are missing from the information space or product, which you should consider adding.

Complementary methods of UX research

Since a data-analytics system can generate data only after you've launched an information space or product, you *must* rely on other types of UX research to inform the definition of the requirements, design, and implementation of a *new* information space or product.

Once you've launched an information space or product, you can use relevant demographic data about actual users and their behaviors in creating new user personas that represent its key target audiences. You should also conduct complementary, user-centered forms of UX research to represent specific segments of your audience, such as user interviews, contextual analyses, observations, and surveys to ensure that you fully understand each of these user groups. Your analytics data can inform that UX research, as well as its analysis. Leverage *all* of your learnings about the information space's key target audiences to inform the creation of personas.

After creating representative personas, you can segment the information space's audience by user groups that correspond to these personas. You can then track specific user groups' behavior on the information space. By recruiting participants who match these personas' essential characteristics for your UX research, you can gain reliable insights into the ways in which different groups of target users engage with the information space.

While usage-data analytics can show you what content users are viewing and how they're engaging with it, they cannot reveal the intentions motivating users' behaviors. To enable your team to understand the *whys* behind users' interactions, it is necessary to triangulate your quantitative analytics data by conducting complementary, qualitative user-centered methods of UX research. For example, to better understand why users are taking certain actions, you could conduct user interviews or a usability study or intercept individual users immediately after they've taken an action and ask them to complete a brief survey about it.

Once you've analyzed an information space's metrics to identify its common user journeys and their painpoints, you can focus an expert review, usability testing, or another method of *evaluative* research on better understanding those painpoints. For example, evaluating a specific task scenario or user journey could help you understand how users are navigating the information space and where they're encountering obstacles. Plus, by identifying the most common user journeys, you can determine what task scenarios you should test during usability studies.

Conduct usability testing to better understand any problematic user journeys you discover through your analytics data—for example, users' failing to find specific information that resides on the information space or product—then optimize the content categories and information hierarchy through iterative design and testing to remedy their painpoints.

You can conduct discrete experiments by iteratively conducting A/B testing—comparing your hypothesis for a redesign either to the previous version or to a different hypothesis—then making small, focused, incremental changes to the information architecture or user interface. Then, following these experiments, you can measure the impacts of your design changes on whatever metrics would be most relevant. By doing multiple experiments, you can extend your learnings across the entire information space or product.

Alternatively, you can use the usage data for the period prior to your implementing design changes as a benchmark against which to measure the impact of those changes. To ensure that it's clear what specific design changes have had what impacts, avoid making too many changes at once. To show the return on investment (ROI) from your design and research work, track improvements to your KPIs over time, demonstrating how the impacts of specific design changes have effected those improvements.

When—You should monitor and analyze usage data for an existing information space or product during *Discovery* or, for a new information space or product, immediately after its launch. Your learnings can then help you determine user requirements for the redesign of its information architecture. You can also monitor and analyze usage data throughout an iterative process of design and testing.

Where—You can analyze usage data anywhere, using an online analytics system.

Participants—*All* the actual users of an information space or product unknowingly contribute to your learnings from usage-data analytics. It is essential that you maintain their privacy by anonymizing all usage data and presenting it *only* in aggregate.

Type of data—Because a data-analytics system, if set up optimally, automatically gathers data on *all* of an information space or product's pages, it can generate large, even massive volumes of accurate, quantitative, statistically significant, summative data. This usage data can give you an overview of all users' actual behavior patterns when using the information space or product and engaging with its content, providing quantitative evidence that is very convincing to stakeholders and enabling you to determine whether the information space is meeting your KPIs. In addition to informing your team about exactly *what* users are doing on the information space, usage data can tell you how much time it takes users to complete specific usage scenarios. However, usage data can tell you nothing about *why* users are taking specific actions. See the section "Metrics that are particularly useful to information architects," later in this chapter.

Data analysis—While analyzing such a large volume of data enables you to identify both the typical behavior patterns of users and outliers, it can be very time consuming. So, you must allocate sufficient time and resources for data synthesis and analysis. While one or more analytics experts might be responsible for analyzing analytics data, an information architect or other UX professional can provide a user-centered perspective.

While usage data can tell you a lot about the relative performance of particular pages, you should try to consider the broader context so you can understand the reasons behind the numbers. Therefore, before analyzing the data, review any existing annotations that might be relevant to your analysis, which could reveal the additional context that is necessary to accurately interpret the numbers. When events coincide, beware of assuming causation.

First, you should quickly get an overview of the available usage data and do a preliminary analysis of that data to learn about the typical levels of daily traffic on the information space, common attributes of its users, what pages are getting the most page views, and what pages or sections of the information space are receiving the least traffic and, thus, performing poorly with users.

Then, begin your analysis by taking a deep dive into the details that matter—the usage data that can inform your team's business strategy and design strategy for the information space or product. To focus your analysis, interpret usage data and related KPIs within the context of the organization's business goals, users' goals, the information architecture and content strategy for the existing information space or product, and their impact on search-engine optimization (SEO). Only once you've considered all of these factors can you determine whether you need to make changes to the information architecture to address any issues that you've discovered.

When you're analyzing usage data to determine how well an information space or product is performing with users, your data should generally represent extended periods of time such as months or years. Looking for trends in specific metrics over time rather than focusing on actual numbers provides greater context. You can compare the data from different years to identify seasonal usage patterns or year-over-year trends—for example, to find out whether key metrics such as page views or conversions are increasing or decreasing.

If you've configured user groups, you can analyze the behaviors or outcomes of the different user groups, or audiences, who are engaging with the information space by segmenting the usage data in the analytics system's reports. For example, since you can segment the data by individual *or* multiple attributes and behaviors of users, you could decide to segment the data by demographic groups to which users belong, then by the pages they're viewing, and learn what content is of interest to certain user groups. Data about how much time users are spending on the information space could indicate whether they're seeking particular information or just browsing. Segmenting users by traffic source provides information about the context from which they came to the information space, including their interests. Direct traffic could indicate users' previous awareness of the organization to which the information space or product belongs.

Your analysis should concentrate primarily on identifying issues that you need to address. Initially, focus on the usage data for specific pages that are either strategically important or receive high traffic. Consider a page's bounce rate and exit rate and the average time users spend on that page to determine whether the page is causing users to leave the information space. A high exit rate for a page that users would view at a midpoint in a user journey indicates a painpoint that has caused users to abandon the journey's goal. Also, look at the page value or total goal value for each page. High page values indicate pages that are part of key customer journeys *and* result in valuable conversions, enabling you to identify the most strategically important pages on an information space. Focus on improving pages that have both high page values *and* high exit rates. They represent painpoints that are preventing users from converting. Low page values indicate underperforming pages.

If you've configured content grouping, you can evaluate the performance of specific content types and page types, using the same metrics as for individual pages. Content groups with high bounce rates or exit rates or low times on page probably indicate issues. Calculate the average page value for each content group, then evaluate groups that have particularly high values or low values. Determine where in specific user journeys users are encountering issues that you need to address. For more information about user journeys, see the next section, "Path analysis."

Analyzing usage data can help you to identify common usage scenarios, or user journeys, the painpoints at which users have encountered obstacles and abandoned their attempt to complete specific user journeys, and identify unexpected actions, revealing the underserved needs of users whose usage patterns differ from those that are typical.

If you've set up event tracking, you can monitor users' behaviors on specific pages, which enables you to understand how users are engaging with the content and interacting with the functionality on those pages. Segmentation of this data by page types or content types *or* users' attributes, device types, or countries would likely to provide additional insights.

You can use analytics data to identify usability issues relating to specific Web browsers on specific *device categories*—desktops, tablets, or mobile devices—and *device types*—such as Android, iPhone, or iPad. This data can obviate the need for extensive, time-consuming, cross-browser testing, which could never be comprehensive anyway. Compare conversion rates or pages per session for specific device categories, whose differences typically depend on the contexts in which users employ particular devices, but could also be the consequence of design issues.

Your analysis of usage data should be actionable and determine your team's next priorities. Analytics can quantify the impacts of the issues that you identify. Because pages that have high page values are usually important pages, addressing any issues on these pages that prevent users from completing their goals is a high priority and might signal the need for additional UX research. Following your analysis, you might decide to do further UX research. For example, you could conduct user interviews to understand the *whys* behind users' navigation decisions or conduct an expert review or usability testing to better understand the obstacles users are encountering in attempting to complete specific usage scenarios.

Usage data represent the actual behaviors of real users, but do not reveal the motivations behind those behaviors. During analysis, you need to consider *why* users have taken specific actions. Therefore, you must first synthesize relevant findings from both the analytics data *and* complementary forms of UX research, enabling you to understand users' behaviors, identify issues they've encountered, propose and test possible solutions for these issues, and determine what actions to take to remedy them—for example, you might address information-architecture issues by moving, renaming, or removing specific categories of information or changing the information hierarchy. For more information about complementary UX research, see the section "Complememtary methods of UX research," later in this chapter.

Level of effort—Once your team has set up the information space or product's analytics system correctly, it provides a quick, easy means of gathering and monitoring quantitative usage data, enabling you to continually obtain answers to your questions about how users are engaging with the information space or product or multivariate. Data analytics offers similar—though much broader—benefits to A/B testing, but at a much lower cost. See the section "Configuring a data-analytics system," later in this chapter.

Deliverables—Periodically deliver brief, easy-to-read data-analytics reports to everyone on your project team and all other interested stakeholders, providing them as frequently as weekly. Structure your reports to convey a meaningful narrative, integrating various usage data *and* findings from user-centered research to support that narrative. For clarity, avoid the overuse of analytics jargon and emphasize visualizations of the data to ensure its easy consumption by stakeholders. The findings in your reports should be actionable. Include relevant, quantitative usage data and your analysis of these metrics and their related KPIs in your presentations to stakeholders, enabling everyone to leverage your findings in business-strategy, content-strategy, and design-strategy discussions and documentation. Broadly distribute your reports and make them available to the entire organization online to ensure that anyone who is interested can access these reports. Provide access to the raw analytics data as well, as appropriate. Elicit feedback on your data-analytics reports from your project team and other stakeholders, then iteratively refine the reports to ensure that they include all the information stakeholders need.

Investment—Usage-data analytics is a low-cost method of UX research that can focus your organization's financial investment on the design and development of improvements to an information space or product for which your team has verified a real need. [43, 44, 45, 46, 47, 48, 49, 50]

Recommended book—Luke Hay. *Researching UX: Analytics*. Collingwood, Victoria, Australia: SitePoint, 2017.

Path analysis

Purpose—As Luke Hay says in his book, *Researching UX: Analytics*, "Looking at pages or groups of pages in isolation will only ever tell you part of the story." An analytics system's user-flow and behavior-flow reports let you track the specific paths that users have taken when using an existing information space or product, enabling you to conduct a path, or *clickstream*, analysis, which provides greater context and helps you to identify and analyze the most common user journeys.

You can determine what paths users have taken when entering the information space and understand how users generally access its content—for example, by starting from the home page or specific landing pages, using the navigation system, or using site search. You can track what links users have clicked when navigating the information space to seek specific information or perform particular tasks. You can ascertain the *drop-off rates* at specific points within user journeys, at which users have abandoned interactions that comprise multiple steps, revealing sources of issues that are preventing users from completing their tasks. Plus, you can see the points at which users have deviated from the expected paths of common user journeys, which could indicate that they became confused.

Description—Before conducting a path analysis, define the questions that you want your analysis to answer. Use clickstream-analysis software to log, monitor, and analyze the paths that users take when using the information space, including any originating sites from which users navigate to the information space, the pathways users follow through the information space, and any destination sites to which users navigate from the information space. Note the amount of time users spend on each page they visit. For pages with high bounce or exit rates, determine whether those pages or the pages that precede them in user flows have issues that might have caused users to leave the information space.

You can set up custom goals for specific user interactions that correspond to particular KPIs. Configuring goal funnels lets you track and analyze entire user journeys for interactions that comprise multiple steps, across several pages, and see the drop-off rate, or *abandonment rate*, at each step of the journey—as both the percentage and the total number of users—versus the users who have successfully advanced to the next step.

Once your path analysis has identified the points at which users are encountering issues that are preventing them from completing their goals, you need to distinguish precisely what is causing each issue, then determine how to remedy it. Focus your efforts primarily on solutions that could deliver an increase in conversions and, thus, provide the highest return on investment. Once you've implemented a solution, you can assess its impact using the same goal funnel.

Follow up your path analysis with user interviews or surveys to learn why users have visited your information space or product, what useful information they found, what barriers they encountered that prevented their finding the information they were seeking, and *why* they navigated away from your information space. Use all of this data to inform the creation of user personas. Once you've created representative personas, you can segment the audience by user groups that correspond to those personas. You can then track what pathways these user groups are taking through the information space and the pages from which they're exiting the information space. You can also compare the behaviors of different user groups.

Your identification and understanding of an information space or product's most common user journeys should inform the definition of the task scenarios that you should test during usability studies. By evaluating a specific user journey, you can understand how users are navigating the existing information space, *why* users are clicking specific links, and where they're encountering obstacles.

When—You should conduct a path analysis for an existing information space or product during *Discovery*, when your findings can help you to determine the user requirements for a redesign of its information architecture.

Where—You can conduct a path analysis anywhere, using either an online analytics system or clickstream-analysis software.

Participants—*All* of an information space or product's actual users unknowingly provide path, or *clickstream*, data and, thus, contribute to your learnings from a path analysis. You must maintain users' privacy by anonymizing their path-analytics data and presenting it *only* in aggregate.

Type of data—A path analysis generates large amounts of automatically logged, reliably accurate, statistically significant, quantitative path, or *clickstream*, data. This quantitative evidence is convincing to stakeholders and helps you assess whether the information space is meeting your KPIs. You can learn exactly *what* paths users are taking through the information space and how much time it takes them to complete specific user journeys, but not *why* users have chosen to take those paths.

Data analysis—The analysis of users' navigation paths through an information space is key to understanding how well that information architecture is performing, so be sure that you schedule sufficient time for conducting a path analysis. Focus on identifying key user journeys, analyzing users' degree of success in achieving their goals, and understanding what issues are preventing them from completing these user journeys—such as common errors that users are making.

By analyzing the funnel reports for specific user journeys comprising multiple steps, you can identify what pages are causing users to abandon those user journeys. Consider each page's context in the overall user journey. Look for the pages with the highest drop-out rates—for example, pages that appear at an early stage of a user journey when users might not yet be fully committed to accomplishing a goal. But identifying pages that have a high number of drop-outs because users are abandoning a user journey at a later stage in the funnel, when they are more invested in achieving their goal, is even more significant because addressing such issues could enable you to recover more value.

Level of effort—Once your team has set up path analysis in your data-analytics system or *clickstream-analysis* software for the information space or product, path analysis provides a quick, easy means of continually obtaining answers to your questions about what paths users are taking on the information space.

Deliverables—Periodically provide brief, easy-to-read path-analysis reports to your team and other interested stakeholders. Your reports should convey a meaningful narrative about an information space or product's most common user journeys, integrating path-analysis data *and* your findings from user-centered research. Avoid the overuse of analytics jargon and emphasize visualizations of the data to ensure its easy consumption by stakeholders. Ensure that the findings in your reports are actionable. Include both the quantitative path-analysis data and your analysis of their related KPIs in your presentations to stakeholders, enabling your team to leverage your findings in business-strategy, content-strategy, and design-strategy discussions and documentation. Broadly distribute these reports and make them available to the entire organization online. Provide access to the raw path-analysis data as well, as appropriate. Elicit feedback on your path-analysis reports from your project team and other stakeholders, then iteratively refine the reports to ensure that they include all the information stakeholders need.

Investment—Path analysis is a low-cost method of UX research and enables organizations to focus their financial investment on the user journeys on an information space or product that deliver the most value. [51, 52, 53, 54]

Recommended book—Luke Hay. *Researching UX: Analytics*. Collingwood, Victoria, Australia: SitePoint, 2017.

Search-log analytics

Purpose—By analyzing an information space or product's search-log analytics, you can discover what *keywords* and phrases, or *search queries*, users are submitting to its search engine in endeavoring to find specific information on that information space. Because users' search queries hold meaning for them, they clearly convey users' *intent* to seek specific information. Thus, search queries can tell you what information is of interest to your users, how they think about the information they're seeking, and what keywords and phrases they typically use in describing the information for which they're searching.

In this way, search-log analysis enables you to learn the vocabulary of your users. Do they use the plain language of the general public or industry jargon and acronyms? Do they know the names of the organization's products or refer to products using generic terms? Are they using the terms you expected them to use when referring to specific concepts or a variety of different terms? Learning your users' vocabulary informs and enables you to improve the labeling of links, page titles, and other user-interface text, as well as the language that authors use when creating the information space's content and the content managers use in its metadata. Search-log analysis can also inform the development of controlled vocabularies, or *thesauri*. By making *all* of this language consistent with your users' actual vocabulary, you can greatly improve search results' *relevance* to users. But you must balance two countervailing tendencies: the search engine's *precision*, or the percentage of results that are relevant to users and align with their needs, and its *recall*, or the percentage of relevant results that the search engine retrieves.

Once you've analyzed an information space or product's search logs and know what information users are seeking, you can assess how well its information architecture and content are meeting users' needs. Are there categories of content that users are either unable to find or have difficulty finding by browsing? Is some content that users need missing from the information space?

By iteratively conducting search-log analyses, you can identify, then remedy any issues that are negatively impacting search information retrieval. To explore and understand your findings from search-log analyses more deeply, conduct complementary methods of qualitative user research such as user interviews or contextual analyses. Consider doing an open card sort to determine common groupings of keywords, as well as the optimal terminology to use in labeling and metadata.

Through search-log analytics, you can greatly increase the quality of the search results that the information space's search engine delivers, which ultimately improves users' satisfaction with the information space, user retention, and conversion rates, and, thus, delivers better business outcomes.

Description—Search analysts are generally responsible for tracking and evaluating the search-log analytics for an information space and determining how to improve them. An information space's search logs capture real users' actual search queries, which describe what the users were seeking in their own words.

Setting up a more sophisticated search query–analysis tool rather than just relying on usage-data analytics enables you to manipulate your data in more ways. For example, to include data from particular user sessions in your report *or* exclude such data from your report, you can filter queries by date and time, by users' location, or by IP address. Thus, you could, for example, exclude data for internal traffic that might skew your numbers. These tools can also identify popular queries that are retrieving *no* results and help you determine whether there actually is no relevant content on the information space *or* users' keywords have been ineffective in finding the content that does exist. You can also use these tools to determine whether specific queries are becoming more or less popular with users. Plus, these tools can tell you what popular queries are retrieving very large numbers of search results and help you to determine more specifically what users were actually seeking. However, the specific metrics that you can track depend upon the capabilities of your search query–analysis tool. For in-depth information about the performance metrics and KPIs that you can track, see the section "Search system–performance metrics," later in this chapter.

Search system–performance metrics and KPIs

Before you conduct a search-log analysis, you should define the questions that you want your analysis to answer. Determine what search system–performance metrics would be most helpful in identifying issues that are preventing users from finding the information they need and would inform your product and content strategy, as well as the redesign of the search system and information architecture. Clearly defining what KPIs you need to measure enables you to understand the organization's specific business goals and verify that the information space is meeting them. Also, consider what search metrics would have the greatest impact on achieving the KPIs the organization has defined for the information space and the business.

Marko Hurst, an expert in information-retrieval systems, has compiled a useful, albeit not exhaustive, list of search metrics for Web sites' or other information spaces' internal search engines that provided the basis for the following list of metrics:

- **Percentage of search queries, or search sessions, for which users clicked a search result**—A *low* percentage typically indicates that the search engine is frequently failing to retrieve relevant search results for users—*unless* the search-results page itself is delivering the information users need, as is becoming increasingly common, making this metric less useful. A *high* percentage indicates that the search engine typically retrieves highly relevant or engaging search results. Teams often define a KPI based on this metric.

- **Percentage of search queries for which users clicked no search result**—A *high* percentage is a key indicator that the search engine is frequently failing to retrieve search results that are relevant to users and, thus, meet their expectations. Determine the keywords for which the search engine is unable to find useful search results—for example, because that content does not exist on the information space or the matching content's metadata does not clearly identify that content. Address issues that are preventing the search engine from retrieving relevant results.

- **Percentage of search queries that retrieved zero results**—If a search query fails to retrieve any results, this likely indicates that either content that would be of interest to your users is missing from the information space; the search engine is delivering low-quality search results because, for example, it does *not* check and, if necessary, correct the spelling of keywords, stem words by reducing them to their root form, substitute keywords with synonyms from a controlled vocabulary, or for overly constrained search queries, provide partial matches; *or* the metadata for any matching content that does exist on the information space does not clearly identify that content. A *high* percentage indicates that the search engine often fails to retrieve *any* search results for users' queries. Determine what relevant search queries are retrieving no results and address the issues that are preventing the search engine from retrieving relevant results. Teams often define a KPI based on this metric.

- **Percentage of search queries for which users exited the information space from a search-results page**—This *search exit rate* indicates that users are leaving the information space immediately after searching, without clicking any search results. While a *high* percentage could indicate that the search-results pages are poorly designed and provide a poor user experience, this typically indicates that the search engine often fails to retrieve any relevant search results that meet users' needs, so users have likely concluded that the information they're seeking does *not* exist on the information space. Teams often define a KPI based on this metric.

- **Percentage of user sessions during which users performed a search**—A *high* percentage indicates that users often rely on searching rather than browsing for information—either because they generally prefer searching over browsing; they have usually been successful in finding the information they need using the search system; *or* they've found the navigation system frustrating. Users' heavy reliance on searching could provide a strong justification for making a greater investment in the information space's search system. Teams often define a KPI based on this metric.

- **Average number of searches per user session**—A *high* number indicates that users rely heavily on searching rather than browsing for information—either because they generally prefer searching over browsing, have usually found the information they need using the search system, *or* found the navigation system frustrating. However, if users *also* frequently refine their search queries, repeatedly search using the same keywords, or search for synonymous keywords, a high number could indicate that the search engine is retrieving poor-quality search results. A *low* number of searches per session likely indicates that users are using the navigation system and browsing successfully, so are searching less frequently.

- **Average time users spent on search-results pages**—This metric is a strong indicator of the usability of the search-results pages. A *high* amount of time spent across *multiple* search-results pages likely indicates that these pages are usable, the results provide sufficient information scent and lead to relevant information, and users were fully engaged in their information-seeking task. However, a *high* amount of time spent on just the first page of results probably indicates that users found the results page confusing or completely unusable—perhaps because each search result includes an overwhelming amount of information. A *low* amount of time spent on the first page of results likely indicates that highly relevant search results appeared early among the results, and users could easily identify relevant results and traverse their links.

- **Average number of search-results pages users viewed for each search query**—This metric is a strong indicator of the relevance *and* quality of the search results. Users' *not* clicking any results on the first page of results, then viewing a *high* number of results pages indicates that the search engine is *not* displaying the most relevant results on the first results page. However, if users are clicking search results on the first *and* other search-results pages, a *high* number likely indicates that the search results are relevant and have good information scent, and users are fully engaged in their information-seeking task. A *low* number usually indicates that highly relevant search results that fully meet users' needs appear early among the results on the first search-results page.

- **Average number of pages users viewed before searching**—A *high* number of pages indicates that users typically rely heavily on browsing rather than searching for information, likely because they generally prefer browsing over searching, the information space provides good information scent, users are finding high-quality content or calls to action that are of interest to them, and they are fully engaged in their information-seeking task. A *low* number of pages indicates that users were unable to find the information they needed by browsing or became frustrated with the navigation system. Compare this number to the *Average number of pages users viewed after searching*.

- **Average time users spent on the information space before searching**—A *high* amount of time probably indicates that users prefer browsing for information over searching. However, while the information space's navigation system might have seemed to provide good information scent, users might ultimately have failed to find the information they sought by browsing and become frustrated, so decided to try searching instead. A *low* amount of time might indicate that users strongly prefer searching over browsing *or* quickly became frustrated with browsing. Compare this amount to the *Average session duration for all user sessions that did not include searches*.

- **Average number of pages users viewed after searching**—A *high* number of pages indicates that the information space is providing good information scent; users are finding relevant, high-quality content or calls to action that are of interest to them; and they are fully engaged in their information-seeking task. A *low* number of pages indicates either that users had a limited need for information at the outset of their user journey, which they've satisfied, *or* they're not fully engaged with the information space. Compare this number to the *Average number of pages users viewed before searching*, as well as the *Average time users spent on the information space after searching*.

- **Average time users spent on the information space after searching**—This metric is a strong indicator of users' satisfaction with the information space and the content they've found on it. A *high* amount of time indicates the information space is providing good information scent; users are finding relevant, high-quality content or calls to action that are of interest to them; and they are fully engaged in their information-seeking task. A *low* amount of time indicates either that users had a limited need for information at the outset of their user journey, which they've satisfied, *or* they're not fully engaged with the information space. Compare this amount to the *Average number of pages users viewed after searching*, as well as the *Goal-completion or conversion rate for users who performed a search on the information space*.

- **Average session duration for all user sessions that did not include searches**—This metric is a strong indicator of the effectiveness of the navigation system and the extent to which users explore when browsing. A *high* session duration indicates that the information space provides good information scent; its navigation system is easy to use and users can find relevant, high-quality content that is engaging and meets their needs or calls to action that are of interest to them; and users were fully engaged in their information-seeking task. A *low* session duration might indicate either that users strongly prefer browsing over searching, but were unable to find the information they needed by browsing, *or* they became frustrated with the navigation system. Compare this session duration to the *Average session duration for all user sessions that included searches*, as well as the *Goal-completion or conversion rate for users who performed a search on the information space*.

- **Average session duration for all user sessions that included searches**—This metric is a strong indicator of the effectiveness of the search system and the extent to which users explore after searching. A *high* session duration indicates that the information space provides good information scent; the search system is easy to use and users can find relevant, high-quality content that is engaging and meets their needs; and users were fully engaged in their information-seeking task. A *low* session duration might indicate either that users strongly prefer searching over browsing, but were unable to find the information they needed by searching, *or* they became frustrated with the search system. Compare this duration to the *Average session duration for all user sessions that did not include searches*, as well as the *Goal-completion or conversion rate for users who performed a search on the information space*.

- **Average number of items users added to their cart from ecommerce search results**—For any information space that provides ecommerce capabilities that enable users to search for products, then add them to a cart or basket, a *high* number of items indicates that users are successfully searching for and finding the products they need, then selecting them for purchase. A *low* number of items could indicate that users are unable either to use the search system successfully to find the products they need—perhaps because they are not available for sale—or to successfully add them to their cart, or users have satisfied their very specific shopping needs. By using this metric in concert with the *Goal-completion or conversion rate for users who performed a search on the information space, Most frequent search queries and their corresponding conversion rates*, and *Average time users spent on the information space after searching*, you can make a strong business case for investing in the search system.

- **Goal-completion or conversion rate for users who performed a search on the information space**—A *high* rate of completion or conversion suggests that the information space's search system provides a good user experience, while a *low* rate suggests that the navigation system and browsing provide a better user experience. However, there are many other factors that could influence goal completion or the conversion rate. Teams often define a KPI based on this metric.

- **Percentage of searches resulting in a conversion**—A *high* percentage of searches that result in conversions indicates the overall value of search to business success. Teams often define a KPI based on this metric.

- **Search-query refinement rate**—Users modify, or *refine*, their search queries when their original queries do *not* yield results that meet their expectations, but they want to try searching again, using a slightly different query, to obtain better results. This metric can help you to measure the quality and relevance of search results for specific keywords and, thus, determine for what keywords the search engine is unable to provide useful results. If there are many such keywords and, thus, a *high* search-query refinement rate, you should conduct user research to improve your awareness of the vocabulary of your users. However, if the search system provides faceted search that lets users narrow their search queries or filters that enable users to narrow their search results, choosing a facet or a filter could constitute a refinement, artificially boosting this metric and making it necessary to adjust the rate accordingly. Teams often define a KPI based on this metric.

- **Most frequent search queries**—Compile a list of the search queries that users most commonly use and keep them top of mind when designing the information architecture and especially the navigation system, and when deciding what content to create. Track whether and how users' top search queries change over extended periods of time. These top search queries are both the most popular with and most valuable to users and, thus, are also the most valuable to the organization. Prioritize addressing any issues that are preventing the search engine from retrieving relevant results for users' most popular search queries. Compare this list of queries to that for the *Total number of unique searches.*

- **Most frequent search queries and their corresponding conversion rates**—This metric is a very strong indicator of users' needs. Using this metric, you can track the performance of users' most popular search queries that actually result in conversions, as well as how the conversion rates for these search queries change over extended periods of time. Prioritize addressing whatever issues are preventing the search engine from retrieving relevant results for users' most popular search queries that result in conversions. Teams often define a KPI based on this metric.

- **Top keyword groupings**—This metric indicates what keywords users most commonly use together in their search queries. Prioritize addressing whatever issues are preventing the search engine from retrieving relevant results for users' most popular keyword groupings or phrases.

- **Top decomposed keywords**—This metric decomposes all search queries into the individual keywords that they comprise and orders these keywords according to their popularity. Prioritize addressing whatever issues are preventing the search engine from retrieving relevant results for users' most popular keywords.

- **Top keywords and phrases that returned zero search results**—This metric indicates the most popular keywords and phrases for which the search engine has failed to retrieve *any* search results. These keywords and phrases that are retrieving no results might indicate content that would be of interest to your users, but is missing from the information space; that the search engine is delivering low-quality search results because, for example, it does *not* check and, if necessary, correct the spelling of keywords, stem words by reducing them to their root form, substitute keywords with synonyms from a controlled vocabulary, or for overly constrained search queries, provide partial matches; *or* that the metadata for any matching content that does exist on the information space does not clearly identify that content. Prioritize addressing whatever issues are preventing the search engine from retrieving relevant results for these popular keywords and phrases.

- **Top pages on the information space from which users most frequently initiated a search**—Initiating a search typically indicates that users have become frustrated with the quality of some element of their current context—whether it's the content on the page they're currently viewing, the page's design template, or the navigation system. However, if searching is the *first* action a user takes on the information space, it's likely because the user prefers searching over browsing, so you can exclude such cases from this metric. If users often initiate searches from multiple pages that are based on the same design template, this is a clear indication that the design template is faulty. Teams often define a KPI based on this metric.

- **Total number of unique searches**—This metric reveals the full breadth of the search queries that users have employed when searching for information on the information space. Because this list of search queries captures every unique search query, it is very instructive regarding both what information users are seeking on the information space and the language that users commonly use in describing specific information, which can inform the design of the information architecture for the information space, including its taxonomy, keywords, and other metadata that describe its content, and the labeling of navigation links and page titles. Thus, this data can help you to improve the findability of the information space's content, the relevance of its search results, and its SEO.

Always consider the search system's performance metrics within the broader context of the overall information space and use these metrics in concert with one another. Their context determines what these metrics actually mean and, thus, their impacts on the organization's KPIs. Test your assumptions. Observe how these metrics change over time as you make design changes to remedy issues. Compare the information space's search metrics from month to month to see how users' interests and information needs are shifting over time.

When—Ideally, you should conduct a search-log analysis for an existing information space or product during *Discovery* so you can leverage your findings in defining the user requirements for your redesign of its information architecture, focusing particularly on requirements for improving search results. For a new or redesigned information space, do a search-log analysis immediately after launch. Continually monitor and periodically analyze the information space's search-log analytics to ascertain how its search system's performance is impacting the KPIs your team has defined.

Where—You can conduct a search-log analysis anywhere, using an online search log.

Participants—As with other types of data analytics, an information space or product's actual users unknowingly provide the data in the information space search logs. Be sure to maintain users' privacy by anonymizing this data and presenting it *only* in aggregate.

Type of data—The search log is a text file and includes all of the search queries that users have submitted during the period of time for which you're analyzing the data. The parameters for each search query include the search terms, or *keywords*, the user has submitted and the user's language. For each search query, the search log also typically captures the number of results the search engine retrieved and a *timestamp*, the date and time at which the user submitted the query. For a Web site, the data might also include the IP address and identity of the user—either from tracking a cookie or the user's signing into the information space; the Web address of the search-results page and of the page from which the user searched, and the user agent—that is, the user's browser or app.

Unlike the other types of automatically logged analytics that you can collect for an information space or product, search logs capture semantically rich, *qualitative* data, comprising users' actual search queries, which accurately convey their intent to search for specific content they need.

Search logs can also provide numeric, behavioral data such as the number of times users have employed specific keywords or phrases during a specific timeframe. Typically, a frequency-distribution graph of users' search queries ranges from a small number of high-frequency queries that they commonly enter to the many lower-frequency queries that are part of a long tail of queries, with few instances of each query.

Data analysis—Understanding users' motivations, or *intent*, is key to learning how and why users are or are *not* successful in searching for and finding the information they need. Your insights from analyzing an information space or product's search logs should be actionable and, thus, help the organization achieve its strategic business goals.

In his book *Search Analytics for Your Site*, Lou Rosenfeld outlines five approaches to the analysis of search-log analytics, as follows:

1. **Pattern analysis**—Search-log analysis lets you identify and understand trends in users' search behaviors, including the types of content for which they're searching and the language, or keywords, they're using to describe that content, as well as trends in where and when users are searching. Identify patterns in the behaviors of the users who search for content on the information space or product, as well as any exceptions to those patterns. Begin with a brief exploratory data analysis to quickly get an overview of the information space's content, then iteratively try different ways of grouping or sorting users' most common search queries to see them from different perspectives. Do users' queries employ common language that everyone knows *or* jargon and abbreviations that only experts know? Which of the synonymous search terms is in prevalent use so should be the preferred term? What time-based patterns exist? Would a longitudinal pattern analysis reveal any seasonal differences in users' search needs? Are users searching for information relating to specific topics or types of content? What outliers exist in the data? What new content would best satisfy users' unmet needs? What language should you use in the taxonomy, the labeling for navigation links and page titles, and the content itself? What metadata would enable the search engine to find relevant content?

2. **Failure analysis**—Identify and endeavor to understand users' failures to find the information they need by searching. What search queries are returning no results or poor search results? By conducting a failure analysis, you can identify what failures are commonly occurring when users are searching for content that is pertinent to the information space's subject matter, but either the search engine is unable to find—perhaps because its terminology is not consistent with that of the audience or the search engine has not indexed it—or that is missing from the information space. These are top-priority issues that you should address as quickly as possible. Be sure to employ the users' terminology when defining the taxonomy, labeling, page titles, and metadata and when creating content. The search results that the search engine retrieves may be less than optimally useful because the best matches don't appear first in the list of results and those that do are not relevant to users' queries. Fixing these issues requires the support of the Development team. If the usability of the search-results pages *and* the quality of the search results are both poor, users may express their dissatisfaction by immediately leaving the information space. Consider defining modes of failure that are specific to the search engine's inability to provide search results that meet particular needs of the users or the organization.

3. **Session analysis**—*If* your search query–analysis tool is capable of identifying individual users, you can analyze users' behaviors during individual search sessions and find out who is searching for what information, where, and when. Users often submit multiple search queries during the same search session, and you can follow an individual's entire user journey throughout a search session. Thus, analyzing users' search sessions provides greater insight into their intent to search for specific information than any of these other approaches to analysis. How do users' needs and behaviors evolve as their familiarity with and understanding of the information space's content and search system increase during a search session? How do users refine their search queries? Are they typically narrowing or broadening their search queries? Focus on search sessions that eventually result in failure so you can better understand *why* they're failing and remedy the causes of failure; *or* on sessions that begin with, include, or end with users' most common search queries, or those having strategic importance to the business.

4. **Audience analysis**—To narrow the scope of your analysis, prioritize analyzing the search behaviors and information needs of users who belong to particular audience segments. Segmenting an information space's audience can help make the insights from your search-log analysis more actionable because you'll frequently need to analyze behavioral metrics that relate to particular groups of users. Depending on the organization's business goals, the audience segments that would be most useful would differ, but a few examples of dimensions along which you might segment the information space's audience include first-time versus repeat users, as well as users' loyalty, conversions, geographic location, and language. Consider how the behaviors and needs of specific audience segments differ from one another. Their behaviors and needs could differ significantly from the behaviors of all users in aggregate. Try applying the other four approaches to the analysis of each audience segment, then assess what findings differ or are consistent across the segments. Use your learnings to inform the definition of a user persona for each audience segment. Establish the desired behaviors for each audience segment, and design solutions that are tailored to their needs and support the desired behaviors. Develop content that addresses the information needs of specific audiences.

5. **Goal-based analysis**—By leveraging search metrics, you can measure, monitor, analyze, and optimize the performance of the information space or product against specific business goals and KPIs—whether goals and KPIs that relate specifically to the performance of the search system or broader performance goals and KPIs for the overall information space or even the organization. A failure to set goals and define and track KPIs for search can result in the loss of revenues. KPIs enable organizations to define business goals in a way that makes them measurable and actionable. Because KPIs are based on an organization's goals, they're very specific to each organization. However, specific types of information spaces and products tend to employ similar KPIs, along with the metrics that support them. For example, the goal for a content-rich, predominantly informational product is to engage its visitors in consuming the content. For a product with an ad-based business model, a useful KPI might be to increase the time visitors are spending on the information space, while the corresponding search metric would be the average time visitors have spent on the information space after searching. Or, for an informational product with a subscription-based business model, a KPI might be to

increase the rate of subscription renewal. Other types of information spaces include those that provide a service—for example Customer Support and commercial information spaces, whose main purpose is lead generation *or* ecommerce. A robust, high-performance search system is essential to ecommerce because search is users' primary means of finding products. Earlier in this section, I discussed how search system–performance metrics can inform KPIs in greater detail.

Search-log analytics provide convincing evidence that supports UX design, information architecture, and content-strategy decisions and enables you to persuade stakeholders to invest in implementing them. For detailed information about leveraging your learnings from search metrics, see the section "Applying your learnings from search metrics," later in this chapter.

Applying your learnings from search metrics

Once you understand users' information needs, their vocabulary, and the ways in which they're using the information space's search system through search-log analysis, you can improve the design of the information space's search system, navigation system, labeling system, and metadata, as well as its content.

Begin by optimizing the information space to ensure the success of a small number of the users' highest-frequency search queries, which constitute a large percentage of all users' searches and have the potential to deliver the greatest value to both users and the organization. Also, identify strategically important searches that are not performing well and optimize the information space to retrieve better results for them. Examine any high-frequency search queries that retrieve zero results to identify the search engine's indexing gaps, as well as content that is missing from the information space.

Evaluate the typical length and form of users' search queries, which can help you to ascertain your audience's degree of technological sophistication. More technology-savvy users tend to submit longer search queries and might use Boolean operators such as AND, OR, and NOT. Make sure the **Search** box is wide enough to accommodate users' longer search queries and that the search engine can support whatever query syntaxes users are employing. Enhance the search user interface by implementing autocomplete, which should ideally leverage data about both the user's own previous searches and the searches of other users to ensure that it presents the best possible matches. Autocomplete should also check and correct the spelling of the user's search terms, as necessary, and list the best matches first.

For a search system that supports faceted search—typically by allowing users to narrow their search by selecting a facet from a drop-down list—your learnings from search-log analytics should inform what facets you include in that list. On information spaces for specialized industry domains such as travel and real estate, for which search is users' primary means of seeking information, users typically constrain their searches up front by filling out several form fields that the search user interface comprises. Search-log analytics can inform what those fields should be.

To refine users' search queries and, thus, improve their results, in addition to checking the spelling of search terms, you might need to stem words by reducing them to their root form or substitute some keywords with synonyms from the information space's controlled vocabulary. Also, consider whether the section of the information space from which a user searches should constrain the scope of the search. Provide partial matches for overly constrained search queries that might otherwise have no or very few results.

To enable users to refine their search queries as necessary to improve their search results—perhaps by narrowing or broadening their scope—be sure to include the **Search** box on the search-results pages and populate it with the user's query. For a faceted search system, include the list of facets as well, preserving the user's selection.

In optimizing the design of search-results pages, consider whether you should create a unique format for displaying search results pertaining to particular types of data—for example, if the user has searched for a place, a person, an organization, a product, an acronym, or a unique identification number. On zero-results pages, suggest some similar search queries the user might want to try, by leveraging synonyms from the information space's controlled vocabulary.

Once you've learned the vocabulary of the information space's users, you should apply your learnings to improve both search information retrieval *and* findability. If different users are using a variety of keywords and phrases in their search queries to describe the same content, define a preferred term for each concept, as well as its variants, and add both the preferred terms and their synonyms to the information space's controlled vocabulary, or thesaurus. Then, use these preferred terms when defining the metadata for the information space's content.

To improve the findability of the content on the information space—whether for search information retrieval or for users who are browsing—make sure that the content, its metadata, labeling for navigation links, and page titles conform to the users' vocabulary. This ensures that links, titles, and metadata clearly identify the content.

To improve the quality of the information space's metadata, you can do the following to identify any gaps in the metadata:

- Compare the keywords from the top search queries to the metadata and look for synonyms in the metadata. Try searching for keywords from the metadata. Then, compare the performance of the top search queries to that of their synonyms in the metadata—particularly the relevance and precision of their search results. Consider replacing underperforming metadata.

- Collect some important *or* new documents, then list the search queries that have successfully retrieved those documents. Determine whether the metadata for each document is using the same terminology as that in users' queries. If not, make the language of the metadata consistent with that of the search queries.

- To maintain the currency of the information space's metadata, determine whether there are any newly trending queries for which users have not searched before *or* whose use is increasing rapidly or significantly. Focus on the search queries for which users are searching most frequently. If the trend continues, consider adding corresponding keywords to the metadata. Also, identify any search queries whose use has diminished significantly and consider adding synonyms for those search keywords to the metadata, leveraging the language that users are currently using in their queries.

When you're working to improve the information space's navigation system, identify any important categories of content for which users are typically relying on searching rather than browsing. Then, restructure the information hierarchy as necessary to incorporate those content categories and add navigation links for those categories to the primary or secondary navigation, as appropriate.

Once you've discovered what content is missing from the information space, create new content that meets the newly identified users' needs. In contrast, you might find that users are no longer searching for particular content, and you should consider whether to remove or update that content or reduce its prominence.

Analyze what search queries users are entering during the same user session, which can provide context that could reveal how users relate different content and indicate what content you should group within sections and what associative links you should create.

Level of effort—Once your team has set up search-log analytics for the information space or product, analyzing your search logs is a quick, easy way of continually collecting and monitoring data about users' search queries, enabling you to answer your questions about how users are engaging with the content on the information space or product. Evaluate users' most frequent search queries at least once each month to understand users' key information needs, how they've changed over time, and whether their search queries are retrieving relevant results or no results at all. To ensure that your search-log analyses have the greatest possible impact, always scale your efforts according to the scope of the most common search queries on which you're focusing. Spending just an hour each month analyzing your search-log data can provide very useful insights.

Deliverables—Periodically disseminate brief, easy-to-read search log–analytics reports to your project team and other interested stakeholders. Your reports should convey a meaningful narrative and integrate quantitative behavioral data, qualitative data from the search logs, *and* user research in support of that narrative. Avoid the overuse of analytics jargon and emphasize data visualizations to ensure the easy consumption of the data by stakeholders. The findings in your reports and presentations to stakeholders should be actionable and include your analysis of the search metrics and their related KPIs, enabling stakeholders to leverage these findings in business-strategy, content-strategy, and design-strategy discussions and documentation. Broadly distribute these reports and make them available to the entire organization online. Provide access to the search logs as well. Elicit feedback on your reports from your project team and other stakeholders, then iteratively refine the reports to ensure that they include all the data stakeholders need.

Investment—Search-log analytics is a low-cost method of research, with a high return on investment. [55, 56, 57, 58, 59, 60, 61, 62]

Recommended book—Louis Rosenfeld. *Search Analytics for Your Site: Conversations with Your Customers*. Brooklyn, NY: Rosenfeld Media, LLC, 2011.

Summary

In this chapter, I've provided overviews of some types of UX research methods that apply specifically to the design of effective information architectures for digital information spaces and products, then described some important methods of each type.

First, I covered three common card-sorting methods that can help you to understand, design, and evaluate information hierarchies, as well as their labeling: open card sorting, modified-Delphi card sorting, and closed card sorting. Then, I discussed another method of UX research that is useful in discovering an information space or product's optimal structure and labeling: free-listing. Next, I described several UX research methods that you can use in evaluating an information space or product's findability: reverse card sorting, card-based classification evaluation, and tree testing. Finally, I explored several types of data analytics that can enable you to learn about users' actual behaviors and, thus, help you understand and improve the information architecture of a digital information space or product: usage-data analytics—for example, Web analytics; path, or clickstream, analysis; and search-log analytics. Synthesize your insights from all of your UX research to inform your information-architecture strategy and design.

Next, in *Chapter 6, Understanding and Structuring Content*, we'll consider content-analysis heuristics and methods and content modeling. Then, in Chapter 7, *Classifying Information*, we'll look at another foundational element of information-architecture strategy: the classification of the information that a digital information space or product comprises.

References

To make it easy for readers to follow links to the references for this chapter, we've made them available on the Web: `https://github.com/PacktPublishing/Designing-Information-Architecture/tree/main/Chapter05`

Further reading: Books

Earlier in this chapter, I recommended some excellent books that focus on the specific methods of generative and evaluative UX research and data analytics that this chapter covers—methods that apply specifically to the design of information architectures for digital information spaces and products. This list recommends some additional books that either discuss specific methods or provide broad information about many methods of UX research and.

Austina De Bonte and Drew Fletcher. *Scenario-Focused Engineering: A Toolbox for Innovation and Customer-Centricity*. Redmond, WA: Microsoft Press, 2014.

Avinash Kaushik. *Web Analytics: An Hour a Day*. Indianapolis, IN: Wiley Publishing, Inc., 2007.

Avinash Kaushik. *Web Analytics 2.0: The Art of Online Accountability and Science of Customer Centricity*. Indianapolis, IN: Wiley Publishing, Inc., 2009.

Bella Martin and Bruce Hanington. *Universal Methods of Design: 100 Ways to Research Complex Problems, Develop Innovative Ideas, and Design Effective Solutions*. Beverly, CA: Rockport Publishers, 2012.

Brad Nunnally and David Farkas. *UX Research: Practical Techniques for Designing Better Products*. Sebastopol, CA: O'Reilly Media, Inc., 2017.

Catherine Courage and Kathy Baxter. *Understanding Your Users: A Practical Guide to User Requirements Methods, Tools, and Techniques*. San Francisco: Morgan Kaufmann Publishers, 2005.

Donna Spencer. *A Practical Guide to Information Architecture*. 2nd edition. Northcote, Victoria, Australia: UX Mastery, 2014.

Donna Spencer. *Card Sorting: Designing Usable Categories*. Brooklyn, NY: Rosenfeld Media, 2009.

Elizabeth Goodman, Mike Kuniavsky, and Andrea Moed. *Observing the User Experience: A Practitioner's Guide to User Research*. 2nd edition. Boston: Morgan Kaufmann Publishers, 2012.

Eric T. Peterson. *The Big Book of Key Performance Indicators*. Self-published, 2006.

James Lang and Emma Howell. *Researching UX: User Research*. Collingwood, Victoria, Australia: SitePoint, 2017.

Louis Rosenfeld. *Search Analytics for Your Site: Conversations with Your Customers*. Brooklyn, NY: Rosenfeld Media, LLC, 2011.

Luke Hay. *Researching UX: Analytics*. Collingwood, Victoria, Australia: SitePoint, 2017.

Mike Kuniavsky. *Observing the User Experience: A Practitioner's Guide to User Research*. San Francisco: Morgan Kaufmann Publishers, 2003.

Peter Morville and Louis Rosenfeld. *Information Architecture for the World Wide Web*. 3rd edition. Sebastopol, CA: O'Reilly Media, Inc., 2007.

Rochelle King, Elizabeth F. Churchill, and Caitlin Tan. *Designing with Data: Improving the User Experience with A/B Testing*. Sebastopol, CA: O'Reilly Media, Inc., 2017.

6

Understanding and Structuring Content

Although digital information spaces and products exist at specific, quite persistent locations, it is actually their content, or the information itself, that defines an information space and its ever-shifting, but hopefully clear, boundaries. Because an information space's content evolves continually, its information architecture must be resilient to remain coherent. Plus, information spaces typically comprise distinct sections that must also have clearly defined, albeit constantly shifting, boundaries.

Before designing an information architecture for a digital information space or product, you must first become familiar with its content. What does the term *content* comprehend? An information space or product's content might include textual information such as articles, profiles of people, marketing copy, product descriptions, customer-support information, news, press releases, or calendars of events; documents such as PDF reports or papers, ebooks, or slide presentations; numeric data such as statistics, financial data, stock reports, spreadsheets, graphs, or charts; images such as photographs, illustrations, or maps; media such as video, audio, or animations; and dynamic, structured-content components.

Understanding the content that an existing information space or product comprises, as well as any new content that is being developed for it, is an essential precursor to devising an effective information-architecture strategy for redesigning its information architecture. Likewise, it is essential that you understand the content strategy for a new information space or product, which informs information-architecture strategy and design.

In this chapter, whose focus is on content, I'll cover the following topics:

- **Content-analysis heuristics**—Fred Leise's eleven heuristics provide one useful, qualitative approach to evaluating an information space or product's content.

- **Content-analysis methods**—These methods include content-owner interviews, content mapping, content inventories, content audits, and competitive content analyses. Through these methods, a team can gain a deep understanding of an information space or product's content, which informs the design of an effective information architecture.

- **Content modeling**—Through content modeling, you can decompose your content into logical components and elements, enabling your content to support contextual navigation, chunked content, personalized content, filterable content, sortable content, and reusable content.

Content-analysis heuristics

In 2006, Fred Leise—now Principal Taxonomy Designer at Contextua—developed eleven content-analysis heuristics for the qualitative analysis of an information space or product's content. You can also use these content-analysis heuristics to provide structure to and, thus, promote the thoroughness of your content-analysis reports. Some or all of the following content-analysis heuristics may be applicable to a given information space or product:

1. **Collocation**—Create collections of similar content that users would want to find in one location—especially content on related topics, but also content by the same author or that is published on the same date, *or* documents of the same type—and make it easy for users to find all of the content that is relevant to their information needs by placing it in well-labeled sections or subsections of the information space or product.

2. **Differentiation**—Create discrete collections of content that users can easily distinguish from one another and whose content they would expect to find in different locations. Ensure that each collection comprises content on related topics that are dissimilar from those in other collections. Enable users to easily differentiate these collections by placing them in well-labeled sections or subsections of the information space or product.

3. **Completeness**—Identify gaps that exist within the scope of the information space or product's content—for example, any content that is available on key competitors' information spaces or products, but missing from the content you're analyzing. Also, make sure that any content to which you refer actually exists on the information space or product.

4. **Information scent**—Ensure that all collections and pages of content provide good information scent, reassuring information seekers that they are on the right path to find what they need, as I've described in the section "Information foraging," in Chapter 2, *How People Seek Information*. Use the language of the target audience in creating the labeling for the navigation links that display the information space or product's sections and subsections and contextual links that display individual content pages, as well as in the destination pages' titles and section headings to clearly identify their content.

5. **Bounded horizons**—Create a well-structured information hierarchy and navigation system that clearly delineate the scope of the information space or product, showing both its breadth and depth. This heuristic, in combination with information scent, indicates the level of effort that information seekers must invest in determining whether the information space or product might provide the information they need.

6. **Accessibility**—Ensure that people having different levels of cognitive and physical abilities can use the information space or product and consume its content. Enable users to access *all* content on the information space or product *both* by using its navigation system to browse its information hierarchy and by searching for content using its search system.

7. **Multiple access paths**—Support the findability of the information space or product's content for users who have different mental models of that content by enabling them to follow various paths in seeking the same content. Leverage faceted navigation and search systems to increase the visibility of the various available paths and make it easier to follow them.

8. **Appropriate structure**—Organize the content on the information space or product in a way that both reflects users' mental models of its information hierarchy *and* supports a variety of information-seeking behaviors, which I've described in the section "Information-seeking behaviors," in Chapter 2, *How People Seek Information*. These include *known-item information-seeking behaviors* such as searching for known items, browsing to find known items, returning to known items, and seeking updated information, and *exploratory information-seeking behaviors* such as exploratory browsing, exploratory searching, and integrated searching and browsing.

9. **Consistency**—Ensure that similar content on similar pages within a particular section of the information space or product conforms to a consistent structure and format, enabling users to more readily form a mental model of the content. Examples of page types whose user experience benefits from the consistency of structured content include product pages, article pages, and profile pages. By creating or adopting a design system, an organization can ensure optimal design consistency across content elements or pages of a given type.

10. **Audience-relevance**—Balance the creation of generally useful content and targeted content that meets the needs of *key* audience segments. Structure the hierarchy of the information space in a way that matches the audience's mental models and helps specific audience segments find relevant content that meets their needs. Design the navigation system using clear labeling that employs the audience's vocabulary. Leverage collocation and differentiation when organizing the content within the information space to ensure that you place it in the most appropriate location within its structure, making it easier for specific audience segments to find all relevant content.

11. **Currency**—Ensure the timely and thorough maintenance of the information space or product's content, whose accuracy is essential to fostering the trust of the people making up its audience. Keeping the content up to date requires a good maintenance plan that schedules content reviews and updates at regular intervals; defines what people in what roles are responsible for them *and* how customers, users, and content authors can request them; and determines what business events should trigger an unscheduled content review.

To increase the precision of these content-analysis heuristics, you can indicate the degree to which either the overall information space, a specific section of the space, or a particular content type conforms to individual heuristics by rating them on Fred Leise's five-point *Likert scale*, as follows:

1. "Strongly deviates from the heuristic"

2. "Deviates from the heuristic"

3. "Neither deviates nor conforms to the heuristic"

4. "Conforms to the heuristic"

5. "Strongly conforms to the heuristic"

If the information space, a section, or a particular content type fails to fully satisfy a specific content-analysis heuristic, describe content and design improvements that would provide better outcomes, focusing on resolving issues with any content that strongly deviates from one or more of these heuristics. [1]

Content-analysis methods

Peter Morville and Lou Rosenfeld characterize content-analysis methods as the foundation of a "bottom-up approach to information architecture." Through content analysis, you can gain a deep understanding of and methodically describe an existing information space or product's current content. You can also identify and remedy any gaps that exist between that content and the content requirements the organization's content strategy has defined. Content analysis enables you to appreciate the overall scope of the information space or product's content, as well as learn about the various types of content it comprises. You can discover common themes and meaningful patterns and relationships between discrete content objects, then by assessing which of these are predominant, identify optimal ways of organizing the content to ensure its accessibility.

When devising a content strategy or information-architecture strategy for a new, digital information space or product, you should assess the organization's existing content to determine what content your team could reuse or adapt for publication on that information space. Determine what content requirements the existing content satisfies and what new content your team needs to create. Again, identify the types of content and the themes, patterns, and relationships between content objects.

By analyzing the content of either an existing or a new information space or product at a more granular level, you can inductively ascertain what patterns exist within unstructured content—and, thus, how you might structure sets of related content objects. You can also learn what language appears on certain page types or within other content objects, compile a list of their common words and phrases—for example, keywords and phrases in titles and subheadings—and tally their numbers of instances. Or you could tally all the instances of concepts that are represented in both language and images. You could then infer what categories of content an information space or product should comprehend, which informs the design of the metadata that describes its content objects, as well as its labeling and navigation systems.

You must also determine what types of metadata the content team should use to describe content objects, as follows:

- **Structural metadata**—This metadata describes the information hierarchy of a content object—from its title through its sections and subsections—or specific chunks of information to which associative links might be useful or chunks that might be reusable.

- **Descriptive metadata**—This metadata describes the content object itself—for example, the type of *media* such as text, document, numeric data, image, video, audio, or dynamic content components; its *category* such as an article, profile, or product description; its *topic*, or subject matter; alt text; information source; target audience; or language.

- **Administrative metadata**—This metadata might describe the content object's authors, owner, publication date, archival date, keywords, or other business-related information.

When preparing to conduct a content analysis, you should first gather and print out a sufficiently representative sample of an information space or product's existing content or, for a new information space, the organization's existing content. To find out what content you should collect, browse the existing information space or product to become generally familiar with its content, then conduct content-owner interviews to gain a better understanding of that content. For a new information space, interview content owners and ask them to provide representative examples of their content, as well as data about the scope of their content. As you gather content, use your judgment to balance the completeness of your sample with your project's time, financial, and staffing constraints. Include a few examples of the most prevalent types of content, as well as some examples of other important, but less common types of content.

In choosing different types of content for inclusion in your sample, consider factors such as the types of media; categories of content; topics, which are typically specific to particular industries or even to a certain information space or product; target audience, which might be a single type of person or, for a large enterprise, perhaps several—such as customers, partners, and employees; information source, which is usually an important factor for an organization's Web site or intranet and comprehends information from all departments, as well as any external sources; and, if an information space or product includes translations, in what languages.

Content analysis is an iterative process through which you'll gradually discover and come to understand the full scope of the content that an information space or product comprises. Your learnings and insights from this process inform your information-architecture strategy and every aspect of designing an information architecture—from its organizational structure to its metadata and to its labeling, navigation, and search systems.

Content-analysis methods that you can employ in learning about an information space or product's content include the following:

- Content-owner interviews
- Content inventories

- Content audits
- Content mapping
- Competitive content analysis

Depending on your project team's available resources, time, and budget and the strategic goals of your organization, choose whichever of these content-analysis methods would enable you to learn as much as possible about an information space or product's content before redesigning its information architecture or rethinking its content strategy. Among these content-analysis methods, conducting content-owner interviews is most universally useful for learning about either existing or new information spaces' content. While these content-analysis methods are *not* mutually exclusive, certain methods should generally precede others. You'll find more information about when to use what content-analysis methods later in this chapter.

[2, 3, 4, 5, 6]

Content-owner interviews

Purpose—To design an effective information architecture, you must first know what content a digital information space or product should comprise and the strategy behind the production of that content. Conducting content-owner interviews is the best way of quickly getting an overview of an information space or product's content. You can then gradually develop a deeper understanding of that content through successive interviews with stakeholders, who include business leaders, content managers, content strategists, content creators, editors, and others.

By conducting content-owner interviews, you can learn about an existing information space or product's current and planned content *or* the content requirements for a new one. You can discover what topics and types of content already exist, what additional content would be necessary to meet all of an organization's content requirements for an information space or product, and what content the organization could acquire or would need to create. You can also learn about both the organization's content-development lifecycle and its content-management process.

Description—To learn about an information space or product's content, you should talk with the people who define the requirements for and share ownership of that content—the organization's *content owners*. You can either interview individual content owners one-on-one *or* facilitate group discussions among several content owners—whether those working in similar roles *or* who are representative of a multidisciplinary content team.

Ask these content owners to prepare in advance to share examples of what content on the existing information space or product is and is *not* effective and explain the reasons for their assessments. Also, ask them to share good examples of effective content from competitors' information spaces or products—and perhaps from some unrelated information spaces or products as well—and explain what makes them successful. Ask them to provide printouts of the examples, as well as links to examples on the Web.

Prior to your content-owner interviews or group discussions, prepare a list of the topics or questions you want to cover. Craft open questions that allow content owners to tell their own stories and communicate their own viewpoints. Some broad categories of questions that you might ask content owners include the following:

Questions about planning an information space or product's content:

- Does a content-strategy team exist within the organization? If so, what is the team's purview?

- What are the vision and key goals for the information space or product? How does it contribute to the organization's competitive advantage?

- Who are the target audiences for the information space or product?

- Who is responsible for identifying content requirements for the information space or product and what process do they follow? Who determines what existing content to include on the information space or product? Who identifies business and user requirements for new content? At what levels and in what parts of the organization do key people defining content requirements reside?

- What business-related events drive the need to create and publish new or updated content—for example, product launches, adding new products to an ecommerce product catalog, news such as quarterly or annual financial results, events such as conferences or workshops, or hiring new leaders or staff?

- What content requirements have the team defined for the information space or product? What new types and topics of content will the information space or product include? What is the business goal for including each type and topic of content? What user needs does each type and topic of content satisfy?

- How does the team prioritize these content requirements? What challenges has the team encountered in determining priorities for content requirements? How does the team plan to address them?

- Who can request or approve the creation of content for publication on the information space or product? Who is responsible for assigning the creation of that content to content authors? How does the team determine what types of content to create to satisfy specific content requirements?

- Who is responsible for planning and maintaining the organization's editorial calendar? How much lead time is necessary to create and publish new content? Does this differ depending on the type or topic of content?

- What types and topics of content will the team need to acquire or license from an external content developer rather than creating them in house?

Questions about an information space or product's existing content:

- What types and topics of content does the information space or product currently comprise?

- Is additional content now under development? If so, what types of content, on what topics?

- Who is responsible for assessing the accuracy, quality, and consistency of the information space or product's content, and how? Has the organization received any complaints from customers regarding the content?

- What content is most effective? What content needs improvement? Is there any content that isn't serving a useful purpose and should be removed?

- What content does and does *not* meet the business needs of the organization?

- What content is effectively meeting the needs of the target audiences for the information space or product?

- What gaps currently exist in the content that prevent the information space or product from fully satisfying the needs of either the organization or its target audiences?

- Which content is receiving the highest and lowest rates of traffic?

- How and when does the organization measure the usefulness of specific types and topics of content to users?

- What is the return on investment (ROI) for specific types and topics of content?

Questions about creating an information space or product's content:

- What is the organization's content-development strategy—whether for an entire organization, a business unit, a department, or a product or service?

- What phases make up the organization's content-development lifecycle? What occurs during each of these phases? Is this lifecycle consistent across the entire organization? If not, what deviations occur within particular departments? Which aspects of the content-development lifecycle are and are *not* working well?

- What tools and technologies do authors who create specific types of content use during their content-development process? Do the authoring tools and technologies provide all of the capabilities their process requires? If not, what capabilities do they need? Has the team defined any new technology requirements? Are the tools fully compatible with one another? Do authors encounter any issues in using them? Can authors view their content in its final form during its development?

- What is the organization's content-development process? Do this process and the standards that exist adequately support the content-development strategy? Do they meet the needs of content authors and other content owners? Is the organization experiencing any challenges as a result of the current process or standards? Is the team planning to make any improvements to them? If so, what changes might improve them?

- What dependencies exist at different points in the content-development process? With whom do content authors interact throughout this process?

- How long does completing each iteration of the content-development process take? How much time does authoring the first draft of each specific type of content take? How long is each editing cycle?

- What are the organization's version-control processes—both for the iterations of a single document and for the multiple variants of content components for different contexts? Are these processes consistent across the entire organization? If not, how do these processes deviate across departments? What aspects of these version-control processes are and are *not* working well?

- Has the publication-design team created visual-design specifications, templates, and guidelines that ensure the consistent presentation of the information space or product's information?

- Are content authors, editors, production managers, and others on the content team familiar with the organization's content-development lifecycle and process, as well as the current content strategy? Are their multidisciplinary teammates and other stakeholders familiar with them? Do effective documentation and training exist for them?

- Do content authors, editors, production managers, and other content creators have access to and know how to use the tools, technologies, visual-design specifications, templates, and style and branding guidelines they need to do their work? Is there effective documentation and training on how to use them? Is there an adequate budget for developing and delivering that documentation and training?

- Who are the subject-matter experts from whom content authors need to obtain the source information for specific types and topics of content? Where does the source information reside for the content the team needs to create?

- Are content authors likely to encounter any obstacles in getting access to subject-matter experts to obtain the source information they need to develop specific types or topics of content or any other challenges in developing that content?

- Where does the source information for content authors' projects reside? Is it in a single repository or scattered across multiple teams' servers?

- How effectively do the departments and teams within the organization share information across projects? Do content authors have adequate information about other projects that might impact their work?

- Do content authors have access to the templates, style and branding guidelines, and legal standards they need? Do these templates and information reside in a central repository? Who is responsible for keeping everything up to date and notifying authors of any changes? How do such notifications occur?

- Who is responsible for creating specific types and topics of content? In what departments do key people responsible for developing content reside? Do they have sufficient resources to hire all the content authors they need?

- Are there currently people or teams within or available to the organization who have the skills and tools that are necessary to create all types of content, including video, audio, and other media?

- Does the organization have adequate staffing to support all aspects of the content-development process? If not, where do deficiencies exist—whether in the existence of specific roles, staff with sufficiently well-developed skillsets, or the need to hire additional human resources?

- Do content authors have sufficient time to iteratively develop the content for which they're responsible?

- With what other disciplines do the authors creating specific types of content need to work—for example, business analysts, content managers, publication designers, content strategists, information architects, graphic designers, editors, production managers, or developers? When is it important to have access to the people in these other roles?

- At what stages of the content-development process does editing occur? What are the stages of the organization's editorial workflow? What is the process for editing content? Where does the content team stage content for editing? Do the editors receive notification that content is ready for editing sufficiently far in advance? How do editors receive notification that content is ready for editing? How much time do editors have to complete their edits? How do they provide their edits to content authors? How do authors resolve issues with edits? Does the editorial process consistently result in high-quality content that meets people's needs? Is the editorial process efficient and effective? What process issues do editors typically encounter? How might the content team improve the editorial process?

- What is the organization's review and approval process for content? At what stages of the content-development process do reviews and approvals occur? Where does the content team stage content for review and approval—and, ultimately, approve the content for publication? Do the reviewers receive notification sufficiently far in advance that they can be available to provide feedback? How do reviewers receive notification that content is ready for their review? How much time do reviewers get to conduct their review and provide feedback? How do they provide their feedback to content authors? Does the review and approval process consistently result in high-quality content that meets people's needs? Is the review and approval process efficient and effective? What process issues do reviewers typically encounter? How might the content team improve this process?

- Who is responsible for reviewing and approving content? Does this depend on the specific type or topic of content? Are some of these reviewers the same people who originally requested the content? On what kinds of content issues should specific reviewers focus—for example, assessing the information's accuracy and currency, providing editorial feedback, or giving feedback on branding or legal issues? Who provides the final approval for publication?

- How do content authors negotiate the resolution of the feedback they receive from various reviewers to ensure the highest-quality outcomes and final approval? Who makes the final determination on what types of issues? Is it easy for reviewers to compare the version of the content they originally reviewed, with their feedback, to the revised version to verify the implementation of the changes they requested? Is this process efficient and effective? How might the content team improve the process?

- Does the information space or product's content require translation or localization? What are the translation and localization processes? Do translation and localization standards and guidelines exist? What tools support this work? Do they separate content from formatting? Is there an effective change-management process to ensure consistency across translations? Do effective documentation and training exist for these processes? Who is responsible for translating and localizing content? Who is responsible for reviewing translations and localized content? Are there any issues the content team typically encounters during the translation and localization processes? How could they improve them?

Questions about the organization's content-management process:

- Does the organization have an influential, centralized content-authoring team that consistently follows the same content-management process and policies? *Or* is content ownership distributed across multiple departments that use different processes and policies?

- What are the organization's content-management and publication processes and workflows? What are the organization's content-management policies? Are these processes and policies well documented? Are they efficient and effective? Does the content team deviate from them in any way? Are improvements necessary?

- How much time does publishing new content or making revisions to existing content take?

- What content-ownership roles exist in the organization? What are the job responsibilities of the people in each role?

- Are specific content owners responsible for particular types and topics of content? If so, who is responsible for what types and topics of content?

- What content-management system (CMS) is the organization using? Does it meet all of the organization's authoring and publication needs or is it deficient in certain respects?

- Do content authors, editors, and other content creators have access to the CMS? How can they obtain accounts for the CMS? Do they know how to use the CMS? Does effective documentation and training exist on how to use the CMS? Is there an adequate budget for developing and delivering that documentation and training?

- Does the CMS support the use of controlled vocabularies and thesauri?

- Does the CMS automate any parts of the publishing process? Can the CMS automate content categorization or index generation?

- Does the information space or product's technology support personalization?

- Who is responsible for entering content and its descriptive metadata into the CMS? Who is responsible for publishing content using the CMS?

- What standards exist for metadata? Who is responsible for assessing the completeness, quality, and consistency of the metadata that describes the information space or product's content?

- Who is responsible for revising, maintaining, sunsetting, and archiving content in the CMS? What processes exist for updating and archiving content? Does the organization regularly schedule reviews, updating, and archiving of content? If so, at what time intervals? Do certain events trigger the updating or archiving of content?

- What capabilities does the information space or product's search engine offer—for example, autocompletion of search queries; search zones; an integrated spelling checker, word stemming, and thesaurus; a query language for advanced searches; or the filtering of search results?

- Do the current content-management and publication processes and policies adequately support the organization's needs? Has the team experienced any challenges because of them? Are there any issues of legal compliance relating to that process? Does the team plan to make any improvements to the content-management or publication processes or policies?

- How well does the CMS handle publishing across multiple platforms and templates?

- Does the information space or product consist of static pages or dynamic content? Does dynamic content comprise reusable textual components and other digital assets? Where does this content reside? Does the CMS adequately support content reuse?

- Do content owners, content strategists, and information architects have access to usage-data and site search–log analytics? How can they obtain accounts to access that data?

During interviews *or* group discussions, you should avoid asking your questions in a rigid order. Instead, these sessions should be flexible and conversational. Introduce new topics of discussion as they occur to you. Because content owners can answer your questions in as much or as little detail as they want, this kind of interview puts them at ease. To ensure that you gain a deep understanding of an information space or product's content—as well as the organization's processes—pursue any important lines of discussion that arise during your sessions by asking probative, follow-up questions to elicit additional information or clarify a participant's previous responses. Never pose leading questions that would bias participants' responses—especially when asking probing questions to elicit more detail.

During group discussions, content owners typically respond to their colleagues' comments—reminding each other about or questioning specific details. Thus, while group discussions let you obtain more information more quickly, your learnings might not represent the consensus of the group. Through close observation during group discussions, you can also learn about the organization's culture and political dynamics. As the facilitator of a group discussion, you must prevent overbearing participants from dominating the dialogue or unduly influencing others and encourage all participants to share their own points of view. When facilitating a group discussion, you'll hopefully have the opportunity to hear the differing opinions of content owners who are working in a variety of roles.

To benefit from both getting individual content owners' unique perspectives and gaining more diverse learnings from group discussions, you could interview some individual content owners in various roles first, then gather content owners in multidisciplinary roles together for a group discussion.

When—Whether you need to learn about the content for a new or an existing information space or product, conduct content-owner interviews during *Discovery*. Gathering this data informs the definition of content requirements, content strategy, information-architecture strategy, and the design of an information architecture. Continue these conversations with content owners throughout *Design*, whenever you have questions they could answer.

Once you've conducted in-depth content-owner interviews and have a deep understanding of both the information space's content and the content-development process, doing brief interviews to find out what content is changing is generally sufficient when preparing for future design iterations. If you conduct content-owner interviews before doing a content inventory for an existing information space, you'll be better able to understand its content and its structure.

Where—Individual content-owner interviews should typically take place face-to-face, in each content owner's workspace. However, if necessary, you could interview content owners remotely, using online-meeting software or on a phone call. Group discussions ideally occur face-to-face, in a conference room—usually at the workplace of the senior people participating in a discussion.

Participants—To learn about an information space or product's content—as well as an organization's content-development and content-management processes—you must obtain information from the people who define the requirements for and share ownership of that content—those who request, inform, create, acquire, review, edit, approve, publish, and govern that content. The roles and titles of content owners may differ across organizations and their responsibilities often overlap. In smaller organizations, one person may fulfill the responsibilities of multiple roles. Common content-ownership roles within organizations could include the following:

- **Business leaders**—In any business organization or unit, executives and other leaders are responsible for defining a business strategy that generates maximal business value, including determining what content would be most valuable to the business's target audience, as well as building a content-management system (CMS) that would ensure its reliable delivery. These business leaders are often responsible for requesting or approving the creation of large categories of content. In a large organization, a C-level executive, VP, or director of content management might hold primary responsibility for content management.

- **Business analysts**—The responsibilities of a content-management business analyst include securing a mandate and defining the business need for a CMS project; gaining the support and cooperation of sponsors on CMS initiatives; devising a strategy that ensures the CMS supports an organization's broader strategic goals; focusing content owners on your organization's business goals; identifying and tracking the progress of all existing efforts that could contribute to the success of the CMS; determining what efficiencies to build into the CMS; assessing the performance of your content against strategic goals and metrics; fully understanding authors' needs in successfully developing content; defining content requirements that improve your content's performance throughout its lifecycle; and analyzing the return on investment (ROI) of the content team.

- **Content managers**—A content manager's responsibilities typically include establishing the discipline of content management within an organization; leading any Content Management team; overseeing the definition and governance of content-management standards, policies, and procedures; planning and executing content-management initiatives; specifying requirements for and implementing a content-management system (CMS) that automates inputting, storing, and publishing the organization's content; acquiring controlled vocabularies and thesauri and adapting them to the organization's needs; defining, maintaining, and applying metadata standards, which dictate the structure and content of the metadata that identifies specific topics and types of content; enabling the metadata-driven publication and reuse of content and automatic structuring of navigation systems; requesting and approving the creation of specific content; reviewing, providing feedback on, and approving the publication of new and updated content; approving the sunsetting and archival of specific content; and acquiring and processing content from external content developers.

- **Publication designers**—The key responsibility of the designers on a publication's staff is ensuring that they publish well-designed, high-quality publications that come out on schedule. Therefore, they must design a standardized, automated CMS publication system that lets the organization publish the content components that reside within its repository and plan what content to publish in specific publications and when. These designers are also responsible for creating visual-design guidelines and specifications for the team's use, designing and implementing templates and style sheets and applying them to each of an organization's publications to ensure the optimal presentation of their content, and reviewing content to ensure it meets all publication-design standards.

- **Content strategists**—A content strategist's responsibility is the planning, creation, delivery, and governance of an organization's content throughout its lifecycle, across multiple channels. The responsibilities of a content strategist include identifying and understanding an organization's content requirements; deciding where to publish specific content and creating a strategy for content reuse; devising a content strategy for an information space or product that defines the overall scope, structure, sources, topics, types, and components of its content; assessing the quality and utility of all existing content; identifying gaps in that content; determining what content to acquire or create to fulfill all of the content requirements for an information space or product; deciding who is responsible for creating, updating, reviewing, editing, approving, publishing, and archiving specific content; defining standards and guidelines against which reviewers, editors, and approvers evaluate content; and ensuring that new and revised content satisfies all strategic requirements.

- **Information architects**—An information architect's responsibilities include understanding the business context for a digital information space or product, as well as its target users' needs; gathering content requirements for topics and types of content that meet both business and user needs; devising strategies for structuring content that ensure findability; creating taxonomies, or classification schemes that define the hierarchical and associative relationships between an information space or product's pages or content objects; creating or adapting controlled vocabularies and thesauri that enable the effective description of pages and content objects and their relationships; defining and maintaining standards for metadata that describes or classifies specific topics and types of content *and* applying the appropriate metadata to pages and content objects; devising content models that define consistent sets of content objects and the interconnections between them, which facilitate contextual navigation, and designing page templates that support these models; designing organization, labeling, and navigation systems that make an information space or product's content browsable; and designing search systems and query syntaxes that deliver useful search results whose content and formatting ensure their usability.

- **Information providers**—These are subject-matter experts who hold the source information that content creators need to develop content that meets an organization's requirements. Depending on the nature of the content, these people could be in any role within an organization, including business leaders, business analysts, product managers, technologists, marketers, salespeople, customer-service representatives, and Human Resources departments. They might have existing documents they can share with content creators *or* convey information to them verbally. Once content creators have developed content, these subject-management experts should review the content to ensure its accuracy.

- **Content creators**—Those who plan, design, create, and revise the content for a digital information space or product include authors of textual content such as those who write articles, marketing copywriters, instructional designers, and technical writers; graphic designers, illustrators, and photographers; producers of video, audio, and animated media; and community managers and creators of user-generated content (UGC) who write or respond to social-media posts and comments. Your organization's content creators are responsible for fulfilling its content strategy by creating accurate, engaging, readable content that meets business and user requirements, reflects the organization's brand and messaging, and helps your audience to accomplish their goals; *and* for adhering to all editorial standards and guidelines. They also determine what illustrations, other digital media, and links to incorporate in textual content. Content creators should also understand the content structure the CMS requires, define and write reusable content components, provide appropriate metadata for pages, be able to input content into the CMS correctly, and optimize content for findability by search engines.

- **Editors**—An editor-in-chief's responsibilities include establishing the editorial discipline within an organization; the creation and governance of all content standards and guidelines; defining the editorial process, content lifecycle, and policies; active involvement in the planning and implementation of all content-strategy and content-management initiatives; and leading the editorial team. The editorial team guides all content development; defines all standards and guidelines that govern textual and visual content—including both style and branding guidelines that specify authors' use of language, tone of voice, and color *and* standards that ensure legal and regulatory compliance; reviews *and* edits all content to ensure that it meets the editorial standards and guidelines; approves content for publication; and is responsible for the editorial calendar and maintenance plan. The editorial team also leads any efforts to translate and localize content for specific audiences worldwide. The team's responsibilities include *developmental editing*, which is substantive editing whose focus is structuring content effectively, enabling its reuse, and ensuring that it clearly and accurately conveys all of the necessary information, at the correct reading level for the intended audience; *content editing*, which ensures content delivers consistent messaging, satisfies readers' information needs, is optimally readable, and uses the appropriate branding and tone of voice; *copyediting*, which ensures that content observes all grammatical and punctuation rules and style guidelines; and *proofreading*, whose purpose is correcting all remaining typos and grammatical, punctuation, and spelling errors.

- **Content processors**—A content-processing staff is responsible for designing and implementing a system that automates the process of converting content any organization has acquired from external sources or created in-house to the structure the CMS requires. Inputting the content into the CMS correctly, using the proper formats, restructuring content or breaking it into components as necessary, and tagging elements with the appropriate metadata ensures that the content retains its meaning across various platforms and templates. Content processors are also responsible for reviewing all acquired and authored content to verify the accuracy of automated conversions and correcting any errors. They might also need to convert and add metadata to some content manually.

- **Production managers**—The primary responsibility of a production manager is publishing—ensuring a steady flow of high-quality content into the CMS. This requires the efficient processing of all the types of content that an organization develops, as well as content acquisitions from external content developers. A production manager's responsibilities also include establishing an effective production process, including deciding what parts of the workflow to automate, managing production staff, planning and scheduling the production of content, monitoring the creation of content, and reviewing and approving the content's production quality across all digital channels.

- **Community managers**—An organization's community managers are responsible for devising an audience-engagement strategy, which defines the interactions of Marketing, Customer Service, Customer Support, Customer Success, and others in the organization with its communities of partners, customers, and users. They converse with these communities through social-media channels, by writing and disseminating email newsletters and press releases, writing blog posts, and responding to requests for information and comments. Community managers often create new content in response to user-generated content (UGC). They also publicize new content—often by engaging specific audience segments.

- **SEO analysts**—A search-engine optimization (SEO) analyst is responsible for monitoring and analyzing the performance of Web pages in search engines and devising a strategy that improves their performance. An SEO strategy considers an organization's business goals, the needs of a Web site's target audience, what content to deliver to meet them, and the scope of the Web site's current content. SEOs endeavor to improve a Web site or page's relevance rankings in search-engine results and, thus, the amount and quality of traffic that it receives from *organic*, or unpaid, search-engine results. Thus, the ultimate goal of SEO is to improve the findability of Web sites or pages by search engines. SEO analysts improve the visibility of Web pages to search-engine algorithms by determining what *keywords* or phrases users would type into search engines to find specific content, then ensuring that content authors incorporate the appropriate terms into both the Web pages' content and their metadata.

Try to interview individuals who are representative of each role on your organization's content team. In a small organization, this might mean interviewing everyone on the content team. The overall number of content owners you must interview depends on what roles exist within your organization and the size of your content team. Try to limit each group discussion to between five and seven participants.

Type of data—Interviewing content-owners typically yields large amounts of qualitative, self-reported data. The data that you gather during your content-owner interviews or group discussions should provide a complete record of these sessions. Whenever possible, capture video recordings—or at least photos and audio recordings—of the sessions. However, don't allow notetaking to distract your attention from participants. Just quickly capture some key quotations or your own thoughts and the times at which they occur during a session. Whenever possible, have a notetaker take detailed notes.

Data analysis—Organizing the unstructured data from your content-owner interviews or group discussions prior to data analysis can take a significant amount of time, making this process more challenging. To build a shared understanding of your findings from content-owner interviews, synthesize the findings, identify common themes and patterns in the data, and have your project team do affinity diagramming immediately following the interviews. Affinity diagramming enables your team to give tangible form to and make sense of your findings. Alternatively, you could rely on the expertise of a team of analysts and collaboratively develop and align on hypotheses and models from your learnings.

Level of effort—Writing good questions to ask the content owners requires considerable skill and effort. While individual, in-depth content-owner interviews typically last between an hour and a half and two hours, they can sometimes take longer. Group discussions could take several hours. The time your interviews require overall depends primarily on how many content owners you must interview and whether you decide to conduct group discussions. Plus, preparing your interview questions and setting up the sessions with individual content owners or groups, then analyzing the resulting data takes significant effort. Depending on the number of content owners who participate in your interviews, completing them could take between two and four weeks. You should devote about half the time to conducting the interviews, a full day to affinity diagramming, then the remainder of the time to analyzing your findings and preparing your report or presentation. Conducting face-to-face content-owner interviews is a labor-intensive, highly immersive method of research and can be very time consuming. However, this is still a very efficient way of obtaining the information you need. Conducting remote content-owner interviews requires a more moderate level of effort.

Deliverables—Once your analysis is complete, prepare your report or presentation to deliver your findings to stakeholders. Ultimately, you'll capture and share the information you've obtained by conducting content-owner interviews or group discussions in content requirements or content-strategy documentation.

Investment—Conducting content-owner interviews requires only a moderate level of investment.

[7, 8, 9, 10, 11, 12, 13, 14]

Content inventories

Purpose—Inventory an existing information space or product's content to learn what content it currently comprises and *where* that content resides in its information hierarchy. You must understand the full breadth and depth of *all* the existing content—as well as any content that is currently under development or planned—to be able to organize the content effectively. This is especially true if you're unfamiliar with the domain of the information space or product.

Only by learning and documenting what content already exists can you identify any gaps in that content, determine what new content your content team needs to create to fill those gaps, decide what existing content requires revision, and justify the development of new content to stakeholders.

To determine the full extent of a digital information space or product's scope, it is necessary to conduct a comprehensive content inventory. However, a content inventory's scope might extend beyond an individual information space or product to an organization's broader or even its entire digital information ecosystem. For example, a content inventory might be necessary to prepare for merging multiple information spaces or products *or* for migrating content from one information space or product to another—for example, moving content from a customer extranet or a corporate intranet to a public Web site. Information that should be on an intranet might reside on various departmental servers. A content inventory might also be necessary if your content team needs to transform one type of content into a different format—for example, to repurpose content from a presentation or PDF document for publication on a Web page.

Understanding the current structure of an information space or product could inform the design of a new information architecture *or* help you to ascertain how to restructure or expand the existing information hierarchy—both to improve findability and accommodate new content. Your stakeholders require this shared knowledge to approve your making extensive changes to the information architecture and when planning and budgeting for information-architecture and content-development initiatives. This information can also be helpful to content authors who must find and refer to existing content as source material and avoid replicating content that already exists.

Description—Before beginning your content inventory, your team should establish the scope of and align on clear strategic goals for the inventory. What does your team need to accomplish by conducting a content inventory? You could inventory either complete pages, the chunks of textual content and media assets that pages comprise, *or* reusable content components. Whatever the level of granularity on which your content inventory focuses, you must determine what information you need to capture about each page *or* whatever other types of content objects you want to record in your content inventory. What data is relevant can vary significantly by project, depending on both the types of content objects for which you intend to gather data and how you intend to use that data. Get input from your entire project team regarding what data you should capture. Be as precise as possible in making these initial decisions, but be aware that, as your content inventory progresses, you'll likely discover some additional types of data that are relevant.

You can manually gather data about an information space or product's structure, the pages or other types of content objects it comprises, and their attributes. Alternatively, you could automate the process to some extent by using a digital crawling tool to generate a complete list of all of its pages or other content objects *or* by exporting information about its content objects from the content-management system (CMS), including attributes such as titles, authors, dates, and metadata such as topics or categories. A very effective approach is to begin by automating the gathering of as much data as possible, then using the information hierarchy to structure your content inventory and capturing additional details manually.

If you're doing a content inventory manually, create a content-inventory spreadsheet in which you can capture data about each page or other content object in the rows and the specific attributes of each content object in the columns. The structure of this spreadsheet should conform to the information hierarchy of the information space or product.

Before you record any data in this spreadsheet, the members of your multidisciplinary project team and other stakeholders should review and provide feedback on its structure, the terminology it uses, the definitions of the attributes that you intend to capture, any attributes you should omit because they would not add business value, and any attributes you've overlooked that you should add. Ask your team's subject-matter experts exactly how you should categorize certain types of content. Once you've created your content-inventory spreadsheet's initial columns, document and train your team on their proper use. Assign the responsibility for capturing specific types of data to particular team members who have the necessary expertise.

When conducting your content inventory, you should catalog *all* of an information space or product's existing pages and any other relevant content objects in your content-inventory spreadsheet. To ensure that you discover all of its pages, begin with its home page, then traverse every link in the main navigation system, working your way through every level of the information hierarchy, as well as all the links in any auxiliary navigation systems—such as the links that appear in any footer.

As you progress through each section of the information space or product, click all links within each page's body content to determine whether there are any internal links that do not replicate links in the main navigation system and would display a document that you could not otherwise capture in your content inventory. However, avoid including any page or document in your content inventory more than once. You could also choose to record any embedded images or other media that reside on your server and appear on pages. Finally, ask your colleagues whether there is any additional content that you would have difficulty discovering—for example, landing pages for marketing campaigns, dynamic promotional content, or product demos.

To increase your efficiency and prevent losing your place when working on a content inventory manually, display the information space or product on one device or computer monitor and your content-inventory spreadsheet on another monitor. Also, try to capture all the information you need about a page or other content object in one pass. Copy and paste whatever information you can.

Descriptive attributes of pages or documents

In the columns of your content-inventory spreadsheet, you should gather whatever data about the information space or product's content would meet the organization's needs. For example, you might choose to record the following *descriptive attributes* of each page or document that your information space or product comprises:

- **Page ID**—Record a page or document's numeric identifier in a **Page ID** column, which should be the first column in your spreadsheet. Refer to the section "Identification of pages or content objects," later in this chapter, for detailed information about assigning page IDs.

- **Page title**—Capture the textual identifier for a page or document.

- **Purpose**—Briefly describe the purpose of the page or document. What value does it provide to target users and the organization? What user needs does it address? How does the page or document help the organization achieve its goals for the information space or product?

- **Content type**—Record a page or document's content type—for example, a page that employs the template for a particular type of content such as a landing page, an article—or more specifically, a feature article, column, review, news report, or editorial—a personal profile, marketing copy, a product description, or company information; a functional page such as a form or shopping cart; or a downloadable document such as a text document, spreadsheet, PDF document, or presentation. These content types might correspond to those of the CMS. Is the page's content static or dynamic? Does it integrate any social media? As you work on your content inventory, compile a separate working list of all the content types you've assigned to pages or documents and define each content type, then use these content types consistently throughout your content inventory.

- **Components**—List the content components that the template for the page's content type comprises.

- **Topics or categories**—Add the topics or categories that describe the content to either a **Topics** or a **Categories** column in your content inventory. In identifying these topics or categories, you could rely either on your own expertise and/or that of your colleagues who are subject-matter experts. Alternatively, you could use an industry-standard classification scheme. As you assign topics or categories to pages or documents, maintain a separate list of the topics or categories and define them, ensuring that you can use them consistently throughout your content inventory. These topics or categories could also be useful in defining a page's metadata.

- **Section**—Document the section within which a page or document resides. You could choose to include just the highest-level section or its subsections as well. You can derive these sections from Web addresses.

- **Location**—Record a page or document's Web address or other data source. The Web address shows the page or document's location within the information space or product's information hierarchy and, thus, the section in which it resides.

- **Linked content**—You could identify each internal link to downloadable content on a page—for example, PDF documents, spreadsheets, or other downloadable documents—and list their Web address or other data source and their type. It's important to gather this information for pages the content team is revising.

- **Embedded content**—For Web pages only, you could identify each instance of a page's embedded content—for example, images, presentations, videos, or audio clips—and list their Web address and, if necessary, their type. Gather this data for pages the content team is planning to revise or rebuild from content components.

- **Crosslinks**—*If* you want to record which of an information space or product's pages link to one another, you could add a column for internal crosslinks and list the Web addresses for all the internal crosslinks that are on a page.

- **Notes**—Capture any notes that would help you understand a page or document—for example, a strategic goal for its content—or details about whatever action you plan to take on the content, such as revising or removing it.

Descriptive attributes of content components

If you're creating a content inventory that comprises the individual content components on pages—for example, chunks of textual content and embedded media such as images, video or audio recordings, or animations—*or* the dynamic content components that you intend to reuse across multiple pages or channels, your content-inventory spreadsheet might include columns for the following *descriptive attributes*:

- **Content ID**—Record the numeric identifier for a content component in a **Content ID** column, which should be the first column in your spreadsheet. Refer to the section "Identification of pages or content objects," later in this chapter, for more information about assigning content IDs.

- **Heading or filename**—Capture the subheading for a chunk of content on a page or a reusable content component in a **Heading** column; *or*, for embedded media, record its filename in a **Filename** column.

- **Purpose**—Briefly describe the purpose of the content component and the value it provides to target users and the organization. What user needs does it address? How does this component help the organization achieve its goals for the pages on which it appears?

- **Content type**—Record the file format of a digital-media asset that is embedded in a page, such as an image or video, *or* a reusable content component. These content types might correspond to those of the CMS. Create a separate list of all the content types you've assigned to content components and define them, ensuring that you use them consistently throughout your content inventory.

- **Data source**—Add the single data source at which a content component resides within a CMS or other repository—that is, where an author or editor can revise the content.

- **Location**—Capture a content object's location on a page. Instances of reusable content objects might appear in multiple locations—either on the same channel or across multiple channels—each of which refers to the same data source. List all the locations at which a reusable content component appears within the information space or product—or across all of an organization's channels.

Alternatively, if capturing reusable content components within the context of the pages on which they appear is more important than listing the locations of *all* instances of a reusable content component together, you could either nest the subheadings of a page's content components under the page's title, in a **Page Title / Components** column, *or* list all of the components on a page in a separate **Components** column.

Governance attributes of content objects

Depending on your needs, you could add columns for the following *governance attributes* of each page or other content object to your content-inventory spreadsheet:

- **Author**—Who was the content's original author? Who most recently revised it?

- **Editor**—Who most recently edited the content?

- **Owner**—Did someone within your own organization create this content or an external content creator? What department or organization owns the content? Who requested, approved, and published this content? These details can be helpful if an author or editor has left the organization and you need to know about prior decisions regarding the content or require approval for revising or removing the content. Did a vendor provide this content? Is this user-generated content (UGC)?

- **Metadata**—What metadata—such as a page title, description, keywords, alt text, or topics or categories—currently exists for a page? Capture any relevant metadata.

- **Publication date**—On what date was the content originally published?

- **Revision date**—On what date was the content last updated?

- **Trademarks and copyrights**—What trademarks and copyrights apply to this content?

- **Status**—What is the current state of this content object? Is the content high quality, accurate, and accessible? Is it consistent with style and branding guidelines? Is this content up to date or obsolete? Does this content require revision? Should this content be removed from the information space or product? Therefore, the status of a page or other type of content object might be, for example, *Active, Inactive, Deleted, Obsolete, Being Updated, Updated, Requires Revision, Under Revision, Revised,* or *Remove.*

Governance attributes for content migrations

If you're doing a content inventory preceding the migration of existing content to a new CMS, be sure to capture the following *governance attributes* in your spreadsheet:

- **Current location**—What is the content's current location in the CMS?
- **New location**—What is the content's projected location in the CMS?
- **Author**—Who was the content's original author?
- **Editor**—Who most recently revised the content?
- **Owner**—What department owns the content? Is this user-generated content (UGC)?
- **Revision date**—On what date was the content last updated?

Identification of pages or content objects

Structure your list of pages or other content objects in your content inventory by leveraging the hierarchy of the information space or product's existing navigation system. Similar to a document outline, you should indent each level in the information hierarchy and assign a unique, numeric page ID or content ID to each page or other content object, respectively. For example, these IDs could have the following formats:

- **Home page**—0.0
- **Sections' landing pages**—1.0, 2.0, 3.0, 4.0, 5.0, 6.0, and so on, for each of the primary sections
- **Sections' subordinate pages**—1.1, 1.2, 1.3, 1.3.1, 1.3.2, 1.3.3, 2.1, 3.1, and so on. Capture each section's subordinate pages, at all levels of the hierarchy.
- **Linked content**—Under any pages containing internal links to other content objects to which users cannot navigate using the navigation system—such as a link to a PDF document or spreadsheet—designate each content object's type and list them in numerical order—for example, 1.3.3-PDF1 or 1.3.3-Spreadsheet1.
- **Embedded images and other media**—For pages that contain images or other embedded media that reside on your own server, list all of the page's embedded media in the sequential order that they appear on the page. List the media in numerical order, indicating each content object's type—for example, 1.3.3-Image1, 1.3.3-Image2, 1.3.3-Video3. Either nest this list under the page's title *or* list all of the media in a separate **Media** column.
- **Content components**—Under pages comprising content components, add a sequential list of all the page's content components, in numerical order, with the letter *C*, for *Component*, preceding each number—for example, 1.3.3-C1, 1.3.3-C2. Either nest this list of components under the page's title *or* list all components in a separate **Components** column.

Page IDs and content IDs help make the relationships between pages and other content objects clear. You can use these IDs when referring to specific content in your other project documentation. Once your team has determined a useful numbering scheme and format for content inventories, you should use them consistently across all related projects.

When—You should ideally conduct a content inventory during *Discovery*, immediately preceding a content audit, in preparation for redesigning an information space or product; changing its existing information architecture; creating new content; aggregating content from multiple information spaces or products; or dividing a large information space or product's content among smaller ones. Doing a content inventory can also be a helpful first step in preparing content cards prior to card-sorting activities.

While conducting a content inventory is useful primarily in understanding the content on an existing information space or product, you could incrementally build up a content inventory as your team develops new content for either an existing information space or a new one.

You could continue using this method iteratively, throughout your product-development process, whenever you need to assess the scope of an information space or product's existing content and ascertain whether there are any gaps in that content.

Alternatively, it might be necessary to conduct a content inventory when your organization is preparing to migrate an information space or product's content to a new CMS *or* to reuse its existing content components across multiple pages or channels. When migrating content to a new CMS, an exhaustive content inventory is absolutely essential.

Where—If an information space or product has a limited scope and you're working alone, you could conduct a content inventory in your own workspace. However, you should ideally conduct a content inventory in a conference room, where you can post screenshots of pages or other content objects on the walls, visually replicating the structure of the information space or product. Having this amount of space is necessary when you're working as a team.

Participants—Determine who should be responsible for conducting the content inventory and maintaining the quality of the data in the spreadsheet. Depending on the scope of the information space or product, one or more teammates could work on a content inventory. Involve everyone working on the content inventory from the beginning of the process. You could assign a different teammate to audit each of a large information space or product's major sections. If you're doing a content inventory prior to migrating a large information space or product to a new CMS, you should automate as much of the process as possible—especially if your team has limited bandwidth.

Type of data—A content inventory comprises qualitative data about each page or other content object on an information space or product, including the type of page or content object and other key attributes.

Data analysis—Rely on your findings from earlier UX research and the subject-matter expertise of your teammates in determining what gaps exist in the information space or product's content and what new content your content team needs to create. The creation of new content determines, to some extent, the changes you would need to make to the information space or product's information architecture. Usually, after completing your content inventory, your next step should be to conduct a content audit, through which you can begin to analyze your content.

Level of effort—Conducting a thorough content inventory can be very time consuming, depending on the amount of content an information space or product comprises. The larger the scope of the information space or product, the more time a content inventory takes—ranging from a few days to several weeks of effort. To complete your content inventory more quickly, you could divide the work among several members of your team or automate the process.

Alternatively, for a large information space or product, you could limit the scope of your content inventory to just one section, *or* you could conduct what Lou Rosenfeld calls a *rolling content inventory*—first completing a content inventory of one part, then another and another, until you've inventoried its full scope, then cycling back to the first part and inventorying the entire information space or product again, over time. Lou says, "Ongoing, partial content inventories are likely to be far more cost-effective than trying to achieve the perfect, all-encompassing snapshot of your content." When you take this approach to a content inventory, your team is more likely to look at the content within the limited scope of part of the information space or product more carefully and more often.

Conducting a content inventory really should be an iterative process. Keep your content inventory up to date as you change the information space or product's structure and your content team adds new content and removes obsolete content. Maintaining your content inventory is worth the effort because it can remain useful for years. Plus, creating a content inventory is a labor-intensive job that your team should have to do from scratch only once.

Deliverables—Your primary deliverables are your content-inventory spreadsheet and a report that distills your key findings and describes how to address them. Be sure to track the dates and scope of each iteration of your content inventory.

Investment—A content inventory is an inexpensive method of content analysis.

[15, 16, 17, 18, 19, 20, 21, 22, 23, 24, 25, 26, 27, 28]

Recommended book—Paula Ladenburg Land. *Content Audits and Inventories: A Handbook*. Laguna Hills, CA: XML Press, 2014.

Content audits

Purpose—By conducting a content audit, or *content assessment*, you can learn what types of content an existing information space or product currently comprises and how you might improve its information architecture. You can also evaluate the content's quality and relevance to the organization's business objectives and target audiences. These learnings inform your organization's content strategy and the definition of requirements for adding new content and making changes to existing content. For example, you might discover content issues such as missing, redundant, or obsolete information; pages that comprise similar content, but have inconsistent structures; inconsistent usage of link text, alt text, Web addresses, and the titles of their destination pages; or broken links. Thus, you can identify what content the content team needs to create, revise, or remove from the information space or product, so you can improve its overall quality and accessibility.

By keeping abreast of forthcoming changes to the information space or product's content, you can understand what changes you must make to its information hierarchy—especially large-scale changes such as adding substantial, new sections. Conducting a content audit can also help you establish the business case for redesigning an information architecture or launching a new content-development initiative and determine the project's scope and cost.

Description—Before beginning a content audit, your team should establish and align on clear strategic goals for the audit and determine its scope and depth. If you're auditing a small information space or product, you can probably complete a comprehensive audit of all the content. However, if you're auditing a large, enterprise information space or product, it might be necessary to identify specific, high-priority sections on which you want your audit to focus—such as those that deliver the greatest business value, impact key user tasks or customer journeys, or receive the highest levels of traffic. When conducting an in-depth content audit, you could narrow the focus of your deep dives to such specific sections. In contrast, to limit the effort your audit requires, you could decide how many levels of the information hierarchy you want to evaluate. You might look at *all* content at just the upper levels of the information space or product—for example, to assess what content attributes matter most in designing a faceted navigation system.

Depending on what you want to learn about the information space or product's existing content and how your team intends to use your learnings from the audit, you should decide exactly what information you need to capture about its pages or other content objects. Why are you conducting a content audit? In addition to whatever content attributes your team captured when preparing the content inventory, you should add columns in which to gather other attributes of content objects that would provide strategic value to what now becomes your content-audit spreadsheet.

Before using your expanded content-audit spreadsheet, ask your project team and other stakeholders to review and provide feedback on the spreadsheet's structure, terminology, attribute definitions, and attributes that you should omit or need to add. Train your team in the proper use of the spreadsheet, and assign specific responsibilities to particular team members. During your content audit, you'll capture additional attributes of the information space or product's pages or other content objects.

When—You'll usually conduct a content audit immediately following the completion of a content inventory, during *Discovery* or perhaps in the early stages of *Design*, when you're devising your information-architecture strategy for an information space or product. You should conduct a content audit whenever your team is preparing to redesign an existing information space; redesign an existing information architecture; aggregate content from multiple information spaces; or divide the content of a large information space among two or more smaller ones. However, you might also conduct a content audit independently of a content inventory if your team is adding or revising a significant amount of content in a way that doesn't necessitate revising the information architecture—for example, when you're preparing to apply revised style, branding, or accessibility guidelines.

Although conducting a content audit is useful primarily in evaluating the content on an existing information space or product, you could incrementally evaluate the content for a new information space as your team develops it. You could use this method continually, throughout your product-development process, whenever you need to evaluate an information space or product's coverage of topics or categories or the quality and usefulness of its content.

Once you've completed your audit of an information space or product's content, consider conducting a competitive analysis of key competitors' content. By assessing the strengths of the information space or product's content against the weaknesses of competitors' content, you can identify opportunities to differentiate its content from that of competitors. To ensure that the information space or product fully capitalizes on key areas of opportunity, you must emphasize the unique value proposition that content on specific topics provides by elevating it in both the information hierarchy and the navigation system and continuing to develop content on those topics. You've also identified gaps in the information space or product's content that may represent competitive weaknesses in comparison to strengths in competitors' content. To remedy these weaknesses in the information space or product's content, you should endeavor to fill the gaps in its content.

Where—When conducting a content audit for a small information space or product on your own, you could work in your own workspace. A remote team could do a content audit, using an online spreadsheet. However, you and your team should ideally conduct a content audit in the same conference room in which you've just completed your content inventory, where you've posted screenshots of pages or other content objects on the walls to illustrate the structure of the information space or product. When you're working collaboratively with your team, having plenty of available wall space is essential because you may need to show what additional content the content team would need to create to fill gaps in the existing content, including what new sections you would need to add.

Participants—Determine who should be responsible for conducting the content audit and maintaining the quality of the data in the content-audit spreadsheet. Involve all teammates who are working on the audit from the beginning of the process. When preparing your team to conduct a content audit, make sure everyone understands the audit process and knows how to audit each type of content the information space or product comprises and provide examples of what data to collect for each content type. The scope of the information space or product and that of the content audit itself determine whether one or several teammates should audit its content. Your team's available bandwidth also constrains an audit's possible scope. For a very large information space or product, consider having a different teammate audit the content in each of its major sections.

Type of data—When you're conducting a content audit, you can gather quantitative data about the numbers of pages or other types of content objects that the information space or product comprises, as well as the numbers of pages or content objects that share particular attributes. You can also collect qualitative data about each page or other content object, including its content type, topic, or category, and other attributes such as quality, business value, and usefulness. Quantitative and qualitative data complement one another and analyzing them in concert results in more meaningful findings. The balance of quantitative versus qualitative data that you should gather depends on your goals for the content audit and the information you need to achieve them.

Data analysis—During your project team's analysis of the data you've gathered during your content inventory and audit, you should endeavor to answer questions such as the following: Are users visiting the information space or product's most important pages? Do its information architecture, navigation system, and search system enable users to easily find the information they're seeking? Are any broken links preventing them from doing so? Does the information space or product's content fully address the business requirements and meet users' needs?

Integrating data analytics with a content audit can enable you to assess how well specific content is performing. Page views and unique visitors are particularly relevant. Once you've added data analytics to your content-audit spreadsheet, you could optionally sort it by the pages that receive the most traffic, helping you to prioritize whatever changes you need to make to the information architecture or revisions to the content by those that would have the greatest impact.

Your project team could do affinity diagramming to identify common themes and patterns in your data. Alternatively, you could rely on the expertise of your project team in collaboratively developing and aligning behind hypotheses regarding what solutions would best address your learnings from your content inventory and audit.

Level of effort—Conducting a detailed content audit requires human judgment, so is labor intensive and, depending on the scope of the content you're auditing and the breadth and depth of your analysis, can be very time consuming. While you could complete a content audit that has a limited scope in as little as one to five days, a large-scale audit could take significantly more time—perhaps many weeks or even a couple of months. To complete your content audit more quickly, you could divide this effort among several people, as for a content inventory—for example, you could assign specific sections of an information space or product to different teammates.

You can employ this method iteratively, whether by taking various approaches to auditing the same content or by focusing on different parts of an information space or product's content over time. You could also choose to capture information about additional content attributes. Consider how frequently you should assess particular types of content.

Try to keep your content-audit spreadsheet up to date as you restructure the information space or product and as your content team adds new content, revises content, and removes obsolete content. If the content changes frequently, the frequency with which you conduct content audits might be more important than the completeness of a single cycle of content auditing. Maintaining your content audit is essential to sustaining its usefulness.

Deliverables—Consider your audience's needs when preparing the deliverables in which you'll present your findings from the content audit. Do they need to inform the design of the information space or product's information architecture, its content strategy, or the design of new templates for specific types of content? Or is their purpose to justify the development of a new information architecture or new content?

In addition to sharing your content-audit spreadsheet with stakeholders, visualize your key findings from the content audit—for example, by creating charts or graphs that show the proportions of the different types of content an information space or product comprises or the content's various attributes—or by creating a content map, as I'll describe in the section "Content mapping," later in this chapter. Prepare a report that indicates what content to keep as it is, what content to move to another location in the information hierarchy, what obsolete content to remove from the information space or product, and what content your content team needs to revise or update. Describe what specific revisions or updates are necessary in detail and indicate who should be responsible for them. Track the dates and scope of each iteration of your content audit.

Investment—Content auditing is a fairly inexpensive method of content analysis, but the scope of each content audit determines its overall cost. [29, 30, 31, 32, 33, 34, 35, 36, 37]

Quantitative content audits

Purpose—Conducting a quantitative content audit lets you determine how much content an existing digital information space or product comprises—that is, the overall scope of the information space or product—and, specifically, the numbers of pages or other content objects on the information space or product or within its specific sections. Your learnings from a high-level, quantitative content audit can inform your project planning—by helping you to define the scope and estimated cost of an information-architecture redesign or a content-development project—and, thus, enable you to persuade stakeholders to support such initiatives.

Description—You could do *only* a quantitative content audit, recording the numbers of instances of various types of content that an entire information space or product comprises *or* those in its specific sections. Alternatively, you could capture quantitative data regarding specific pages or other types of content objects when conducting a quantitative content audit as the first stage of a more thorough auditing process that also comprehends a qualitative content audit. Depending on the organization's goals for a content audit, you could gather quantitative information for an entire information space or product, only one of its major sections or each of them over time, or consider only the content at the highest levels of the information hierarchy. You should consider most quantitative data within the context of the information space or product's sections, look at the distribution of various types of content across those sections, and assess whether that distribution is consistent with stated strategic priorities.

Capturing Quantitative Content Attributes

When conducting a basic, quantitative content audit, you could determine the amounts of certain types of content on an information space or product and capture the numbers of instances of content that have specific attributes, in columns such as the following:

- **Page count**—Count the overall number of pages that an entire information space or product comprises or those in individual sections to which users can navigate from the primary navigation system.

- **Topics or categories**—Count the number of pages regarding specific topics or belonging to specific categories and add this information to either the **Topics** column or the **Categories** column in your spreadsheet. Use the topics or categories that you assigned when working on your content inventory.

- **Target audience**—Each page or document has one or more target audiences. An information space or product's content strategy should define the target audiences for each page or document. Ideally, a persona should exist for each target audience. Count the numbers of pages for each target audience and record this information in either an **Audience** column *or*, if there are multiple target audiences, a column for each specific audience you've defined.

- **Content type**—Count and record the number of pages an information space or product comprises or those in its individual sections; the numbers of each type of document such as PDFs or spreadsheets; the number of functional pages; *or* the numbers of each type of digital-media asset such as images or videos. Use the content types that you assigned when working on your content inventory.

- **Links**—Count the numbers of outbound links on each page—including all internal links, internal links to pages in the same section, internal links to pages in other specific sections of the information space or product, and external links to other domains—as well as inbound links to each page.

- **Template**—List the templates that provide page layouts for specific types of content and record the numbers of pages that employ each of these templates.

- **Author**—Count and record the numbers of pages by particular authors.

- **Owner**—Count and record the numbers of pages belonging to specific internal departments, external organizations, or users.

- **Last revision date**—Count and record the overall number of pages whose last update occurred either prior to or since a particular date. Pages that have not been revised recently may contain content that is obsolete or does *not* observe current style or branding guidelines or comply with current trademarks.

Depending on the focus of your project, you might want to gather other quantitative data about the information space or product. Such quantitative data could help you to define the project's scope by determining how much of the information space or product's information hierarchy requires revision *or* the approximate number of pages or other content objects that your content team needs to revise or develop.

Capturing Quantitative Attributes for a Qualitative Content Audit

When conducting a quantitative content audit as a precursor to doing a qualitative content audit, add columns for whatever quantitative information you need to capture about each page or other type of content object such as the following:

- **Content type**—Count and capture the numbers of instances of each different type of digital-media asset such as images or videos, on a page or within another content object. Use the content types that you assigned when working on your content inventory.

- **Words**—Count and record the numbers of words on each page or in another type of content object. Ascertain whether pages—particularly those in the same section of an information space or product—or content objects of similar types have word counts that are roughly comparable.

- **Analytics**—Capture any relevant data analytics that would demonstrate specific pages' performance, adding the data to a separate column for each type of data. For example, you might capture page views, direct traffic, or unique visitors, which could demonstrate the usefulness of pages to users.

- **Type of data**—As the name of this type of content audit indicates, when conducting a quantitative content audit, your focus is on gathering quantitative data.

- **Level of effort**—You can quickly conduct a quantitative content audit, which requires significantly less effort than a qualitative content audit. [38, 39, 40, 41, 42]

Qualitative Content Audits

Purpose—Conduct a qualitative content audit to assess the overall quality of each page or other type of content object, the content's usefulness to an information space or product's target audience, and the business benefits that accrue from each page or content object. By collectively examining a subset of the information space or product's constituent parts, you can both evaluate its overall quality and usefulness and begin to identify some of its most significant information architecture and content issues. Your learnings from an in-depth, qualitative content audit inform the scope of the changes the content team must make to an information space or product's content, which in turn affects the scope of the changes you need to make to its information architecture.

Description—Evaluate the quality and other attributes of a representative sample of each type of content the information space or product comprises. When assessing the content on a page or in another type of content object, obtain answers to whatever questions would be relevant to your organization, then capture the resulting qualitative data in your content-audit spreadsheet, in columns such as the following:

- **Topics or categories**—To what topics or categories does this content belong? This information is in the **Topics** or **Categories** column of your content-inventory spreadsheet. Use the topics or categories that you assigned when working on your content inventory.

- **Target audience**—What target audiences does this content serve? What are the information needs and tasks of each target audience? How relevant is the content to each audience? Rate how well this content meets the needs of each audience on a scale from *1* to *5*, from the lowest to the highest satisfaction of their needs. Record this data in either an **Audience** column *or*, if there are multiple target audiences, a column for each specific audience.

- **Quality**—Is textual content well structured, well written, clear, concise, easy to read, engaging, and consistent with style and branding guidelines? Are grammar, punctuation, and spelling correct? Are the voice and tone of the content consistent with the brand? Are all page titles and subheadings meaningful? Does link text make all links' destinations clear? Are images well-rendered and legible? Are other types of media meaningful and well-produced? Rate the content's quality on a scale from *1* to *5*, from the lowest-quality to the highest-quality content. Sort the spreadsheet by **Quality** to identify content that the content team should prioritize revising. If you know what specific revisions are necessary, note them briefly in the **Notes** column.

- **Currency**—Is the information up to date? Rate the content's currency on a scale from *1* to *5*, for the least to the most current content. Sort the spreadsheet by **Currency** to see what content most needs updating. If you know of specific updates that are necessary, note them briefly in the **Notes** column.

- **Accuracy**—Is the information accurate? Are all links functional? Has the author included all appropriate trademark and copyright information? Rate the accuracy of content on a scale from *1* to *5*, from inaccurate to completely accurate content. Sort the spreadsheet by **Accuracy** to see what content is most in need of revision. Note any necessary revisions briefly in the **Notes** column.

- **Credibility**—How credible is this content? Does the content inspire trust among users? Rate the content's credibility on a scale from *1* to *5*, from the least to the most credible content.

- **Business goals**—What are the business goals for this content—for example, engaging visitors' interest, converting visitors to customers, providing essential information to create awareness, making a sale, or generating leads? Record these business goals. Rate how well this content meets these goals on a scale from *1* to *5*, from the lowest to the highest satisfaction of these goals.

- **Business value**—What is the business value of this content? Rate the content's business value on a scale from *1* to *5*, from the least to the most valuable content.

- **Usefulness**—How useful and valuable is this content to users? Does the content meet *all* target users' needs? Is this content helpful to users? Rate the content's usefulness on a scale from *1* to *5*, from the least to the most useful content. Data analytics can demonstrate a page's usefulness to users, including page views, direct traffic, or unique visitors.

- **Accessibility**—How accessible is this content to users with different physical abilities and levels of reading skills? Does the content meet *all* target users' needs? Rate the content's accessibility on a scale from *1* to *5*, from the least to the most accessible content.

- **Insights**—What insights have your team derived from your content audit and analysis?

Depending on the organization's goals, determine whether you should gather any of the following qualitative information about specific pages' metadata and SEO performance during your content audit:

- **Metadata**—How well does the page's metadata ensure the findability of the page through the information space or product's navigation *and* search systems or with a Web-search engine? Does the metadata support the CMS's effective organization and display of the page's content objects? What is the quality of the page's metadata? Include each page's metadata in a **Metadata** column. Plus, you could rate the quality of each page's metadata on a scale from *1* to *5*, from the lowest-quality to the highest-quality metadata.

- **Keywords**—List the page's keywords.

- **SEO**—How well is the page's content optimized for search engines? Are the keywords in the page's metadata and in the page's title, subheadings, and body content performing well in search engines? Does the use of keywords on a page ensure that the page is both findable *and* readable? Rate the content's SEO on a scale from *1* to *5*, from the worst to the best SEO.

Gathering such qualitative data can be helpful in determining the scope of an information-architecture redesign or content-development project—such as when you're deciding either how much of an information space or product's information hierarchy requires revision *or* the approximate number of pages or other content objects that your content team must revise or develop.

Type of data—As the name of this type of content audit indicates, when conducting a qualitative content audit, your focus is on gathering qualitative data.

Level of effort—Although a qualitative content audit is much more labor intensive than a quantitative content audit, its scope need not be comprehensive. The content audit's scope determines the overall level of effort. [43, 44, 45, 46]

Recommended book—Paula Ladenburg Land. *Content Audits and Inventories: A Handbook*. Laguna Hills, CA: XML Press, 201

Content mapping

Purpose—Through content mapping, you can represent the content inventory for an existing digital information space or product visually, enabling you to see the relationships between different pages or other types of content objects. For example, you could create a conceptual, visual representation of an information space or product's content that depicts its overall structure, the relationships that exist between various sections or specific topics or categories of content, or detailed layouts of the content components on specific types of pages.

You can create a content map in the form of an information hierarchy, mapping the full scope of a relatively small information space or product, which enables you to assess its breadth versus its depth; mapping only the upper levels of a larger information space or product, allowing you to see its full breadth; mapping certain individual sections, letting you assess their full depth; *or* depicting an information space or product's key navigation paths.

Creating a detailed content map showing the pages that an information space or product comprises can help your team and other stakeholders to better understand the information space or product's information hierarchy, where specific content resides within it, and what content already exists or is lacking. The primary goal of this content-mapping process is to optimize the information space or product's structure and navigation system to improve the findability of its content.

You could create a content map to demonstrate the business case for making revisions to an information space or product's information hierarchy or navigation system or to show gaps or other deficiencies in the existing content that would justify the creation of new content.

You could use your content map in conjunction with usage-data analytics to determine how an information space or product's structure impacts the performance of specific pages, as well as the findability of particular information. For detailed information about usage-data analytics, refer to the section "Usage-data analytics," in Chapter 5, *UX Research Methods for Information Architecture*.

Alternatively, you could create very detailed, visual representations of the specific types of pages that an existing information space or product comprises *or* design layouts for new page types, showing the content components that each of them comprises. Each content component should represent the smallest logical chunk of content that would be useful in mapping content to page layouts. One specialized variant of a content map is the *content-reuse map*, whose purpose is ascertaining what content components might be reusable and where each instance of a reusable content component should ultimately reside.

Description—Once you've conducted a thorough content inventory and audit and really understand the content that an information space or product currently comprises, create whatever visual representations of the existing content would help you address your team's strategic needs. Both the scope and the purpose of content maps can differ greatly, so the type of content map you should create depends entirely on your business needs. You could render a content map by hand on a whiteboard or on paper, or using one of the many digital tools that are available—some of which are general-purpose, graphic-design, or wireframing applications, while others' specific purpose is the creation of content maps.

A content map depicting an information space or product's information hierarchy might consist of boxes that represent pages and arrows that indicate the relationships between them—and perhaps lines or larger boxes for logical groupings of pages such as sections. Figure 6.1 depicts a content map showing the information hierarchy for *UXmatters*.

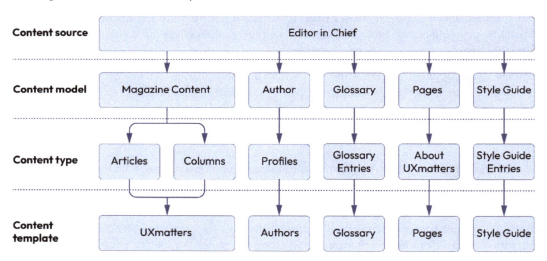

Figure 6.1—An example content map depicting an information hierarchy

Alternatively, to illustrate an existing information space or product's structure in greater detail, a hierarchical content map could comprise screenshots of its pages that your team members post on the walls of a conference room. Or you could represent an information space or product's hierarchy using Post-it notes. In both of these cases, photograph your representation of the information space or product's current state to create a permanent record of your work. Later, during your design process, you could rearrange Post-it notes to try out potential changes to the information hierarchy. If you might need to make major structural changes to an information hierarchy, place the Post-it notes representing each section or subsection on a poster board or a sheet from a Post-it Easel Pad, so you can easily move them around. This approach could also save you a lot of effort if you need to move the entire hierarchy to another location.

Content maps that depict the content components that make up particular types of pages could comprise sketches, wireframes, or screenshots of the individual content components. When chunking existing content into components to enable its reuse, you must divorce each instance of source content from its original container and ensure that all the content components you create remain free of any association with containers at their destination. As you decompose the source content into logical content components, assign a unique identifier to each component and record its source in a content-mapping table. Later, during *Design*, when mapping content components onto the destination pages of an information space or product, record all of their destination pages in the content-mapping table, as shown in Figure 6.2. You could create this table in a document or spreadsheet or as an HTML table, which provides the benefit of letting you link to any destination Web pages.

Source: Print Brochure	Destination: Web Page
Page 1	Product Page: Book cover
Page 2	Product Page: Title and description
Page 3	Product Page: About the author
Page 4	Product Page: Product details
Page 5	Product Page: Pricing information

Figure 6.2—An example of an HTML content-mapping table

Figure 6.3 shows an example of a content map, indicating the content components' roles on a product page. For anything but the smallest information space or product, you'll likely need a content-management system (CMS) and its underlying database to manage your content components and publish your content.

	Awareness	Consideration	Decision
Intent: The user wants to buy a book on a particular topic so visits an ecommerce book store.	The user searches for the topic and sees the cover and title for a book on that topic in the search results.	The user goes to the product page and reads the book's description, information about the author, and notes the publication date and price.	The user decides the book meets his needs and to purchase it.
Content components:	Book cover and title on the Search Results page	Product page: Book cover, title, description, author information, product details, and pricing information	Product page: **Buy Now** button

Figure 6.3—An example content map of a product page's content components

Create such content components when you're gathering content of various types from disparate sources—whether physical or digital—in preparation for designing a new digital information space or product. This type of content mapping is especially helpful when working with various content authors from different departments across an organization, making it important to set standards for identifying content components and facilitate collaboration.

When creating a *content-reuse map*, you should consider whether similar, but somewhat different content components should actually be identical, *or* there are good reasons for those differences to exist. If content components that are different should actually be identical, determine how you should revise the content to make all instances consistent and, thus, fully reusable. Also, determine whether specific content components should be identical across different audiences, media, or channels or tailored to specific contexts. Having this knowledge enables the content-management team to configure each reusable content component's reuse logic in the CMS. Figure 6.4 shows an example of a content-reuse map.

Home Page, with 10 Article Excerpts

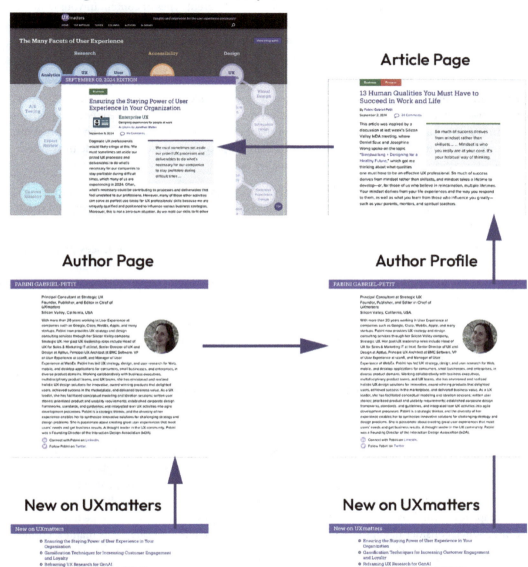

Figure 6.4—A content-reuse map for *UXmatters*

Regardless of what you're representing in your content map, consider using color-coding—for example, to show the levels of an information hierarchy; indicate the different types of content in your content map; or differentiate the direction of internal links, whether down, up, or across the hierarchy. Be sure to include a legend that indicates the meaning behind your use of color or any shapes that you might use in depicting different types of content.

When—You should ideally create a content map immediately following the completion of a content inventory and audit, during *Discovery* and the early stages of *Design*, when you're devising your information-architecture strategy for an information space or product. Create a content map when your team is preparing to redesign an information space or product; change its existing information architecture; aggregate content from multiple information spaces or products; or divide the content of a large information space or product among smaller ones.

While content mapping is useful primarily in visually conveying your findings from a content inventory on an existing information space or product, you could incrementally map a new information space as your team develops and structures its content.

You could use this method continually, throughout your product-development process, whenever you need to evaluate the scope of an information space or product's existing content and ascertain whether there are any gaps in that content, its coverage of specific topics or categories, or the quality and usefulness of its content.

Where—Although remote teams could create content maps using online, content-mapping tools, your project team should ideally work in the same conference room in which you've completed your content inventory and audit. If you've posted screenshots of pages or other content objects on the walls to illustrate the structure of the information space or product, they already effectively constitute a content map. Your team requires plenty of available wall space to depict any additional content the content team would need to create to fill gaps in the existing content, including space for new sections you might need to add.

Participants—Determine who should be responsible for creating and maintaining the quality of the content map. Make sure everyone on the team knows how to represent each type of content the information space or product comprises. The scope of the information space or product and that of your content inventory and audit determine the number of your teammates who should work on the content map. Your team's available bandwidth also constrains its scope. For a very large information space or product, consider having a different teammate map the content in each of its major sections.

Type of data—When creating a content map, you can represent both qualitative data about each page or other type of content object—for example, its content type, topic or category, or other attributes such as quality, business value, or usefulness—and quantitative data—such as the numbers of pages or other types of content objects that the information space or product comprises or the numbers of pages or content objects with particular attributes.

Data analysis—When analyzing your content map, leverage usage-data analytics to elicit answers to such questions about the information space or product as the following: Are users visiting its most important pages? Do its information architecture, navigation system, and search system enable users to easily find the information they're seeking? Are any broken links preventing them from doing so? Does the information space or product's content fully address the business requirements and meet users' needs?

Assess how well specific content is performing, then prioritize whatever changes you need to make to the information architecture or the content according to those that would have the greatest impact.

Your project team could do affinity diagramming to identify common themes and patterns in your data, create clusters of pages or other types of content objects by their themes, and assign a category name to each cluster.

Level of effort—Mapping an entire information space or product is a time-consuming, labor-intensive process that requires great attention to detail. However, you could limit your content map's scope by focusing on specific sections of an information space or product; divide the effort among several teammates to accelerate the mapping process; or employ this method iteratively, either by taking various approaches to mapping the same content, by focusing on different parts of an information space or product's content, or by looking at different content attributes over time.

Try to keep your content map up to date as you restructure the hierarchy of the information space or product or your content team adds new content, revises content, and removes obsolete content. Maintaining your content map is essential to sustaining its usefulness.

Deliverables—While it is the *process* of content mapping that offers the greatest value, the content map is itself a valuable deliverable. Prepare a report that captures your key findings from your team's analysis of the content map, indicating what content to keep, what content to move, what content to remove, and what content to revise or update. Describe what revisions or updates are necessary in detail and indicate who should be responsible for them. If you've taken an iterative approach to content mapping, track the dates and scope of each iteration of content mapping.

Investment—While content mapping is an inexpensive method of content analysis, the scope of the content map determines the actual cost of the process.

[47, 48, 49, 50, 51]

Competitive content analysis

Purpose—The best way to discover any weaknesses or gaps in your information space or product's content—and, thus, learn what topics, or categories, may be missing from your information architecture as well—is to conduct a *competitive content analysis*. You can also discover how best to differentiate your information space or product's content from that of your competitors and, thus, achieve competitive advantage. Plus, you can take inspiration from your competitors' content to potentially improve your content. Therefore, conducting a competitive content analysis can inform the design of your information architecture, as well as your content strategy.

You may already have conducted a broader competitive analysis to evaluate competitors' information spaces or products, target market and audience, product strategy, and business strategy. However, a competitive content analysis focuses specifically on evaluating your key competitors' existing content, as well as on determining their content strategy. Compare the readability, findability, and usefulness of your competitors' content *and* the terminology they use in their content and labeling with those of your own information space or product. Is their terminology consistent with the language of your information space's target audience?

Description—A competitive content analysis might comprise the following steps:

1. **Identify key competitors.** Compile a complete list of the information spaces and products that compete directly or indirectly with your own. Your direct competitors offer information spaces or products that are similar to your own, while the information spaces or products of indirect competitors may solve the same problems or meet the same user needs—although in a different way. To learn about businesses that offer competitive information spaces and products, ask your stakeholders—such as your company's executives, product managers, and Sales and Marketing teams—to identify your key competitors. Prioritize your list, then focus your competitive content analysis primarily on the content of your *first-tier competitors*—up to five direct competitors who provide similar information spaces and products to the same target audience and compete within the same marketplace. To increase your knowledge of competitive businesses and their information spaces and products, search the Web for useful sources of publicly available information about them, their content strategy, and their audience. Also, search for the keywords your business is targeting on Google to ascertain what other businesses are competing for the same keywords.

2. **Conduct a content inventory.** Gain access to each key competitor's information space or product, then conduct an inventory of its content. Fully explore the content on each information space or product, using its navigation system, to determine its scope and learn where specific content resides in its information hierarchy. You can learn how frequently they publish new content, how much content they publish, what content formats they're using, and what topics they're covering. To capture information during your competitive content analysis, add columns to your information space or product's content-inventory spreadsheet for **Company** *and* **Information Space** or **Product**. For detailed information about conducting a content inventory, see the section "Content inventories," earlier in this chapter.

3. **Conduct a content audit.** Determine what sampling of content to assess—for example, you might look at the most popular or the most recent content—then evaluate the quality of the content on each competitive information space or product by conducting a content audit. To capture information during your competitive content audit, add columns for **Company**, **Information Space** or **Product**, and **Overall Quality** to your content-audit spreadsheet. Rate the overall quality of particular types of content as *low*, *medium*, or *high quality*. For Web properties, you should also determine each company's SEO strategy and assess the optimization of each competitive information space or product's content for Web-search engines. When evaluating a page's SEO, look at the use of keywords in its Web address, metadata—particularly the page

description—the page's title, section headings, alt text for images, internal links, and keyword density. Compare your competitors' organic search traffic with that of your own information space or product. For detailed information about how to conduct a content audit, see the section "Content audits," earlier in this chapter.

4. **Analyze content topics.** For each page or component of content that you've sampled, analyze its title or subheading, the description in its metadata, and its actual content to determine what topics are its focus. Then, assign one or more topic tags to each page or component. This step is essential to identifying what topics are missing from your competitors' information spaces or products, as well as your own. Gaps in your key competitors' content present opportunities for competitive differentiation.

5. **Analyze trends in your data.** Identify the most prevalent, high-quality content topics, *or* categories, on your competitors' information spaces or products, and analyze the trends in the data from your competitive content analysis. Identify the popular themes that are common across your competitors' content.

When—Periodically, conduct a competitive content analysis to learn how your competitors' content is evolving over time in comparison to your information space or product's content. You should ideally conduct competitive content analyses on a quarterly basis, but do them at least annually. You might need to do competitive content analyses more frequently if key competitors often make changes to their content strategy or frequently publish new content, *or* when new competitors enter the marketplace. By conducting competitive content analyses on a regular basis, you can benchmark the performance of your information space or product's content against that of your competitors' content.

Where—An individual expert analyst could conduct a competitive content analysis from anywhere, but if multiple analysts are doing a competitive content analysis, they should ideally gather in one place to organize, compare, and synthesize everyone's insights prior to completing their analysis.

Participants—Ideally, a team of expert analysts should conduct a competitive content analysis. The number and scope of the competitive information spaces or products whose content you're analyzing determine the right number of analysts—typically, two or three, but no more than five. Consider involving expert analysts with different areas of expertise, because they may be able to identify different content issues. First, each analyst should independently evaluate the content on your competitors' information spaces or products. Then, the entire team of analysts should work collaboratively to analyze all of their findings. However, if necessary, a single expert analyst can conduct a competitive content analysis.

Type of data—A competitive content analysis generates rich, qualitative data, comprising one or more analysts' insights regarding the content on your competitors' information spaces or products.

Data analysis—If a sole expert analyst is conducting a competitive content analysis, no further analysis may be necessary. However, if several expert analysts have individually gathered information about competitors' content, you'll need to consolidate, organize, compare, synthesize, and analyze the findings of *all* these analysts. When analyzing the content on your key competitors' information spaces and products, try to ascertain the underlying content-strategy decisions behind their content.

Then compare their content strategies with that for your own information space or product. Define the appropriate scope of the content on your information space or product. Determine the content that differentiates your information space or product from those of your competitors. Prioritize the content that is core to your information space or product. Identify any gaps that exist in your content and determine how to remedy them. Decide what content to exclude because it's outside the scope you've defined. Assess the strengths and weaknesses of your content relative to that of your competitors. Analyze trends in your competitors' content over time.

Level of effort—Conducting a competitive content analysis requires significant effort because its scope encompasses the analysis of multiple competitive information spaces and products' content and, thus, can be a time-consuming process. Use this method iteratively as your digital information space or product's content or your competitors' content evolves.

Deliverables—Prepare a report that summarizes each key competitor's content strategy and compare those strategies with that for your information space or product. Identify their differentiators and content gaps, as well as opportunities to remedy your own information space's content gaps. If you've already conducted a competitive analysis to understand your key competitors and created a repository for the competitive information you've gathered, add the learnings from your competitive content analysis to that repository. If not, create such a repository. Share your findings with your team, as well as the broader organization. Maintaining a repository of competitive information lets you track how your competitors' content changes over time.

Investment—Although the investment that conducting a competitive content analysis requires depends on the number of competitive information spaces and products and the scope of the content you need to analyze, how many expert analysts are conducting the analysis, and the thoroughness of their analysis and documentation, doing a competitive content analysis typically requires a moderate level of investment. [52, 53, 54, 55, 56]

Content modeling

Before modeling an information space or product's content, you must first understand that content, its meaning and value to target users, and its business value to the organization developing and publishing the content.

Therefore, during *Discovery*, conduct UX research to understand both the purpose of and the relationships between an existing information space's various types of content and the ways in which target users are consuming that content *or* to understand the content you're planning for a new information space. For example, you could study your target users' perceptions and consumption of the content by creating a paper prototype, or its digital equivalent, comprising various types of content objects, then showing research participants specific content objects and asking them to indicate what information they would want to see next. If a participant asks to see content that your paper prototype doesn't include, create a placeholder for that content on the fly.

To understand the full scope and different types of content that an existing information space or product comprises, you should conduct a content inventory and audit during *Discovery*. Identify the most valuable types of content, considering the content's value to both target users and the organization. Once you've completed your UX research and a content inventory and audit for an existing information space during *Discovery*, you'll have all the information you need to model its content, decide what existing content to eliminate or revise, and plan what content to add.

Content modeling can be a means of either analyzing the structure of an information space or product's existing content *or* designing the structure for new content.

When you're creating a content model, always focus on an information space or product's most valuable types of content, especially those for which large amounts of content already exist or that constitute essential elements of the content strategy for a new information space. Identify pages that are likely entry points to the information space. The content objects that a content model comprehends typically include both pages and components of pages. Content modeling provides the foundation for both content strategy and information architecture.

Typically, various groups within an organization create, manage, and use its content, so content modeling usually requires a collaborative effort across an entire organization to ensure the content model's completeness. Constructing a comprehensive content model ensures that both an information space or product's content-management system (CMS) and its page templates accommodate all the necessary content and that no gaps exist in the content that would represent missed opportunities to meet either the users' or the business's needs.

Types of pages

Pages are the basic building blocks of any digital information space or product. However, content-management systems enable organizations to create and publish digital content independently of its presentation. Therefore, a digital page can be any one of the following:

- **Static**—The content on static pages is typically *unstructured* and is unique to a specific page. Static pages are created manually and remain unchanged unless a content manager, content creator, or production manager updates them.

- **Dynamic**—The content on dynamic pages is *structured*. All pages of a specific type conform to a consistent structure, or template, that comprises a specific set of content *components* and elements. Software processes automatically generate dynamic pages and navigational links to them.

- **Hybrid**—Hybrid pages comprise both static and dynamic content.

Advantages of dynamic pages

Generating consistently structured, dynamic pages that comprise specific sets of content components enables your information space or product to support the following:

- **Contextual navigation**—These navigational components provide deep links that can interconnect similar pages or related collections of content that a user might find of interest, allowing users to move laterally within an information space or product rather than relying on its hierarchical structure and navigation system. Contextual navigation typically provides access to supplementary information or related products. Ideally, your CMS should automatically generate links that provide contextual navigation to similar or related pages, which may be based on rich usage data, but more commonly on pages' metadata.

- **Chunked content**—Pages that contain large amounts of homogeneous content are hard to skim or scan, provide less information scent, and, thus, make it difficult for users to determine whether they've found useful content. Decomposing a page's content into well-defined, clearly labeled chunks, elements, or components—each with a specific purpose and residing in a consistent location on a structured page—makes it easier for readers to consume that content. Structural metadata both enables a CMS to consistently display specific types of chunked content on a page and to link to specific types of chunked content on other pages.

- **Personalized content**—By creating chunked content components, storing them in your CMS, tagging content components to target their display to specific classes of users, and defining conditional personalization rules that control the display of these content components, you can support the personalization of content for specific users, classes of users, or even certain user behaviors or choices.

- **Accessible content**—To reduce the cost of creating accessible content for use across multiple media, devise a unified content strategy that lets you separate content from its formatting. You could then automatically display the content in whatever format is appropriate for people with specific disabilities. For example, you might display text in a larger font size for people with poor vision.

- **Filterable content**—To enable users to filter multiple content objects of the same type on the same page, create one or more separate content components for data that uniquely identifies all instances belonging to a subset of that content type, store that data for all instances of that content type in a consistent format in your CMS, and provide filtering controls on the page displaying that type of content. This capability can be useful when you're implementing faceted search. For example, users might want to filter content or search results by topic, author, type, price, ratings, or location.

- **Sortable content**—To enable users to sort multiple content objects of the same type on the same page, create one or more separate content components for data that uniquely identifies all instances belonging to a subset of that content type, store that data for all instances of that content type in a consistent format in your CMS, and provide sorting controls on the page displaying that content. For example, you could support sorting by date or price.

- **Reusable content**—Creating and storing reusable content components in a CMS provides flexibility in the use of that content and lets you display the same content on multiple pages of various types on the same or a different information space or product; across various business units, functional teams, or publications; or across different content-distribution channels such as the Web or a mobile app. You could automatically adapt the presentation and scope of reusable content to a specific channel. For example, on a mobile device, you might include a brief version of an author's profile on an article page and provide the full profile on an author-profile page. Content modeling is an essential part of planning content reuse and defines both the structure and granularity of reusable content components, as well as the contexts in which they're reused.

Key elements of a content model

A *content model* comprises the following key elements:

- **Content types**—Identify all of the distinct types of content objects that an information space or product comprises. Depending on your industry, typical content types might include, for example, articles, columns, reviews, editorials, author profiles, tutorials, product descriptions, or press releases. All instances of a particular content type share common attributes and serve the same well-defined purpose. A content type could comprise a single, coherent content object that constitutes all the content on a page *or* consist of multiple, related content objects that together comprise the content on a page. Specific content types might appear either directly on static pages *or* within the components of structured pages. A list comprising multiple content objects of the same type might appear on a page, and these content objects could all focus on the same topic or on different topics and be grouped by topic.

- **Content components**—For any information space or product whose scope is sufficient to warrant the effort, decompose each of its content types into its logical components. Each of these content components must have a unique purpose and a specific, well-defined meaning. These content components are the constituent parts of structured pages and determine how your CMS stores and displays the content. Each template comprises a specific set of content components and provides a consistent page structure. Shared metadata defines the relationships between these content components. The optimal approach to defining the rules that govern the decomposition of content into components depends on your ability to balance the semantic chunking of information in ways that provide value against the complexity and cost of implementing the components. Prioritize implementation of the most valuable structural components.

- **Access structures**—Create structural metadata to define the interrelationships between the pages and content components that an information space or product comprises. For pages, access structures are ways of organizing and linking them—including hierarchies, indexes consisting of keywords and phrases, cross-references, and sequences—and, thus, support the navigation systems that enable users to find the content they need. For detailed information about these and other structural patterns, refer to Chapter 4, *Structural Patterns and Organization chemes*. The templates for dynamic pages define the logical relationships between their content components and elements and, thus, the structure of the pages. For content components, their access structures determine how the CMS identifies, organizes, stores, and creates the relationships between the components and their elements, enabling templates to locate the components and elements and integrate them into pages. Creating and maintaining effective access structures is essential, challenging, and time-consuming work.

Creating content models

Depending on a project's scope, a content model might represent just the structure of and relationships between the content components and elements that individual content types comprise *or* the complex interrelationships between the content types that make up an entire cross-channel information ecosystem. As part of your content-modeling process, you'll create diagrams that depict high-level content models. For example, the diagram in Figure 6.5 shows the content model for an individual content type on my Web magazine *UXmatters*, an author profile:

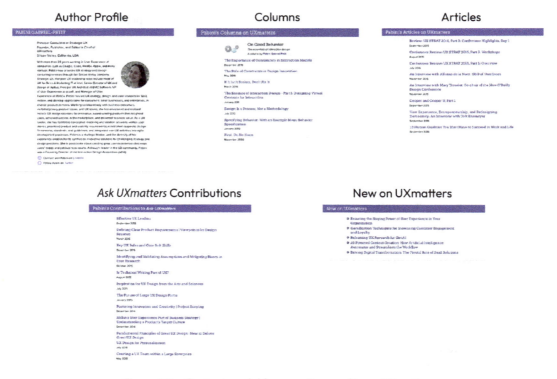

Figure 6.5—Content model for an author profile on *UXmatters*

Figure 6.6 depicts a high-level content model that represents key types of pages on the *UXmatters* Web site:

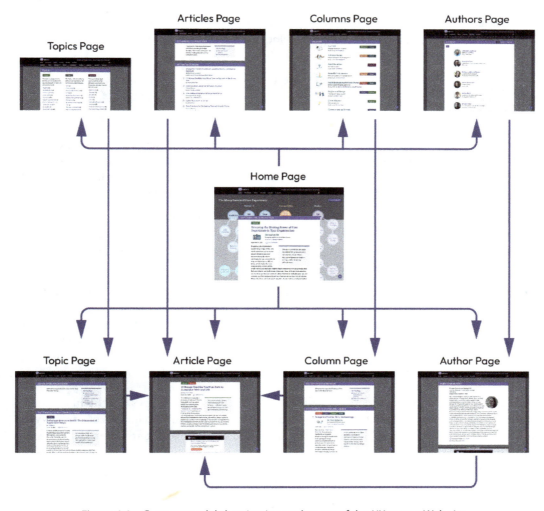

Figure 6.6—Content model showing integral pages of the *UXmatters* Web site

Since the content on most information spaces and products is continually expanding and evolving, a content model is a living, working document that is never really complete rather than a project deliverable. The purpose of content modeling is to aid your understanding of an information space or product's content. Content modeling can help you to decompose content objects into their structural components at an optimal level of granularity and to determine which of these structural components are essential.

To capture the details of your content model, you could create a table or spreadsheet that comprises the following columns or a bulleted list that includes the same information for each content object:

- **Content object**—List the unique, descriptive identifier for each type of content object that the information space or product comprises—for example, a page type such as *author page* or a component class such as *author profile*.

- **Type of object**—Indicate each content object's type. Possible values include *page, component,* or *both.*

- **Description**—Describe the purpose of each content object.

- **Content elements**—List the unique identifiers for all of the semantic, structural content components and elements that each content object comprises. Content elements are the most fine-grained units of information. A page could comprise both content components and elements, while a content component consists only of content elements.

- **Linked content objects**—List all of the interrelated types of content objects for each content object, indicating whether a type of linked content object constitutes the entire contents of a different type of page—for example, an *author page*; or is embedded in another type of page—for example, *author profile, on article page.*

- **Structural metadata**—Specify metadata that establishes the relationships and access structures between the content object and its linked content objects, which defines the content object's *external* structure. This structural metadata supports navigation between these linked content objects. Specify metadata that establishes the relationships between the components of the content object, which defines its *internal* structure. This structural metadata supports the presentation of content components by linking and sequencing related content.

As you construct your content model, you'll be able to ascertain what types of content are most important and, thus, are essential elements of the content model, and which types of content you should exclude from the model. Some considerations to take into account are the value that each content type delivers to both users and the organization and whether that value justifies the added complexity and cost of implementing that content type in the CMS. Plus, although decomposing your content into more content components provides greater flexibility in the CMS's ability to store and display the content, it also complicates entering the content into the system. The extra cost and effort might be worthwhile if you need to create different versions of the content for display in various contexts or on different platforms *or* the components would enable personalization, faceted search, filtering, or sorting.

Only once you've created a content model for an information space or product's content and clearly documented that model can the content-management team implement that model in its CMS and publication designers create templates for the content types that the model comprises. Much of the work of information providers, content creators, editors, content processors, and production managers can then proceed in parallel. Your deeper understanding of the information space or product's content informs its content strategy and all aspects of designing its information architecture and page layouts.

Examples of content models for specific page types

On my Web magazine *UXmatters*, the key types of content pages are article pages and author-profile pages. Let's look at the content models for these pages.

Content model for an article page

The content model for *UXmatters* defines a consistent structure for all article pages, comprising a specific set of content components and elements, each of which has a well-defined purpose. Figure 6.7 shows an example of an article page on *UXmatters*.

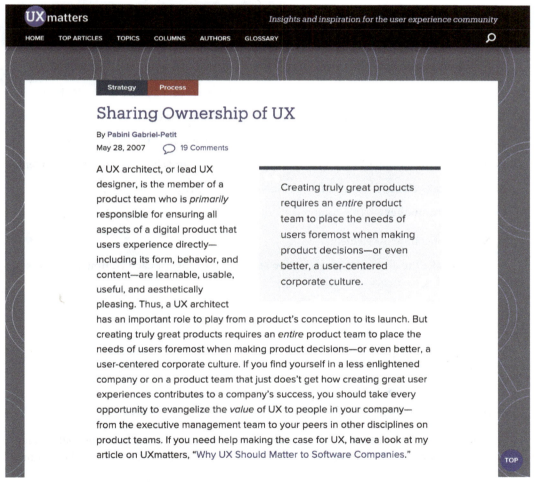

Figure 6.7—Article page on *UXmatters*

An article page comprises the following content components and elements:

- **Excerpt component**—This is a reusable content component that appears on an article page *and* on the home page as well. This component comprises the following elements:

  - Primary topic–tags element—Each of these topic tags links to the *Topics* page, which lists all of the subtopics under each topic.

  - Article-title element

  - Byline element—The author's name links to the author-profile page for that author.

  - Publication-date element

  - Comments-link element—This link displays the comments component lower on the article page.

  - Article-intro element—This element contains the first chunk of the article's content.

- **Advertising component**—This reusable component appears on all pages on the Web site.

- **Body-content component 1**—This component contains the second chunk of the article's content.

- **Advertising component**—This reusable component appears on all pages on the Web site.

- **Body-content component 2**—This component contains the third chunk of the article's content.

- **Subtopics-links element**—Each of these links displays the related subtopic page.

- **Comments component**—This component comprises the following elements:

  - Number of comments element

  - A listing of all readers' comments on the article; the listing for each comment comprises the following elements:

    - Commenter's name

    - Date and time of comment

    - Content of comment

    - **Reply** link

  - **Join the Discussion** form—This form lets users comment on the article.

- **Author-profile component**—This reusable component appears on all pages that display articles, columns, reviews, or editorials by the author. It reuses the content on the author's author-profile page.

- **Links component: author's recent columns**—This reusable component displays a contextual list of links to the author's most recent columns, on all pages displaying content by the same author.

- **Links component: author's recent articles**—This reusable component shows a contextual list of links to the author's most recent articles, on all pages displaying content by the same author.

- **Links component: recent articles on same topic**—This reusable component displays a contextual list of links to recent articles on the same topic, on all pages displaying content on that topic.

- **Links component: new articles**—This reusable component shows a contextual list of links to the five newest articles on *UXmatters*, on most pages on the site, including all article pages.

Content model for an author-profile page

The content model for *UXmatters* also defines a consistent structure for all author-profile pages, comprising a specific set of content components and elements. Figure 6.8 shows an example of an author-profile page on *UXmatters*.

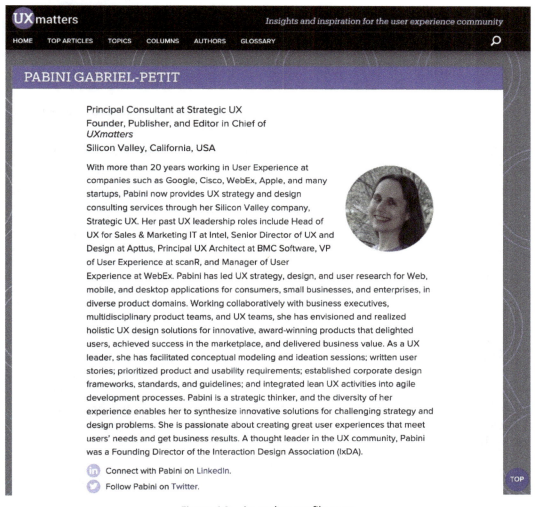

Figure 6.8—An author-profile page

The content model for an author-profile page comprises the following components and elements:

- **Author-profile component**—This is a reusable content component that appears on an author's author-profile page, as well as on all pages displaying content by the same author, and comprises the following elements:

    - Author's name

    - Author's title and employer

    - Author's location

    - Author's profile

- **Author's social media–links element**—Each of these links displays the author's profile on the corresponding social-media site.

- **Links component: all of the author's columns**—This component displays a contextual list of links to *all* of the author's columns.

- **Links component: all of the author's articles**—This component displays a contextual list of links to *all* of the author's articles.

- **Links component: author's contributions to UXmatters**—This component displays a contextual list of links to *all* of the author's contributions to the column *Ask UXmatters*.

- **Links component: new articles**—This reusable component shows a contextual list of links to the five newest articles on *UXmatters*, on most pages on the site, including *all* author-profile pages.

[57, 58, 59, 60, 61, 62, 63]

Summary

In this chapter, you learned several useful approaches to understanding and structuring an information space or product's content. First, I covered Fred Leise's qualitative, content-analysis heuristics for evaluating content. Then I explored a variety of content-analysis methods that enable teams to gain a deep understanding of an information space or product's content to inform the design of an effective information architecture. These content-analysis methods included content-owner interviews, content mapping, content inventories, content audits, and competitive content analyses. Finally, I discussed content modeling, through which you can decompose content into its logical components and elements, enabling the content to support contextual navigation, chunked content, personalized content, filterable content, sortable content, and content reuse.

Next, in Chapter 7, *Classifying Information*, I'll describe some challenges and goals of classifying information. I'll explore various approaches to creating controlled vocabularies and explore creating metadata in greater depth.

References

To make it easy for readers to follow links to the references for this chapter, we've made them available on the Web: `https://github.com/PacktPublishing/Designing-Information-Architecture/tree/main/Chapter06`

Further reading

Books

These books provide useful information about analyzing and structuring an information space or product's content.

Margot Bloomstein. *Content Strategy at Work: Real-World Stories to Strengthen Every Interactive Project.* Waltham, MA: Morgan Kaufmann, 2012.

Bob Boiko. *Content Management Bible.* New York: Wiley Publishing, Inc., 2002.

Kristina Halvorson. *Content Strategy for the Web.* Berkeley, CA: New Riders, 2010.

Ahava Leibtag. *The Digital Crown: Winning at Content on the Web.* Waltham, MA: Morgan Kaufmann, 2014.

Bella Martin and Bruce Hanington. *Universal Methods of Design: 100 Ways to Research Complex Problems, Develop Innovative Ideas, and Design Effective Solutions.* Beverly, MA: Rockport Publishers, 2012.

Lisa Maria Martin. *Everyday Information Architecture.* New York: A Book Apart, 2019.

Peter Morville and Louis Rosenfeld. *Information Architecture for the World Wide Web.* 3rd ed. Sebastopol, CA: O'Reilly Media, Inc., 2007.

Ann Rockley, Pamela Kostur, and Steve Manning. *Managing Enterprise Content: A Unified Content Strategy.* Indianapolis, IN: New Riders, 2003.

Donna Spencer. *A Practical Guide to Information Architecture.* 2nd ed. Northcote, Victoria, Australia: UX Mastery, 2014.

Sara Wachter-Boettcher. *Content Everywhere: Strategy and Structure for Future-Ready Content.* Brooklyn, NY: Rosenfeld Media, LLC, 2012.

Christina Wodtke. *Information Architecture: Blueprints for the Web.* Indianapolis, IN: New Riders, 2003

Articles and papers

You can read these articles and papers on analyzing and structuring an information space or product's content on *UXmatters* and other respected Web sites.

Garenne Bigby. "Why You Need to Map Your Web Site's Information Architecture." *DYNO Mapper,* January 24, 2018. [`https://alistapart.com/article/content-modelling-a-master-skill/`]

Colleen Jones. "Content Analysis: A Practical Approach." *UXmatters*, August 3, 2009. [https://www.uxmatters.com/mt/archives/2009/08/content-analysis-a-practical-approach.php]

Colleen Jones. "Conversing Well Across Channels." *UXmatters*, January 26, 2009. [https://www.uxmatters.com/mt/archives/2009/01/conversing-well-across-channels.php]

Colleen Jones. "Testing Content Concepts." *UXmatters*, December 21, 2009. [https://www.uxmatters.com/mt/archives/2009/12/testing-content-concepts.php]

Colleen Jones. "Toward Content Quality." *UXmatters*, April 13, 2009. [https://www.uxmatters.com/mt/archives/2009/04/toward-content-quality.php]

Anna Kaley. "Content Inventory and Auditing 101." *Nielsen Norman Group*, September 27, 2020. [https://www.nngroup.com/articles/content-audits/]

Fred Leise. "Content Analysis Heuristics." *Boxes and Arrows*, March 12, 2007. [https://boxesandarrows.com/content-analysis-heuristics/]

Margarita Loktionova. "The Ultimate Guide to a Competitive Content Analysis." *Semrush*, June 21, 2021. [https://www.semrush.com/blog/competitive-content-analysis/]

Rachel Lovinger. "Content Modelling: A Master Skill." *A List Apart*, April 24, 2012. [https://alistapart.com/article/content-modelling-a-master-skill/]

Robert Mills. "Creating Good User Experiences by Focusing on Content." *UXmatters*, November 9, 2015. [https://www.uxmatters.com/mt/archives/2015/11/creating-good-user-experiences-by-focusing-on-content.php]

Ellie Mirman. "How to Do a Competitive Content Marketing Analysis." *Content Marketing Institute*, October 23, 2917. [https://contentmarketinginstitute.com/articles/competitive-content-marketing-analysis/]

Nielsen Norman Group. "Content Inventory and Audit Template." *Nielsen Norman Group*, undated. https://www.uxmatters.com/mt/archives/2015/11/creating-good-user-experiences-by-focusing-on-content.php

Matthew Speiser. "Why You Should Perform a Competitive Content Analysis Before Writing Any Content." *Knotch: Pros & Content*, July 23, 2020. (No longer available)

Corey Wainwright. "How to Conduct Competitive Analysis to Step Up Your Content Strategy." *HubSpot*, March 1, 2012. Updated January 20, 2021. [https://blog.hubspot.com/blog/tabid/6307/bid/31619/how-to-conduct-competitive-analysis-to-step-up-your-content-strategy.aspx]

Patrick C. Walsh. "A Map-Based Approach to a Content Inventory." *Boxes and Arrows*, September 6, 2007. [https://boxesandarrows.com/a-map-based-approach-to-a-content-inventory/]

7

Classifying Information

"There is nothing more basic than categorization to our thought, perception, action, and speech."—George Lakoff [1]

Classifying and organizing things is human nature. This characteristic helps us to understand our environment and identify and find the things we need.

The systematic classification of an information space or product's content enables you to group similar content into categories and, thus, structure the information space. By clearly labeling the categories that represent the concepts, objects, and entities an information space comprises, you can facilitate users' ability to browse and search for the information they need. The way you organize and communicate the relationships between content objects influences users' understanding of the information space and the scope of its content, as well as their perception of the organization's brand.

This chapter provides additional foundational knowledge that is essential to designing effective information architectures for digital information spaces, as follows:

- Principles, goals, and challenges of information classification
- The roles of various types of metadata and some examples of metadata schema
- A variety of practical approaches to creating controlled vocabularies, including synonym rings, authority files, taxonomies, thesauri, ontologies, semantic networks, and faceted classification schemes

You'll learn how metadata and controlled vocabularies define the relationships between an information space's content objects to support its navigation and search systems. [2]

Principles of information categorization

In the book *Cognition and Categorization*, Eleanor Roach describes the following two basic principles of categorization:

- **Cognitive economy**—The purpose of a system of categorization is "to provide maximum information with the least cognitive effort." By assigning a content object to a particular category, you're indicating that you consider it to be similar to other content objects in the same category *and* different from content objects that do *not* belong in that category. When creating a category, you must determine which of its attributes are salient in distinguishing it from other categories. Thus, categorization decisions should generally incline toward limiting an information space's categories to *only* those that have highly relevant attributes rather than the creation of many categories, each with very distinctive attributes.

- **Perceived world structure**—People perceive the tangible objects that surround them in the natural world as structured information, whose attributes are neither arbitrary nor unpredictable. Our ability to organize information plays a key role in our understanding of the world. Therefore, people expect an information space's content to be well structured. [3]

Together, these basic principles of categorization can help you to determine the level of abstraction that a particular culture might assign to a given category, its inherent attributes, and the structure of the categorization system. You can achieve the first principle, cognitive economy, either by mapping a system's categories closely to the natural attributes of the structured world *or* by defining or redefining the attributes of categories to match that structure.

Different levels of categorization are *not* all equally useful. *Basic-level categories* are the broad, tangible categories that people can most readily identify and understand from their childhood. The members of basic-level categories have a strong resemblance to one another and a low degree of resemblance to the members of all other categories. These categories represent the highest level at which a single mental image can depict an entire category—for example, *dog* or *cat*—and the members of a category share many attributes in common. More abstract, even broader, *superordinate categories*, such as *animal*, comprehend various basic-level categories, while more specific, *subordinate categories* represent subtypes of a basic-level category.

When defining categories, consider each category's level of inclusiveness—for example, *Golden Retriever*, *dog*, *mammal*, or *animal*—and the distinctiveness of the attributes that are *most* representative of the objects or entities within the category—for example, the attributes for a dog might be furry, four-footed, lolling tongue, and wagging tail—and *least* representative of objects or entities in a different category—for example, a dog's wagging tail differentiates it from a cat, monkey, or horse.

For an information space, the *basic level of categorization* should be the most inclusive level—that is, the most abstract level at which its categories map to the structure of the attributes that people perceive in the real world. To maximize the distinctiveness of its categories, define prototypical categories that exhibit the most representative attributes while excluding the least representative attributes. [4, 5]

Key goals of classifying information

The ways in which we classify and, thus, label content objects influence users' perceptions and understanding of the information they impart. This enables users to more easily, efficiently, and effectively satisfy their information-seeking needs, whether by browsing or searching an information space or product. Therefore, the key goals of classifying an information space's content are as follows:

- **Supporting findability and navigation**—Information architects classify and thereby structure and label an information space's content to ensure that it is findable, enabling users to browse the information space and use its navigation system to locate the content they're seeking. The organizational structure of an information space defines the relationships between the content objects that the space comprises and, thus, its primary navigation paths.

- **Improving the retrieval of search results**—Information architects and others index an information space's content by manually tagging content objects with metadata that defines the categories to which they belong. Information-retrieval systems are most effective when you've defined narrow categories of homogeneous content. However, by using a controlled vocabulary to identify particular search terms' synonyms or related terms (RTs), you can enable a search system to retrieve additional results whose relevance it might not otherwise recognize. [6]

Challenges of categorizing information

Classifying information requires making myriad small decisions about what content objects are similar and, thus, belong in the same category. To make these decisions, you must first clearly define an information space's categories and understand their characteristics, which is typically an iterative process of refinement that happens over time. Then, when assessing each specific content object, you must determine whether its characteristics match those of a particular category.

In their book, *Information Architecture for the World Wide Web*, Peter Morville and Lou Rosenfeld identify four challenges of organizing an information space or product's content:

- **Ambiguity**—Because many words have multiple meanings and can even represent several completely different concepts depending on the context, human language is inherently ambiguous—as are the classification systems that are based on human language. The organization of abstract concepts into categories can be especially problematic. The boundaries between categories are often unclear, so categories may overlap with one another. There are always some core content objects that fit their category better than others that belong to a category only peripherally. Plus, as an information space's content changes over time, you'll probably need to add new categories and rethink the definitions of some existing categories.

- **Heterogeneity**—Most information spaces comprise content objects that are highly *heterogeneous*— that is, dissimilar or unrelated to one another. Therefore, while the content objects within a category might resemble one another in some ways, they might *not* all share the same attributes. Plus, there are usually some content objects that don't really fit into any of the categories you've created. Significant dissimilarities between content objects include different levels of *granularity*, or specificity, and different types of media. Such differences make it difficult to devise a single classification system that you can use in organizing all of an information space's content.

- **Differences in perspective**—People have different perceptions of categories, depending on their knowledge and prior experience. Therefore, the people indexing an information space's content might disagree about what language to use in their labels, the definitions of terms, or the categories in which to place specific content objects. Such decisions are always subjective. To overcome your individual biases and preconceptions in making decisions about categorization, it's important that you rely on your actual knowledge of your target users and their needs, which you've gained through UX research. However, you should avoid creating categories for specific audiences, and it's usually best to create categories that are based on topics rather than tasks.

- **Internal politics**—The differing perspectives, priorities, and goals of the people, departments, and business units within an organization can cause conflicts to arise when teams are making decisions about how best to organize or label content. Because of the shared experiences and knowledge of colleagues working for a specific organization, they often adopt the same language for concepts and may create similar categories. However, there is no single correct way of organizing content, so colleagues often disagree about categorization decisions. Resolving these conflicts often requires making compromises.

As you attempt to classify information in ways that are meaningful to your target users and, thus, can help them find the information they need—whether by browsing or searching—all these factors come into play, making information organization quite challenging. [7, 8, 9, 10]

Metadata

"Well-structured metadata helps publishers to identify, organize, use, and reuse content in ways that are meaningful to key audiences."— *Kristina Halvorson [11]*

Metadata identifies and describes the attributes of the individual content objects on a digital information space and, thus, both provides the basis for their categorization and defines the relationships between them. For information spaces that comprise extensive collections of content, it is essential that content authors and other indexers consistently apply metadata to their content objects to improve information retrieval using the information space's navigation and search systems and to facilitate the reuse and dynamic display of content objects.

Thus, by applying metadata to an information space's content objects, it becomes easier to categorize, store, and retrieve them. The use of metadata makes the purpose and meaning of the content objects clear. The use of metadata standards facilitates the management and sharing of digital content across different organizations, databases, information spaces, and applications, as I'll discuss in the section "Metadata schema," later in this chapter.

Using a content-management system (CMS) *and* a controlled vocabulary lets you take a metadata-driven approach to organizing an information space's content by describing its content objects rather than by placing them within a predefined structure. This enables your CMS and controlled vocabulary to do that organizational work for you automatically.

Whenever you're creating a new information space, adding new content to an existing information space, or revising its existing content, you should devise a *metadata strategy* that defines the type and structure of the metadata you'll create for that content—that is, its *metadata schema*. When conducting a content analysis for an existing information space, always evaluate its metadata schema and consider the quality of the metadata your team has already implemented for its content objects. [12, 13, 14, 15]

Types of metadata

"Metadata has many forms and purposes. ... We employ a word or phrase to describe the subject of a document for the purposes of retrieval. We try to concisely encapsulate its aboutness now to support findability later."—Peter Morville [16]

To ensure the success of an organization's content strategy, you must define the types of metadata you'll use for particular types of content. Some types of metadata are applicable globally to *all* content objects, while others may apply *only* to specific categories of content objects—for example, those belonging to a specific business unit *or* a specific content type or data type.

In their book, *Information Architecture for the World Wide Web*, Peter Morville and Lou Rosenfeld define the following three types of metadata for content objects:

- **Descriptive metadata**—This metadata describes the attributes of individual content objects and, thus, facilitates users' ability to discover, access, and use an information space's content. Many different attributes could potentially describe the nature of a particular content object— for example, a meaningful description of its content, facet, filename, title, topic, related topics, keywords, audience, content type, or data format. Once you've captured all of the possibilities, define whatever metadata would be most useful for the specific information space and the type of content it describes. This type of metadata distinguishes individual content objects from one another so has the greatest impact on navigation and searching.

- **Structural metadata**—This type of metadata describes content objects' internal structure, or information hierarchy—which should be consistent across content objects of the same type—and, thus, supports single-source publishing and the flexible display of reusable content. Possible structural attributes might include subsections and their headings, reusable chunks of content, fields and their length in characters, or the rows and columns of a table. Again, the nature of the content determines what metadata would be most appropriate.

- **Administrative metadata**—This metadata enables content management and workflow routing and describes the business context of specific content objects—for example, the business unit or department to which they relate; their owner, creator, and editor; the dates of their creation and most recent revision, and any expiration date; and tracks their status—for example, *draft, ready for review, in review, reviewed, approved, published, needs revision*, or *archived*. [17]

In their book, *Managing Enterprise Content: A Unified Content Strategy*, Ann Rockley, Pamela Kostur, and Steve Manning define two types of metadata that support a *unified content strategy*, through which you can consistently identify *all* content requirements across an organization, create structured content for reuse, store all content objects in a content-management system, and display content to meet users' needs, as follows:

- **Categorization metadata**—This metadata categorizes your information space's content objects. Group your information space's content objects by category, then define a metadata taxonomy comprising those categories to enable authors and other indexers to consistently assign the appropriate categorization metadata to content objects that belong in particular categories.

- **Element metadata**—Authors use three key types of element metadata to identify, retrieve, and manage specific content objects throughout the publication process: reuse metadata, retrieval metadata, and tracking metadata. *Reuse metadata* identifies a reusable content object, each instance of its reuse, the content types in which its reuse is permissible, and its target audiences. *Retrieval metadata* enables the retrieval of content objects and includes a content object's title, subject, author, date, keywords, and security level. *Tracking metadata* describes the workflow status of a content object and its review status—*accepted* or *rejected*. [18]

Bob Boiko discusses five types of metadata in his *Content Management Bible*, as follows:

- **Structure metadata**—This metadata defines the structural components of content at various levels of granularity—for example, *elements*, such as titles, subheadings, and paragraphs; *components* that comprise coherent collections of elements, or the *content objects* of which pages consist; *nodes*, or pages—content objects, that comprise coherent collections of components; *publications* consisting of coherent collections of pages; and publication groups that comprise coherent collections of publications. Each of these structural components represents a discrete unit that can either stand on its own or be part of a greater whole.

- **Format metadata**—This metadata specifies the rendering of any type of structural component or content object and, other than the formatting of words or phrases in bold or italics, should be separate from the content itself.

- **Access metadata**—This metadata facilitates information retrieval by organizing content objects into access structures—typically, into hierarchies or sequences.

- **Management metadata**—This metadata enables the administration and tracking of content objects and, thus, serves the same purpose as Morville and Rosenfeld's administrative metadata.

- **Inclusion metadata**—This metadata represents external content objects and specifies their inclusion at specific locations within internal content objects. Thus, you can refer to external content and embed it in your organization's content. [19]

Some additional types of useful metadata include the following:

- **Intrinsic metadata**—In addition to descriptive and administrative metadata, Christina Wodtke's book, *Information Architecture: Blueprints for the Web*, also discusses *intrinsic metadata*, which is a subset of a content object's descriptive metadata and comprehends its data format—for example, an HTML page, a Word document, or a GIF or JPG image file—and its file size and dimensions. [20]

- **Social metadata**—Social-media platforms use social metadata to define the structure of users' social-media posts and support sharing and cross-posting them across various services and information spaces. Such metadata typically includes the service that is the source of the post, its subject matter—typically communicated through one or more tags from a user-generated folksonomy—the user's identity, the date and time at which the user posted the information, and whether a post is part of a thread or is a response to another user's post. [21]

- **Embedded metadata**—This metadata identifies a photograph or other image file, a video file, an audio file, or another digital-media asset in a way that is integral to the file and prevents the metadata from being removed from the file. The metadata fields include a headline and caption, the creator and the address of the creator's Web site, a creation date, a credit line, the copyright owner and notice, and the location that is depicted in an image. [22]

- **Faceted metadata**—Faceted search integrates very well with keyword search, increases the findability of content, and improves the organization of search results. Supporting faceted search requires the creation of a faceted classification system and the application of hierarchical, faceted metadata to an information space's content objects. In a faceted classification system, *facets* are sets of categories that describe the various attributes of content objects. Facets are typically hierarchical. Although the use of hierarchical, faceted metadata simplifies the organization of content objects, it does *not* explicitly represent their relationships. [23, 24]

Thus, you can use metadata to support effective information retrieval by search engines and by your users; enable the systematic reuse and dynamic display of content; automatically control your content-development workflow; and track the status of content that is under development or requires revision. Through the consistent use of metadata, you can reduce inefficiencies such as the production of redundant content, identify digital assets or source materials that have changed, identify content that is ready for translation, automatically manage your content-development workflows, and reduce the cost of creating, managing, publishing, and reusing content. Achieving the necessary level of consistency in defining and applying metadata requires the use of a controlled vocabulary. [25]

Metadata schema

A *metadata schema* defines a collection of labeled attributes that describe the content objects on a particular digital information space. Using a metadata schema enables content authors or owners to more efficiently and consistently provide rich descriptions of each of the content objects on an information space, thus enabling users to more easily find the content by browsing or searching. A metadata schema may include both a required, or core, set of attributes *and* optional attributes, as well as provide the capability of extending the schema to meet an organization's future needs.

For a given information space, your alternatives are either to adopt an existing metadata schema—whether a standard metadata schema or one your organization has previously devised for use on all of its information spaces—adapt an existing metadata schema as necessary for your project, or create a new metadata schema. [26]

A common example of a metadata schema is the Dublin Core Metadata Element Set, which was originally devised to describe content on the Web and has been widely adopted for all digital content. This metadata schema comprises the following elements:

- **Title**—The name of a content object.

- **Subject**—The topic of a content object.

- **Description**—A more detailed representation of a content object.

- **Creator**—A person or organization who authored or made a content object.

- **Publisher**—A person or organization who made a content object available.

- **Contributor**—A person or organization who contributed to a content object.

- **Date**—Typically, the publication date for a content object, but possibly another date on which an event that is associated with its content-development lifecycle occurred, such as its creation date, completion date, or revision date.

- **Type**—The kind of content object—for example, a product description, article, author profile, case study, or review.

- **Format**—The medium of a content object—or its file format or dimensions.

- **Identifier**—The unique reference, identifier, or ID, that distinguishes a content object on a specific information space.

- **Language**—The natural language of a content object.

- **Source**—A related content object from which a content object derives.

- **Relation**—A specific content object that is associated with a content object.

- **Coverage**—The organizational, geographic, or temporal scope of a content object's relevance.

- **Rights**—Typically, the copyright information for a content object. [27, 28]

Other useful attributes of content objects that often occur in metadata schema include the following:

- **Keywords**—One or more index terms—words or phrases—that describe the essential characteristics of a content object and that the author assigns to aid its retrieval. [29]

- **Category label**—A term from a controlled vocabulary that describes a category in which a content object belongs, ensuring that the content object appears under the proper label in the information space's navigation system.

- **Web address**—The URL of a content object.

- **Product**—A product's brand name.

- **Product type**—The category of product to which a product belongs.

- **Industry domain**—The industry of the organization to which an information space or product belongs *or* to which its content pertains.

- **Audience**—The target audience for a content object.

- **Education level**—The level of education that the target audience for a content object should have achieved.

- **Security level**—One of several predefined permission levels that constrain who can access a content object to view it or create, modify, or publish a content object.

- **Document type**—The type *and* format of a content object.

- **Bibliographic citation**—The formal citation for the content object to use when another author is quoting it in another document. [30, 31]

The *Embedded Metadata Manifesto* from the International Press Telecommunications Council (IPTC) defines the following specialized metadata schema:

- **Headline**—A brief summary of the content of an image or other type of digital-media asset for publication with the asset.

- **Caption**—A description of the content of an image or other digital-media asset that indicates, for example, what the image depicts, who appears in it, or its location.

- **Creator and Web site**—A person or organization who created an image or other digital-media asset and their Web address.

- **Date created**—The date on which an image or other digital-media asset was captured.

- **Credit line**—Text that describes the person or organization who supplied and should be credited for an image or other digital-media asset on publishing it.

- **Copyright owner**—The name and ID of the organization or person who holds the copyright for an image or other digital-media asset.

- **Copyright notice**—The copyright information for an image or other digital-media asset, which indicates its date, current owner, and the rights reserved.

- **Location**—The geographical location that is depicted in an image or other digital-media asset—for example, a city, state or province, and country. [32]

Schema.org provides a diverse collection of metadata schemas for structured data and shared vocabularies that are in use by many organizations.

Controlled vocabularies

"It's essential to use the language of your users and to do so in a consistent fashion. The tool we use to enforce that consistency is called a controlled vocabulary."—Jesse James Garrett [33]

A *controlled vocabulary* is a well-defined subset of a specific natural language that comprises the standard terms to use for particular concepts on an information space or product. Many controlled vocabularies comprise words and phrases that pertain to a particular industry domain or branch of knowledge, whose terminology would be useful in identifying, indexing, organizing, and labeling the content objects on an information space, as well as for users' retrieving content by browsing or searching.

Two characteristics of all natural language necessitate the disambiguation of an information space's terminology using a controlled vocabulary:

- **Ambiguity**—Different terms that have the same spelling, or *homographs*, can have different meanings and represent different concepts.

- **Synonymy**—A variety of different terms—for example, synonyms and near synonyms—can be equivalent and represent the same concept.

The primary purpose of most controlled vocabularies is to define an authoritative list of preferred terms—for use in categorizing specific concepts or objects and in defining the metadata for content objects—to collect synonyms and other variant terms for those concepts or objects, and to define the semantic relationships that exist between those terms. Thus, a controlled vocabulary facilitates the alignment of the natural language of internal content creators, managers, and indexers, as well as an information space's target users, enabling users to retrieve *all* the content that is relevant to their searches, including any content that uses different terminology.

You could choose to use a standard controlled vocabulary that another organization has published; derive your own controlled vocabulary from one or more previously published controlled vocabularies, which might be either broader in scope *or* more specialized; or create your own controlled vocabulary from scratch.

In a controlled vocabulary, each preferred term must be unique, have only one meaning, and represent only a single concept or entity. Therefore, a preferred term's meaning should *not* overlap with the meanings of other terms. The consistent application of a controlled vocabulary to an information space's content objects enables its CMS to automatically define links between them. Equivalent terms should have equal granularity, or specificity, and, within a hierarchical classification scheme, should reside at the same level in the hierarchy. Depending on the type of controlled vocabulary your information space or product requires, it might comprise the following types of terms, which have specific semantic relationships to one another:

- **Preferred terms**—These indexing terms represent specific concepts, and their use is generally accepted within an industry domain, a branch of knowledge, or a specific organization. *Always* use a preferred term in lieu of *any* of its variant terms. Add *scope notes* to constrain the definitions of any natural-language terms that have multiple meanings by specifying their permitted usages as preferred terms, as well as any rules for assigning the preferred terms to content objects. Defining the scope of preferred terms both reduces their ambiguity and provides increased precision in search results. In a controlled vocabulary, you can designate *preferred terms* using the acronym *PT* and *scope notes* using the acronym *SN*. You can use the following syntax to list all the variants of a preferred term: [PT] USE FOR (or UF) [VT, VT, and so on]—for example, *favorite* UF *favourite, fave.*

- **Variant terms**—These include synonyms, near synonyms, alternative spellings, or other terms that are roughly equivalent to the preferred terms for specific concepts, *not* the terms of choice for those concepts. Variant terms have an *equivalence* relationship to their corresponding preferred term. Variants might include common misspellings or different spellings of preferred terms *or* different names for the same person, place, or thing. Rather than ever using variant terms, content creators and indexers should *always* use their corresponding preferred term. By enabling the retrieval of content that uses different terms for the same concepts, defining variants for preferred terms improves recall in search results. In a controlled vocabulary, indicate variant terms using the acronym *VT*. Use the following syntax to indicate the preferred term for a variant: [VT] USE [PT]—for example, *UX practitioner* USE *UX professional.*

- **Broader terms**—These include related terms that describe a more general concept that comprehends the narrower concept a particular preferred term describes. Broader terms, or categories, thus have a *hierarchical* relationship to the preferred term. In a controlled vocabulary, indicate broader terms using the acronym *BT*.

- **Narrower terms**—These include related terms that describe a more specific concept or a subset of the concept that a particular preferred term describes. Narrower terms, or subcategories, thus have a *hierarchical* relationship to the preferred term. In a controlled vocabulary, indicate narrower terms using the acronym *NT*.

- **Related terms**—These terms represent less closely connected concepts that have only a loosely *associative* relationship to a particular preferred term rather than a hierarchical relationship to that preferred term. In a controlled vocabulary, you can indicate related terms using the acronym *RT*. Although defining related terms for a preferred term facilitates users' browsing and berrypicking behaviors by providing a more complete picture of an information space's content, it can be difficult to decide what associative relationships would actually help users find the information they need and when to reject the inclusion of more obscure associative relationships.

In determining whether to choose specific preferred terms to represent certain concepts in your controlled vocabulary, refer to a variety of sources to identify and gather potentially useful words and phrases, then assess them against three different criteria or *warrants*, as follows:

- **Literary warrant**—Consider whether particular terms are natural language that authors and people, in general, would typically use in describing specific concepts or content objects. To determine terms' literary warrant, review a content domain's literature, textbooks, and existing controlled vocabularies, as well as general references such as encyclopedias and dictionaries and particular organizations' or domains' glossaries. For an existing information space or product, determine the vocabulary that the organization has used in indexing its primary-level *and* secondary-level content objects. Also, as you create new content, consider the appropriateness of specific preferred terms for indexing that content and, if necessary, define new preferred terms. Any preferred terms that you add to your controlled vocabulary should be as consistent with their prevailing usage in the literature as possible.

- **User warrant**—Consider whether a particular term is consistent with the language of your information space's target audience. To discover the vocabulary of your users, analyze the language they employ for specific concepts during UX research sessions, when requesting information from Customer Support, and when searching for information.

- **Organizational warrant**—Consider whether a term conforms to the language of the organization, its subject-matter experts, and its industry domain, as well as the importance of using that language. Determine what form of particular terms the organization uses, as well as their formatting.

When *indexing*, or categorizing, an information space's content objects, or pages, content managers, or indexers must consistently assign preferred indexing terms from a controlled vocabulary to the content objects. These preferred terms describe the specific concepts or attributes of the content objects. The purpose of indexing terms is to aid search and retrieval, regardless of whether the terms actually appear *within* the content objects.

Benefits of controlled vocabularies include the following:

- Providing a better understanding of the scope and subject matter of an information space or product

- Facilitating the efficiency of content management through the consistent indexing, organization, and structuring of information

- Defining categories that are clearly distinguishable from one another and, thus, improve findability

- Aligning the natural language of content creators, indexers, customers, and users and, thus, facilitating effective business communications with customers and users

- Promoting consistency in indexers' use and formatting of preferred terms when describing content about the same person, organization, place, or thing

- Improving users' effectiveness in retrieving information by both browsing and searching

- Increasing the speed of a search engine's information retrieval over that which it could achieve using full-text search *and* improving both the *precision*, or relevance, and *recall*, or completeness, of search results

- Improving the labeling and hierarchical structure of navigation systems

- Enabling users to find content about the same specific concept by using different natural-language search terms—for example, synonymous, more generic, or more specific terms

Some challenges of using controlled vocabularies include the following:

- Both an organization's content creators and indexers and its customers and users need to be familiar with an information space or product's controlled vocabulary to make effective use of it. For example, customers and users need to know what terms to use when searching for products or information.

- Although the value of using a controlled vocabulary is significantly greater for large-scale information spaces, the cost and effort of manually tagging their content objects can become too great to bear.

- As the scope of an information space or product's content grows and its domain evolves, a controlled vocabulary can quickly become outdated.

- Developing and maintaining a controlled vocabulary requires a greater investment than using a natural-language vocabulary and can become a challenging, labor-intensive endeavor.

The different types of controlled vocabularies in common use range from the very simple—such as synonym rings and authority files—to the highly complex, including taxonomies, thesauri, ontologies, semantic networks, and faceted classification schemes. In the following subsections, I'll describe each of these approaches to developing controlled vocabularies. [34, 35, 36, 37, 38, 39, 40, 41, 42]

Synonym rings

When conducting information searches, people use different terms to describe the same things. Therefore, search engines provide the capability of configuring synonym rings. A *synonym ring* is a simple list of *equivalent terms* for a specific concept that enables a search engine to broaden its retrieval of search results. The individual terms in a synonym ring could be either words or phrases and include synonyms, broader terms, and narrower terms. Unlike other types of controlled vocabularies, synonym rings are *not* used in indexing content. Thus, synonym rings are especially helpful in cases where no more sophisticated controlled vocabulary exists to support indexing or for full-text searches, which do not rely on indexing.

True synonyms are words that are exactly equivalent to one another, including terms and their abbreviations, acronyms and abbreviations, internal jargon, slang, and even common misspellings; scientific names and their corresponding common names—for example, the Latin name for a plant or animal's genus and species and its popular name; and American and British spellings of English words, as well as synonymous words in other languages. However, *not* all the terms in a synonym ring must be true synonyms. They typically also include roughly synonymous terms, or *near synonyms*—for example, generic terms that describe closely related classes of products, the brand names for specific products belonging to those classes, *and* misspellings of these brand names that have appeared in users' search strings. Typically, synonym rings for broad information spaces for the general public include many near synonyms.

Synonymous terms can be difficult to distinguish—especially in cases where terms have multiple meanings. To determine what terms are synonyms, it's necessary to consider their use within the context of a particular industry, a specific information space or product, and its audience. Therefore, the people who manually create synonym rings should be subject-matter experts on an industry or topic and familiar with an information space or product's content, the needs of its audience, and the terms for which they are likely to search. To keep synonym rings up to date with users' current language, review the search logs for the information space, note any new terms or acronyms for which users are searching, and add them to the appropriate synonym rings.

The use of synonym rings expands the scope of users' search queries to comprehend both synonyms and near synonyms for their search keywords. While *query expansion* improves a search engine's recall significantly, it can also reduce the precision, or relevance, of search results. To balance precision and recall when employing synonym rings, order the search results by displaying the exact matches to the user's actual search terms first. (For more information about precision versus recall, refer to Chapter 13, *Designing Search*.) You could optionally enable the use of synonym rings only once an initial search has failed to retrieve any useful results and the user has expanded the search to include related terms. [43, 44, 45, 46, 47]

Authority files

An *authority file* comprises an index of *authority records*, which define authoritative terms, or *preferred terms*, for the concepts within a body of knowledge such as a library or digital knowledge-organization system (KOS) and is the simplest form of controlled vocabulary. Each *authority record* comprises a unique preferred term for a particular concept; a list of *cross-references*, or *variant terms* for that concept—typically in alphabetical order; and notes that validate the authority of the preferred term, which cite the title and publication date of one or more sources. Because an authority file's purpose is to organize information rather than to inform, it should include *only* information that is necessary to disambiguate terms.

By referring to an authority file when authoring, categorizing, or indexing content, you can prevent the use of synonyms or loosely equivalent terms in lieu of preferred terms; ensure the consistent use of terminology in an information space's content and navigation system; and, incidentally, educate readers about the proper terminology for an industry or branch of knowledge.

There are two common types of authority files: name authority files and subject-heading authority files. A *name authority file* comprises an alphabetical list of unique, *authoritative names* for particular entities and objects within a specific industry or branch of knowledge. Examples include the names of people, organizations, species of plants or animals, events, places, products, and publications. Because name authority files typically do *not* indicate any relationships between the names, lists of authoritative names tend to be shorter than the lists of preferred terms for other types of controlled vocabulary.

A *subject-heading authority file* typically comprises an alphabetical list of the standard subject headings, or *preferred terms*, for either particular subjects, or topics, within a body of knowledge or for an entire field of knowledge, indicating each of their *variant terms* and *related terms* and their *cross-references*; and providing *scope notes* that constrain the usage of the preferred terms. These preferred terms appear in subject headings *and* in cross-references as the preferred terms to which synonymous or related terms should refer. Use an authority file when classifying the subject matter of articles, books, or other documents within an information space or product, thereby grouping them into collections of related content.

Organizations have published domain-specific name and subject-heading authority files for many fields of knowledge, including the arts, humanities, geographical names, agriculture, social and behavioral sciences, and medicine—for example, the *Medical Subject Headings* (MeSH). Some examples of comprehensive, general-purpose authority files in common use include the following:

- *ISO 639-2 Codes for Representation of Names of Languages*
- *Library of Congress Name Authority File* (LCNAF)
- *Library of Congress Subject Headings* (LCSH)
- *Virtual International Authority File* (VIAF)
- *Sears List of Subject Headings*

Authority files usually include both preferred terms *and* synonyms. Thus, such an authority file is simply a synonym ring in which one of the preferred terms has been designated the preferred term, which functions as the unique identifier for that synonym ring, making it easier to find collections of information and add, delete, or edit terms. Synonym rings and authority files are very useful means of controlling the vocabularies of large information spaces for the general public, which are often created and maintained by decentralized teams and for which searching is typically the dominant information-seeking behavior.

In addition to the benefits of controlled vocabularies in general, benefits that are specific to authority files include the following:

- Promoting consistency in the use of names and subject headings
- Increasing the specificity of subject headings through scope notes
- Connecting diverse names for a single entity or subject
- Disambiguating the use of the same name for different entities or subjects
- Reducing errors such as misspellings
- Improving the efficiency of searching for preferred terms, names, and subject headings
- Increasing the relevance of search results through the use of cross-references and related terms
- Easing error detection and correction [48, 49, 50, 51, 52, 53, 54, 55, 56]

Taxonomies

"The organization of ideas and objects into categories and subcategories is fundamental to human experience. We classify to understand."—Peter Morville [57]

A *taxonomy* is a top-down, hierarchical *classification scheme* that comprises the preferred terms for particular concepts. When classifying preferred terms to form the classes, categories, or groups of such a controlled vocabulary, indexers divide general concepts into progressively more specific concepts, forming parent-child relationships. Following the principle of *inheritance*, each class, category, or group inherits the attributes of its antecedents. Within a taxonomy, the preferred terms have *one* of the following logical relationships:

- **Broader/narrower relationship**—This very common type of taxonomy defines the hierarchical relationships between *broader terms* and *narrower terms*—for example, a genus or class and the species or members it comprises. Other than the *top term*, or *root node*, in a hierarchy, which is its broadest term and comprehends *all* of an information space's content, all preferred terms in this type of taxonomy have broader *and* narrower relationships with one or more other preferred terms. The hierarchical relationships of these broader and narrower terms let an indexer classify the preferred terms into categories and subcategories using a content-management system (CMS) or another indexing application. An information space or product's categories inform the design of its navigation system and demonstrate the scope of its content. You can express this relationship as: *[Narrower term] is a [broader term]* or *[Narrower term] is a kind of [broader term]*—for example, *an oak is a tree* or *a petunia is a kind of flower*.

- **Instance-of relationship**—In this type of taxonomy, a hierarchy comprises preferred terms that have *instance-of relationships*, in which a common noun represents a general category of objects, entities, or events that comprise specific instances of that category and often have proper names. Express this relationship as *[Term] is an instance of [term]*—for example, *Zoom is an online-meeting application* or *Bangalore is a city*.

- **Whole/part relationship**—Regardless of context, the logical hierarchy of this type of taxonomy consists of preferred terms for objects or entities that have *whole/part relationships*, in which a broader whole comprehends its inherent parts. Express this relationship as *[Term] is part of [term]*—for example, *a page is part of a Web site* or *California is part of the United States of America*.

Some concepts and the terms that represent them belong logically to multiple categories, forming *polyhierarchical* relationships—for example, *a piano is a kind of percussion instrument* and *a piano is a kind of stringed instrument*. While polyhierarchy is common in complex taxonomies because it can improve the findability of content objects whose categorization is ambiguous, it is not typical of information spaces of limited scope that have simple, browsable hierarchies. Plus, polyhierarchies can be difficult to maintain. *All* taxonomies comprise preferred terms that have either strictly hierarchical or polyhierarchical relationships.

Note—Refer to the section "Hierarchy, or taxonomy," in Chapter 4, *Structural Patterns and Organization Schemes*, where I discussed the differences between strict hierarchies and polyhierarchies. While each term can appear in only one location within a strict hierarchical taxonomy, the practicalities of organizing information often require greater flexibility. Thus, complex taxonomies may actually need to be polyhierarchies, in which the same term can appear in multiple locations.

The *top term* in any hierarchical classification scheme, which is designated by the acronym *TT*, is the most comprehensive term in that hierarchy. While taxonomies usually lack the *synonyms* and *related terms* that are typical of thesauri and subject-heading authority files, any synonyms or other equivalent terms that they do include should reside at the same level within the hierarchy.

When classifying content objects to form a taxonomy, you must consider the cognitive aspects of categorization. In *Principles of Categorization*, Eleanor Roach posits two basic principles of categorization:

- The primary goal of categorization "is to provide maximum information with the least cognitive effort."

- "The perceived world [comprises] structured information." Therefore, to achieve the primary goal of categorization, categories should map to "the perceived world structure as closely as possible." [58]

The most cognitively basic levels in a hierarchy reside neither at the top nor the bottom of the hierarchy, but are its mid-level categories. These are the first categories that children can understand, for which they learn the names. A single image can represent an entire category. We organize most of our knowledge at this level. This is also the level at which people find it easiest to identify the members of a category. The top-level categories are more abstract—for example, *animal*—while the mid-level categories are more concrete—for example, *dog, cat, horse*, and so on. The categories at the bottom of the hierarchy are very specific and distinguishing them requires more knowledge on the part of the user—for example, knowing cat breeds such as *Himalayan, Ragdoll*, or *Maine Coon*.

Multiple taxonomies support some complex information spaces, each serving a particular purpose—for example, enabling users to filter or sort content or conduct a faceted search.

Information architects commonly create spreadsheets to document taxonomies, often with the first worksheet providing an overview of all the information space's taxonomies, with columns for listing, defining the purpose of, and describing the use of each taxonomy; then, separate worksheets that detail each individual taxonomy's list of terms. [59, 60, 61, 62, 63, 64, 65, 66]

Thesauri

"[A] thesaurus is a semantic network of concepts, connecting words to their synonyms, homonyms, antonyms, broader and narrower terms, and related terms. … Its most important goal is synonym management—the mapping of many synonyms or word variants onto one preferred term or concept—so the ambiguities of language don't prevent people from finding what they need."—Peter Morville and Louis Rosenfeld [67]

The type of controlled vocabulary that is in most common use across digital information spaces is the information-retrieval thesaurus. A *thesaurus* resides in a database and is a highly structured controlled vocabulary that is fully integrated into an information space. It typically comprises an alphabetical list of preferred terms that denote specific concepts, and the structure of each of its records derives from a term's semantic relationships with other terms that might describe the same concept. You could alternatively express a thesaurus as a hierarchy or visual map of terms. The purpose of a thesaurus is to improve the indexing of an information space's content *and* information retrieval through an information space's navigation *and* search systems.

An information-retrieval thesaurus can comprehend all the various semantic relationships that characterize the simpler types of controlled vocabularies, including *preferred terms*, *variant terms*, *broader terms* and *narrower terms*, and *related terms*. Thus, a thesaurus defines the equivalence, hierarchical, *and* associative relationships that exist between other terms that could describe an information space's content objects and maps them to specific preferred terms. Terms having these relationships provide a variety of access points to users when they are searching for information, both to the exact content they are seeking and to related content.

In a thesaurus, the semantic relationships between terms must always be reciprocal, so the standard relationship indicators you would use to clearly show the semantic relationships between terms are always used reciprocally. The *reciprocal relationships* between terms can be either *symmetrical*, with each term having the same relationship to the other—for example, with each term being a variant term (VT) *or* a related term (RT) to the other; or *asymmetrical*, with each term having a different relationship to the other—for example, with one term being a broader term (BT) and the other, a narrower term (NT). A thesaurus includes scope notes (SN) that define the meaning, or sense, of each preferred term and its use within an information space.

When compiling an information-retrieval thesaurus, you must first clearly define its overall scope, which could be either narrow or broad, depending on the scope and complexity of the subject matter of an information space's content, the various types of content objects it comprises, and the specific needs of its users. Therefore, a given thesaurus could apply to a specific content domain with a very limited scope *or* correspond to the information-retrieval system for an entire information space that comprises many topics or types of content. The implementation of a thesaurus could benefit the users of any large-scale information space by increasing the speed of information retrieval.

In their book *Information Architecture for the World Wide Web*, Peter Morville and Lou Rosenfeld describe three types of thesauri that are in common use, as follows:

- **Classic thesauri**—This is a standard, manually constructed, subject-specific thesaurus that supports keyword searches, synonym management, hierarchical navigation, and associative linking. The functionality of a classic thesaurus comes into play both when information architects and others index an information space's content objects by consistently mapping variant terms to preferred terms *and* when users retrieve content objects by browsing or searching for information they need. Indexers and users employ the same terms in expressing the same concepts. When performing a search, the search engine searches the metadata of content objects on the information space rather than the full text of their natural-language content. Developing and maintaining a classic thesaurus is highly specialized, labor-intensive, time-consuming work, and requires a significant investment. However, you might be able to use software that automates the categorization of content to facilitate the creation of a classic thesaurus.

- **Indexing thesauri**—This type of thesaurus defines a controlled vocabulary that supports the indexing of an information space's content objects, but does *not* have a synonym-management capability that would support its search engine by mapping users' variant terms to preferred terms. Although an indexing thesaurus does not offer all the power of a classic thesaurus, it does structure and, thus, improve the efficiency and consistency of the indexing process across all indexers. It also lets you build a browsable index of preferred terms that enables users to find *all* the content objects on a specific topic in a single place. Implementing an indexing thesaurus is less costly.

- **Searching thesauri**—This type of thesaurus defines a controlled vocabulary comprising synonyms, near synonyms, broader and narrower terms, and other variants of preferred terms, and even some carefully selected associated terms, enabling users to search for information about the same concepts using a variety of search terms. A searching thesaurus would typically be useful for searching a free-text database—for example, one that comprises news that updates frequently, collections of third-party content, or voluminous content for which the cost of manual indexing would be too great—rather than when searching for indexed content. Using a thesaurus to find more search results could be valuable, but this approach could increase recall at the cost of precision. A searching thesaurus can also support greater flexibility in browsing, by enabling users to navigate a preferred term's equivalence, hierarchical, and associative relationships—perhaps by using predefined search queries. The cost of implementing a search thesaurus does not depend on the amount of content on the information space. [68, 69, 70]

Ontologies

"An ontology [specifies] what exists, or what may exist, in a particular domain … and the vocabulary used to describe what exists."—David L. Poole & Alan K. Mackworth [71]

An *ontology* is a formal specification for a machine-readable controlled vocabulary that can represent large numbers of very specific semantic relationships between the concepts, objects, and entities that are characteristic of a particular domain of knowledge. While an ontology's primary function is the semantic classification of objects, it also defines the attributes of its object classes and models the complex relationships between classes. Thus, an ontology comprises definitions of classes and their attributes, relationships, and functions, including inference rules and axioms that determine their use.

Massive, sophisticated, digital-information systems such as the Semantic Web rely on ontologies to understand the meaning of users' information requests and improve the findability of information. The Resource Description Framework (RDF)—the World Wide Web Consortium (W3C) standard for defining metadata using XML tags—is a machine-readable language whose expressions take the form of a *triple*, a three-part data-storage format. For example, in an ontology, this could be an *individual–property–value* triple *or a* *subject–predicate–object* triple. A *subject–predicate–object* triple describes the *subject* and *object* in a relationship *and* the *predicate*, or relationship, between them—perhaps an action that the subject can take on the object. Using *subject–predicate–object* triples, you can express *any* conceivable type of relationship, including the equivalence, hierarchical, and associative relationships of other types of controlled vocabularies. Table 7.1 shows some examples. OWL, a more recently developed ontology language for the Web, offers similar capabilities.

Relationship	Subject	Predicate	Object
Equivalence	Feline	Is synonymous with	Cat
Hierarchical	Cat	Is a type of	Pet
Associative	Dog	Is another type of	Pet

Table 7.1—Examples of subject–predicate–object triples

RDF works in combination with the RDF Schema (RDF-S), enabling the definition of resources as subclasses of other resources and subproperties as subclasses of other properties; restriction of the domain and scope of properties; and definition of sets, sequences, and alternatives.

An ontology's *data model* enables organizations to define, share, extend, and customize taxonomies and comprises two parts, as follows:

- **Schemas**—One or more taxonomic *schemas* define the ontology's permitted objects and their attributes and relationships, specifying their meaning at a level of abstraction.

- **Domain ontology**—This is typically an ontology for a specific, narrow domain that comprises one or more taxonomies, flat lists, or other collections of objects, each having its own attributes and relationships, *and* the data elements that populate them—for example, preferred terms and their relationships with equivalent, hierarchical, and associated terms.

To communicate shared knowledge, you must create a common vocabulary and agree on the meanings of the terms in that vocabulary. Thus, ontologies specify the meaning of data, disambiguate terms, manage synonyms, and support the customization of the relationships between terms. If, over time, an ontology's schema expands or otherwise changes, the data forming its domain ontology must also change accordingly. Thus, ontologies are extensible and customizable.

When building a large-scale information space, an organization can either adopt an existing ontology to ensure its interoperability with others using the same ontology, customize an existing ontology by extending or modifying it to better meet their needs, or, as a last resort, develop their own ontology. However, for the organization's ontology and databases to be syntactically and semantically interoperable with those of other organizations and, thus, be useful to a larger audience, they should adopt or expand an existing ontology.

Communities that are focusing on a specific knowledge domain should unite in creating a shared ontology specifically for their domain within the Semantic Web rather than relying on a standards organization to define a massive, universal ontology. The goal should be for the ontologies of all organizations working in the same domain to converge on a standard, domain-specific ontology—or at least for them to create mappings between their organizations' ontologies. When different organizations that are working within the same domain share the same ontologies and use their classes, properties, and terminology consistently, people working for all those organizations can have meaningful exchanges with one another's computing devices and gain access to their databases. It's also important that your ontologies integrate well with adjacent and high-level ontologies and use the same terminology for the same objects and concepts.

Ontologies are also an essential element of knowledge-based systems for intelligent agents, not only because they're useful in representing an intelligent agent's beliefs about specific states but also because they store the knowledge that is necessary for them to act or solve problems appropriately when encountering specific situations within their domain. [72, 73, 74, 75, 76]

Semantic networks

In natural languages, *semantics* convey meaning, so the mappings between words and phrases and the concepts that they represent should be clear and unambiguous.

Semantic information is a component of any information model and typically takes the form of some type of markup or metadata. *Semantic tags* are machine readable and define the structural elements of an information model, identify their specific content, and clearly impart their intrinsic meaning. When designing an information model, identify all of its semantic content elements and assign a unique tag name to each element. Tag names are meaningful to content authors and others on your content team. They define the meaning of the content separately from its formatting. You can define as many tag names as are necessary.

A *semantic network*, or *knowledge graph*, represents the semantic relationships between an information system's objects or concepts, typically using both linguistic information and a visual graph to describe them. Thus, the meaning of an object or concept derives from its relationships with and dependencies on other objects or concepts. A semantic network typically takes the form of an oriented, *directed graph*, which comprises the following elements:

- **A set of nodes**—These labeled nodes represent content, knowledge, data, or metadata. Each node can represent either a category or an individual instance and can function as a starting point, an ending point, or another part of a path.

- **A set of arcs**—These labeled links connect the nodes. Each arc in a directed graph connects an ordered pair of nodes, terminates in an arrow, and has a label. [76]

A semantic network can consist of any number of nodes and arcs. As shown in *Figure 7.1*, semantic networks diagrammatically represent the structure of information or knowledge, precisely mapping a knowledge graph to its corresponding knowledge base. Thus, people can easily perceive the relationships that the semantic network represents. Semantic networks are also machine readable.

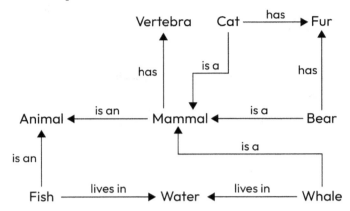

Figure 7.1: A semantic network recreated with reference to a diagram from Wikimedia Commons
Image source: https://en.m.wikipedia.org/wiki/File:Semantic_Net.svg

In semantic networks, arcs represent the relationships between the objects and concepts that certain nodes represent and can connect the nodes either vertically or horizontally. Triples provide alternative representations of arcs—for example, an *individual–property–value* triple—in which the *individual* corresponds to the subject, the *property* to the verb, and the *value* to the object—or a *subject–predicate–object* triple. The predicates of some common types of arcs include the following:

- **Is-a links**—This type of subset arc indicates a class relationship, in which an object node is a subclass, or child, of a broader, parent class *or* a subcategory of a broader category that the subject node represents. These *Is-a* links can form a chain, or hierarchy, in which object nodes represent classes that inherit the attributes of all the broader, subject nodes that contain them.

- **Has-a links**—This type of arc indicates attributes or characteristics of an object node.

- **Inst links**—In this type of arc, an object node represents an individual instance of a broader category that the subject node represents.

- **Part-of links**—This type of arc indicates that an object node is a part of a subject node.

The semantic network's architecture is easy to use. You can clearly and precisely define the meaning of the arcs that a semantic network comprises. You could use a semantic network to represent a knowledge base that comprises a set of interrelated objects or concepts. *Inheritance* is a central feature of the semantic network so is easy to implement. You can organize nodes and arcs into a taxonomic hierarchy of parent-child relationships, in which child nodes inherit the attributes of *all* the parent nodes that contain them. You can express these attributes as *property–value* pairs.

Natural language–processing applications can employ semantic networks for word-sense disambiguation, the semantic parsing of human language for the purposes of ontology induction or conversion to machine-readable logic, and the analysis of large collections of textual content—for example, to identify key themes and topics, uncover biases, or map an entire domain of knowledge. Semantic networks support flexibility in cross-referencing and the sorting of information. [78, 79, 80, 81, 82, 83, 84]

Faceted classification

"In a faceted classification system, each resource is described using properties from multiple facets, but a person searching for resources does not need to consider all of the properties (and consequently the facets) and does not need to consider them in a fixed order...."—Robert J. Glushko [85]

Facets represent discrete attributes or aspects of an information space's content objects *or* the categories of a domain's faceted classification system. They are a form of descriptive metadata that is applicable to an information space's content objects *and*, thus, are part of the metadata that describes those content objects. The collection of facets describing a particular information domain must comprehend that domain's entire scope, and all the facets must pertain to that domain. A facet must be homogeneous, readily distinguishable from all other facets, and mutually exclusive of other facets.

Facets provide a variety of useful ways of classifying an information space's content objects. For example, depending on an information space's domain, facets could classify categories of objects, entities, or concepts, and their attributes—for example, topics, authors, formats, document types, audiences, people, places, objects, products, or prices. Topics represent categories into which certain content objects fit. Each facet isolates a discrete attribute that, in combination with other facets, can express compound concepts that more clearly convey complex ideas.

Facet analysis is an eight-step process for decomposing an information space's content into its semantic components and comprises the following steps:

1. Collect a representative sample of the information space's content objects.

2. Write a brief description of each content object—a phrase or a sentence at most—endeavoring to use consistent terminology for the same concepts.

3. Compile a list of indexable concepts that you've derived from these descriptions, choosing preferred terms—words or phrases—to represent them.

4. Decompose any compound concepts, separating them into their constituent parts, again choosing preferred terms to represent them.

5. Create, evaluate, and, as necessary, revise a collection of categories, or facets, that represents the most frequently occurring concepts you've isolated, ensuring they're mutually exclusive *and* comprehend the entire domain of the information space.

6. Create, evaluate, and revise ordered lists of the preferred terms that belong to each category, or facet, ensuring both that the lists include *all* the original concepts and *all* the preferred terms fit somewhere.

7. Organize all of the information space's concepts and their preferred terms into a taxonomy, faceted classification scheme, *or* a taxonomy comprising categories that contain facets, finalizing its controlled vocabulary.

8. Revise the information space's content objects, as necessary, to ensure that they adhere to both the classification scheme and the controlled vocabulary.

Facet analysis considers multiple attributes, or facets, of an information space's content, so could provide the foundation for a faceted classification scheme. You can use facet analysis to identify the attributes of specific content objects. It can also be helpful in identifying their topics. Facet analysis typically takes a bottom-up approach to information classification, in which you might, for example, initially collect content about, identify, and name narrower subtopics, then form larger collections of content on broader topics by gathering together related subtopics and naming these collections. However, if you're analyzing the facets for a large controlled vocabulary, you might first identify its broader categories, then determine what terms belong in each category.

Facet analysis is especially useful in the following situations:

- **New and emerging domains**—Facet analysis can enable you to assess content about poorly understood domains, whose scope, vocabulary, and the relationships between its content objects remain undefined.

- **Interdisciplinary fields**—Facet analysis can help you to understand the varied perspectives of those working in interdisciplinary fields, whether you're considering the classification of specific content objects or the need for compound concepts.

- **Large, complex information spaces**—For digital information spaces that comprise diverse collections of content and have large controlled vocabularies with complex interrelationships and the need to represent compound concepts, a strictly hierarchical taxonomy or a polyhierarchy might lack the necessary flexibility and expressiveness to clearly convey the controlled vocabulary. In such cases, a facet analysis could produce a more useful organizational structure and a more flexible controlled vocabulary by informing the design and development of a thesaurus or a faceted classification scheme. In contrast, a faceted classification scheme would be unnecessary for a small collection of content objects with a limited controlled vocabulary.

- **Multiple taxonomies**—If classifying *all* of an information space's content would require multiple taxonomies, but the boundaries between some taxonomies are not clear, conducting a facet analysis would inform the design of a faceted classification scheme.

A facet analysis could provide the basis for a taxonomy or some other type of controlled vocabulary. Identifying the primary facet for a specific collection of content could indicate what type of controlled vocabulary would be most suitable for that content. For example, an emphasis on topics might suggest the development of a thesaurus or a taxonomy. If a single dimension of classification would be sufficient for organizing an information space's content objects, implement a hierarchical taxonomy in which each content object inherits all the attributes of its antecedents. However, if many content objects could belong in several discrete categories, create a faceted classification scheme, which allows objects or concepts to appear in multiple locations *and* can express compound concepts much more effectively than a polyhierarchy. The flexibility of a faceted classification scheme enables it to transcend the limitations of a single top-down taxonomy, but can also result in greater complexity in the expression of its concepts.

If you're creating a controlled vocabulary for a very large digital information space—for example, an extensive product catalog for an ecommerce application—implementing a hierarchical taxonomy in combination with subordinate facets can provide greater power and flexibility than a taxonomy alone. For example, an ecommerce store can use facets to suggest other similar or related products.

You could create a *faceted classification scheme* that would enable you to systematically organize anything in the world. To organize an information space's content objects, you can devise a faceted classification scheme by creating generic or domain-specific semantic categories and combining them as necessary to express new concepts. Thus, rather than attempting to define a comprehensive set of categories that exhaustively covers *all* of a domain's concepts at the outset, you can discover new objects and concepts and learn about their relationships over time, and create a highly scalable classification scheme by combining existing facets to create new facets. A faceted classification scheme can readily accommodate new facets without causing any disruption to the existing scheme.

A faceted classification scheme comprehends whatever facets are necessary to collectively describe *all* attributes or aspects of an information space's domain and comprises multiple taxonomies, each focusing on a specific attribute or aspect of its content. Each collection of facets could form either a *flat list* of facets—with alternative paths at a single level, enabling users to quickly narrow search results to one facet—or a *hierarchy* of facets—with nested attributes or aspects providing alternative paths to a variety of relevant information. However, for an information space that has a broad domain, the number of facets could proliferate, making its faceted classification scheme overly complex or even unmanageable.

On an information space that employs a faceted classification scheme, each content object comprises both its actual content *and* descriptive metadata, including one or more facets that authors or indexers have assigned in classifying it, to indicate which of its attributes would be most useful in facilitating information retrieval by searching or browsing. Because faceted classification cross-references information in various ways, a content object's facets can be useful both in constraining searches and sorting and filtering search results.

Once a team has defined an information space's faceted classification scheme and assigned the appropriate facets to its content objects, the team can implement a search system that combines full-text search with faceted search, supporting both discovery and exploratory search, and allows the refinement, or filtering, of search results by facets. Because a faceted classification scheme allows content objects to reside in multiple categories—each of which represents a different perspective on a content object—and thus, provides multiple paths to the same content, it also facilitates faceted navigation and browsing. [86, 87, 88, 89, 90, 91, 92, 93, 94]

Developing a controlled vocabulary

To facilitate the organization of an information space's content, develop a controlled vocabulary that enables your team to control which terms you use for specific concepts, distinguish between *homographs*—that is, terms that have the same spelling, but different meanings—and specify hierarchical and associative relationships that exist between terms.

To determine what type of controlled vocabulary would be most effective for a particular information space or product, then design and develop that controlled vocabulary, you can follow the process that I outline in this section or another similar process.

When you're preparing to develop any type of controlled vocabulary for an information space, the first few steps are part of the *Discovery* process, which informs your design strategy. Many steps in the process of developing a controlled vocabulary can overlap with one another or occur iteratively.

Step 1: Conducting and analyzing user research

The design of a controlled vocabulary requires a user-centered design (UCD) process. What terms would your target audience use to describe your information space's content? What terms would help your users find the information they're seeking? Conduct user research to understand your target audience's information needs and learn the vocabulary they use in describing the subject matter of the information space's domain. Refer to the section "Card-sorting methods," in Chapter 5, *UX Research Methods for Information Architecture,* to learn card-sorting methods that can be useful in discovering users' terminology for specific concepts.

Step 2: Learning from data analytics

Using data analytics, you can analyze the actual information-seeking behaviors of your existing users. In Chapter 5, *UX Research Methods for Information Architecture,* the section "Data analytics" describes several methods that are helpful in assessing whether the navigation system's terminology is providing good information scent. Plus, by analyzing an information space's *search-log analytics,* as the section "Search-log analytics" describes, you can discover what *keywords* and phrases users are employing in their search queries and, thus, learn your users' vocabulary.

Step 3: Analyzing your content

Before devising a design strategy for a controlled vocabulary, you must first understand the organization's existing content, including the content that any existing information space comprises. What concepts does the information space convey? What terminology does the organization's content currently use to describe those concepts? Chapter 6, *Understanding and Structuring Content,* explores several methods of content analysis. To learn about the content and its terminology from the perspectives of the people who create and manage it, as well as what additional content they're planning to create, conduct interviews with them, as the section "Content-owner interviews" describes. Conducting a *content inventory* is an essential precursor to developing any controlled vocabulary and helps you understand the full scope of an existing information space's content *or* an organization's other existing content and the terminology it uses, as the section "Content inventories" describes. When conducting your content inventory, collect all of the terms that are characteristic of the information space's domain.

Step 4: Defining your controlled-vocabulary design strategy

Strategizing and planning are essential parts of any design process, including for the design of an information space's controlled vocabulary. By leveraging the various methods that I've described in Steps 1–3, you've gained a deep understanding of the information space or product's industry domain and content and learned what terms best describe that content. All of these learnings inform your strategy for designing one or more controlled vocabularies for the information space, as well as its content strategy and information architecture (IA) strategy. In defining your design strategy for a controlled vocabulary, you should answer the following questions:

- What type of controlled vocabulary would be most appropriate for the information space and its content? What level of vocabulary control should it impose?

- What are your goals for the controlled vocabulary? Improving navigation, search, or both?

- Does the information space require a complex, hierarchical classification scheme, or taxonomy?

- How much content does the information space comprise? Is the content broad or deep, or is its breadth and depth fairly well balanced?

- Are the distinctions between topics subtle or obvious? Is it difficult to disambiguate different categories of content because there are many similar categories? If so, you must define an exhaustive list of very specific terms to distinguish between them.

- How stable are the concepts and vocabulary of the information space's industry domain? Is it a well-established domain that experiences little change or a new domain that is evolving quickly? If the latter, maintaining the controlled vocabulary will require greater effort.

- Do people agree on what terminology to use for specific concepts or use many variant terms for the same concepts?

- What tool should you use in developing your controlled vocabulary? A spreadsheet *or* a dedicated tool for authoring and managing thesauri and taxonomies?

- Will you use the controlled vocabulary in indexing your content to ensure that your search engine can successfully retrieve the content and improve findability?

- Will you use the controlled vocabulary in categorizing content objects within the information space's content-management system and, thus, improve its navigation system?

- What roles within the organization are responsible for creating and maintaining the controlled vocabulary? Does their work require sophisticated capabilities and tools? Would they need specialized training? What level of effort is necessary to do this work successfully? Are there time constraints that would limit their ability to do this work effectively?

The answers to these questions could change over time, as a domain or information space evolves, so it is essential that a controlled vocabulary be well maintained.

Creating a metadata matrix

When devising your strategy for an information space or product's controlled vocabularies, create a *metadata matrix* to facilitate your team's discussions regarding prioritizing the subject-matter areas for which you might develop controlled vocabularies. This metadata matrix should comprise the following columns:

- **Controlled vocabulary**—List the name of each potential controlled vocabulary, which should be descriptive of its subject matter—for example, topics, document types, organizations, people, roles, target audiences, industries, companies, product types, product brands, places, geographic regions, countries, locales, or languages.

- **Description**—Describe the types of terms or names that you should include in that controlled vocabulary.

- **Example terms**—List a few examples of terms or names that might constitute preferred terms in that controlled vocabulary.

- **Scope of content**—Describe the actual or projected scope of the content relating to that controlled vocabulary—for example, **Very large**, **Large**, **Medium**, **Small**, or **Very small**.

- **Priority level**—Considering the value that the controlled vocabulary would provide to both users and the organization, determine the organization's level of priority for developing that controlled vocabulary—for example, **Committed**, **High**, **Medium**, **Low**, or **Deferred**.

- **Cost**—Indicate the level of the proposed budget for developing that controlled vocabulary—for example, **Very high**, **High**, **Medium**, **Low**, or **Very low**.

- **Time constraints**—Describe the time constraints for developing that controlled vocabulary—for example, **Severe**, **High**, **Medium**, **Low**, or **Trivial**.

- **Maintenance**—Indicate the degree of difficulty of maintaining that controlled vocabulary—for example, **Very easy**, **Easy**, **Moderate**, **Difficult**, or **Very difficult**.

Using this metadata matrix can aid your decision-making process when choosing which controlled vocabularies to develop.

Step 5: Gathering terms from diverse sources

As you conduct your user research and analyses, begin gathering potential terms for the information space's controlled vocabulary from these and other sources into a *terminology matrix*. In addition to the terms you've gleaned from the information space's users—both directly through user research and from usage data—*and* the terms authors, editors, and content managers are using, it could be helpful to interview other subject-matter experts to learn what terminology they use for the same concepts. Conduct a competitive analysis to ascertain what terminology other information spaces, products, and publications focusing on the same industry domain use. Also, collect relevant terms from any thesauri that have been published for the information space's industry domain. Using these other thesauri as references can help you to familiarize yourself with the domain. Through this process, you'll likely gather a large variety of terms that describe your information space's various topics.

Creating a terminology matrix

As you collect terminology belonging to the information space or product's domain from all these diverse sources, compile an alphabetical or otherwise logically ordered list of terms that you might add to a controlled vocabulary in a *terminology matrix*, which comprises the following columns:

- **Term**—In the first column, list all the *entry terms* or names that you're considering for possible inclusion in the controlled vocabulary.

- **Meaning**—Describe the meaning of the term, capturing a key concept for possible inclusion in the controlled vocabulary.

- **Source**—Provide the source of the term—for example, an interview or contextual inquiry with a specific user, survey data, the results of a card-sorting exercise with a specific participant, data analytics, or search-log analytics; a specific author, editor, content manager, or subject-matter expert; an organization's or information space or product's existing content; or a particular published thesaurus.

- **Preferred term**—Indicate whether the term or name is that which a user is most likely to use for a concept or an entity, respectively. If *Yes*, you should include this term in the controlled vocabulary. If *No*, indicate the term or name that is the preferred term.

- **Variant terms**—List *all* terms that have the same meaning as the preferred term. These equivalent terms might include the following:

  - **Synonyms**—These terms might include synonyms and near synonyms of the preferred term.

  - **Abbreviations or acronyms**: These terms could include abbreviations or acronyms for the preferred term.

  - **Alternative spellings**—These might include alternative spellings of the preferred term in the same or a different language.

- **Broader terms**—List any terms that have a hierarchical relationship to the preferred term and reside at a higher level in the hierarchy.

- **Narrower terms**—List any terms that have a hierarchical relationship to the preferred term and reside at a lower level in the hierarchy.

- **Related terms**—List any terms that have an important associative relationship to the preferred term.

- **Notes**—Capture the reasons for your specific decisions regarding what terms to include in or exclude from a controlled vocabulary and extrapolate rules from them. Compile these rules in your documentation of standards and guidelines for creating the controlled vocabulary, then consistently apply them when you're making similar decisions regarding the use of terminology.

Refine your terminology matrix iteratively as your understanding of the information space's content and terminology evolves. Ultimately, you can use this matrix to document most of the decisions you make when developing the controlled vocabulary.

Step 6: Deciding what type of controlled vocabulary to develop

Once you've become completely familiar with your target audience's terminology and needs and the information space's content and domain, you should make your final determination regarding the type of controlled vocabulary that would best meet the overall requirements for the information space. Consider what level of complexity the scope of the information space and the time available for the project would warrant. For a large project, in addition to the information space's primary controlled vocabulary, you might also need to create one or more specialized controlled vocabularies to address specific needs. Once you've decided what type of controlled vocabulary you need to create, you'll know whether to emphasize equivalence, hierarchical, or associative relationships between terms when developing that controlled vocabulary. As appropriate to the type of controlled vocabulary, identify variant, broader and narrower, and related terms, and add them to the terminology matrix for that controlled vocabulary.

Step 7: Defining and grouping preferred terms

A *preferred term* is a term that a particular organization has established for accepted use in describing a particular concept within that organization and on its information spaces and products. Define a preferred term for each unique concept within the information space's domain and add it to your controlled vocabulary. All other semantic relationships between the terms in a controlled vocabulary exist in relation to a preferred term.

When defining preferred terms, always consider what terminology the information space's target audience would use in describing the information space's content. You can conduct user research to learn what terms your target users typically use for specific concepts. You can analyze any existing information space's search-log analytics to identify the keywords that actual users are using when searching for information about specific concepts. Using the vocabulary of the information space's target audience greatly improves the findability of its content.

Also, consider what terminology the organization and other organizations in the same industry domain are already using for particular concepts. Where the terminology of your target users and the industry coincide, your choice of preferred terms is clear. Otherwise, you should usually choose the terminology of the target users over that of the organization. However, when you must choose among various terms that different target users are using for particular concepts and those that are prevalent in the organization's industry domain, choose whichever terms are least ambiguous as your preferred terms and add them to the controlled vocabulary.

To evaluate whether an existing information space's terminology matches your target users' terms for specific concepts, you could try conducting keyword searches using various terms for the same concept and see what results you get. This can be helpful in deciding which of several similar terms should be the preferred term for a given concept.

When your team is defining preferred terms, you must make tough choices in determining the meaning and format of each preferred term you include in the controlled vocabulary. Making these decisions is essential to achieving consistency in the use of terminology, which benefits both content managers and users. Consider the following factors when defining *any* term for inclusion in a controlled vocabulary:

- **Meaning**—A single term could have multiple meanings. For example, homographs have the same spelling, but can have completely different meanings. What is the specific sense of a term you're including in a controlled vocabulary? Only by clearly defining a term's meaning can you determine its appropriate semantic relationships to other terms. You can constrain the meaning of a term to a specific concept by providing *scope notes*. To specify a particular meaning for a homograph, you can use a *parenthetical term qualifier*—for example, *lead (transitive verb)*, *lead (intransitive verb)*, *lead (role)*, *lead (electrical)*, or *lead (metal)*.

- **Specificity**—In a controlled vocabulary, each term should represent only a single concept. Terms should be highly specific and easy to disambiguate from all other terms. Among the terms that target users typically use, choose those that are least ambiguous as preferred terms. Choose completely unambiguous terms whenever possible. If necessary to convey a specific concept, a term can comprise multiple words—for example, *controlled vocabulary* is an example of such a *compound term*. The use of compound terms is common on large information spaces, both to convey the precise meanings of these terms and to increase the precision of search results.

- **Lexical classification**—To what category of word, *or* part of speech, should a term belong? You should generally include nouns rather than verbs or adjectives, but you might encounter situations in which you should include verbs or adjectives. The use of adjectives is especially common in controlled vocabularies for product descriptions.

- **Acronyms or abbreviations**—Do your standards and guidelines permit the inclusion of acronyms or abbreviations in a controlled vocabulary? Terms should generally be full words, but popular usage might dictate the inclusion of certain acronyms, even as preferred terms, but probably as variant terms. It is much more likely that a controlled vocabulary would include acronyms than abbreviations.

- **Spelling**—Determine the correct spelling and capitalization of the term, as well as any internal spaces or punctuation such as hyphenation. For example, should the term be in title case or all in lowercase? Using the proper spelling is particularly important for brand names. An organization might employ a specific dictionary, domain glossary, or internal glossary as the authority in making spelling decisions.

- **Singular versus plural**—For nouns that represent objects or entities that you can count, use the plural form of the noun—for example, readers, books, libraries, departments, or companies. For conceptual nouns such as truth or character, use the singular form. Again, refer to whatever source you've deemed the authority in making these decisions.

- **Stemmed form**—If the information space's search engine supports stemming, you can rely on this capability instead of choosing the singular or plural forms of words. Stemming offers the advantages of handling gerunds and verb tenses as well.

When your team is making formatting decisions regarding the terms you're adding to controlled vocabularies, you should generally follow the *ANSI/NISO Z39.19-2005. (R2010) Guidelines for the Construction, Format, and Management of Monolingual Controlled Vocabularies*. However, the needs of a particular organization might require that you deviate from them in some cases.

When preparing to organize similar terms into logical groupings, conduct closed card sorts to learn how your target users would typically group them. Then, based on your learnings, gather preferred terms and other semantically related terms into groups, according to the subject matter to which they relate. Arrange the preferred terms alphabetically *or* in some other logical order.

Defining preferred terms for the various concepts that pertain to your information space's domain provides a means of uniquely identifying each of your collections of variant, broader and narrower, and related terms. Once you've defined a preferred term for a concept, you can more easily add or delete variant terms or modify your collections of related terms, which facilitates the management of these terms.

By defining preferred terms, you can ensure that an organization uses clear terminology that is both meaningful to its target users *and* characteristic of the information space's domain. Defining preferred terms promotes consistency and clarity in an information space's content *and* labeling.

Step 8: Identifying variant terms

A *variant term* is equivalent in meaning to an organization's preferred term for a particular concept and may be synonymous with that preferred term or just equivalent for purposes of information retrieval. Variant terms are the semantic equivalents of *See* cross-references in a book index, which refer to preferred terms.

Consider what terms the information space's target users might use in lieu of your preferred terms. Identify variant terms for each preferred term, including synonyms, near synonyms, abbreviations, acronyms, alternative spellings, and common misspellings of terms. Analyze your search-log analytics to identify what variant terms your actual users are using in their keyword searches when looking for information about specific concepts for which you've defined preferred terms.

Depending on the level of specificity you want to achieve in the controlled vocabulary, the equivalence relationship could comprehend broader and narrower terms. For example, in addition to the common variants of a product's brand name, its equivalents could include general product categories and the names of specific competitive products. Extending the equivalence relationship to include more general and more specific terms can minimize the levels of hierarchy that are necessary in a taxonomy *and* allow users to use a rich variety of keywords in searching for content, improving the findability of your content.

Step 9: Identifying broader and narrower terms

When creating a taxonomy, or hierarchical classification scheme, for an information space, identify *broader*, or more general, terms and *narrower*, or more specific, terms that have hierarchical relationships to the preferred terms for particular concepts. Another type of hierarchical relationship is the *whole/ part relationship*, in which the hierarchy for an object or entity comprises the parts of that object. These hierarchical relationships support the classification of preferred terms into categories and subcategories, which determines where each preferred term should reside in the hierarchy and informs the design of a hierarchical navigation system. To accurately represent all the relationships between terms, it might be necessary to place the same term in more than one location within the hierarchy, forming a polyhierarchy, *or* to cross-reference a preferred term in another part of the hierarchy.

One effective way of determining what terms target users would use in describing broader concepts is conducting open card sorts, during which participants sort cards representing terms into groups, then label the groups they've created.

Step 10: Identifying related terms

In contrast to terms that have hierarchical relationships to the preferred terms for particular concepts, *related terms* describe concepts that have meaningful associative relationships to the preferred terms for specific concepts. Related terms are the semantic equivalents of *See also* cross-references in a book index and should lead your target users to additional information that might be of interest to them.

Typically, you would identify related terms *only* for the preferred terms in a controlled vocabulary for a very large information space—for example, to connect related products within an ecommerce app.

The use of related terms aids users' discovery of related content. If you intend to include related terms in your controlled vocabulary, you need to determine what content might prompt users to become interested in other, loosely related content. When identifying other concepts that are related to preferred terms, limit these to only the most meaningful relationships. Think about where users might want to go next, and create a list of related terms that describe those other concepts. Use the related terms in contextual-navigation links to the related content.

Avoid overwhelming users with too many options for viewing related content, limiting these options to only those that would be most valuable to both users *and* the organization. To discover what related content target users would find valuable, show research participants particular content objects and ask them what content they would want to view next.

Step 11: Implementing, evaluating, and refining the controlled vocabulary

The implementation of an information space's controlled vocabulary is highly dependent on context, including the goals of the organization to which the information space belongs, the information space's content, and the organization's tools.

If your goal for creating the controlled vocabulary is to improve the information space's navigation system, configure its content-management system (CMS) to enable content managers or indexers to apply the controlled vocabulary in classifying its content objects. Try applying the controlled vocabulary to some new content objects in your CMS. Are you able to easily categorize the content using terms in the controlled vocabulary?

If your goal for creating the controlled vocabulary is to improve search results, design the controlled vocabulary to balance recall versus precision. To improve the precision of search results, you should also configure the search engine to leverage Boolean operators. Conduct some preliminary testing to evaluate the search system. Try searching for various types of information using keywords from the controlled vocabulary. Are you able to easily find the information you're seeking? Do your expected results appear on the first page of the search results? Would these results deliver the content that the organization wants to ensure users see? Would the results satisfy users' information needs?

Once your team has implemented the controlled vocabulary, conduct usability testing with actual or target users to validate the preferred terms that you've added to the controlled vocabulary. Does the terminology the information space uses for specific concepts match participants' expectations? Are the terms sufficiently broad or narrow? Do participants get the search results they expect? Based on the results of your testing, refine the controlled vocabulary.

Creating a controlled vocabulary should be a careful, iterative process. For each iteration, avoid trying to accomplish more than would be possible in the available time. As the scope of the information space's content expands, you'll need to add new preferred terms to your controlled vocabulary.

Step 12: Documenting the controlled vocabulary

Finalize your documentation for the controlled vocabulary. Refine your *terminology matrix* for the controlled vocabulary, ensuring that it is as complete as you can currently make it and the decisions you make regarding the use of specific terminology are consistent across team members. Relying on memory almost guarantees that terminology inconsistencies would creep into the information space's content.

Fully document the *standards and guidelines* that your team established when creating and using the controlled vocabulary, especially if you're creating a large, complex controlled vocabulary. This documentation should include the rules you've extrapolated from specific decisions about what terms to include in or exclude from the controlled vocabulary and the reasoning behind them, as well as some examples of these decisions. Good documentation ensures that your controlled vocabulary is easy to learn and use and simplifies the training of your content team.

Creating effective documentation is essential to ensuring that your team can expand and make changes to your controlled vocabulary as necessary. It also enables your team to use terminology consistently over time—regardless of whether the people who originally made terminology decisions remain on the team. For future team members to use your controlled vocabulary effectively, they must understand the rationale behind those decisions. [95, 96, 97, 98]

Social classification

"The advantage of folksonomies isn't that they're better than controlled vocabularies, it's that they're better than nothing...."—Clay Shirky [99]

The social classification of content on the Web results from social tagging by people with shared interests rather than the classification of content objects by indexers who reliably have expertise in a topic. Thomas Vander Wal coined the term *folksonomy* to describe the user-generated, ad hoc taxonomies that result when regular folks tag content objects using one or more shared social tags, on the Web or other large, shared information spaces that comprise diverse collections of content.

Although del.icio.us, a social-bookmarking site that launched in 2003, and Flickr, a photo-sharing site that launched in 2004, represented early successes of social classification on the Web, the general popularity of this approach to the classification of information has since waned—*except* on social media applications. For example, on Twitter, the use of hashtags (#) in tweets was integral to users' being able to share information of interest with like-minded people. However, the use of social tagging on LinkedIn has been less successful because users' skills-endorsement tags on other people's profiles are often annoyingly inaccurate. Of course, the value of social tags depends entirely on how well they meet the needs of users. Some do, some don't.

Folksonomies are flexible, so people can adapt them to their needs. When tagging content objects, people may use different tags for the same things or the same tags for different things. These tags express people's diverse perspectives on an information space's content. Because social tags can provide various access paths to the same content, they facilitate exploration and discovery.

Social-tagging analytics—particularly, the numbers of tags that people have assigned to specific content objects—indicate people's level of interest in particular topics and show what content is trending.

The biggest deficiencies of social tags are their inability to express semantic relationships—whether equivalence, hierarchical, or associative relationships—and the findability issues that result from their lack of vocabulary control. However, it is possible to use folksonomies in combination with controlled vocabularies to gain those capabilities. Plus, the social tags that people use in creating folksonomies can inform the design of taxonomies and other controlled vocabularies. [100, 101, 102, 103]

Summary

In this chapter, I first discussed some principles, key goals, and challenges of classifying, or categorizing, information. Then, I considered several perspectives on types of metadata that can provide the foundation for information categorization, as well as the use of metadata schema to describe the content objects on a digital information space.

I covered controlled vocabularies in some depth, exploring the need for them, the types of terms they typically comprise—including preferred, variant, broader, narrower, and related terms—and their semantic relationships. I considered how to choose such terms, as well as the benefits and challenges of using controlled vocabularies. I described different types of controlled vocabularies—including synonym rings, authority files, taxonomies, thesauri, ontologies, semantic networks, and faceted classification schemes. Then, I provided a detailed, step-by-step process for developing a controlled vocabulary. Finally, I briefly described social classification.

Next, in Chapter 8, *Defining an Information-Architecture Strategy*, you'll learn how to align your IA strategy with business strategy and synthesize your learnings from conducting UX research, analyzing an information space's content, and classifying its content. Then, I'll discuss a variety of IA concerns that you must consider in devising an IA strategy that meets the needs of both the organization and users. I'll describe how to envision, communicate, and validate conceptual-design solutions for labeling, navigation, and search systems. Plus, you'll learn how to document your IA strategy and create an IA project plan.

References

To make it easy for readers to follow links to the references for this chapter, we've made them available on the Web: `https://github.com/PacktPublishing/Designing-Information-Architecture/tree/main/Chapter07`

8

Defining an Information-Architecture Strategy

"An information-architecture strategy is a high-level conceptual framework for structuring and organizing an information environment. It provides the firm sense of direction and scope necessary to proceed with confidence into the design and implementation phases. It also facilitates discussion and helps get people on the same page before moving into the more expensive design phase."—Peter Morville and Lou Rosenfeld [1]

An *information-architecture (IA) strategy* is a specialized form of UX design strategy that defines the outcomes you intend to achieve through an IA program or project whose goal is creating or improving a digital information space or product. When devising an IA strategy for an information space or product, you should define its outcomes along the following four dimensions:

1. The business value it should create
2. The user needs it must satisfy
3. The scope and structure of an information space's content
4. The technologies its implementation should leverage

Throughout *Discovery*, you've derived valuable insights that should inform your IA strategy and design decisions for an information space or product. By doing stakeholder interviews and competitive research, you've developed in-depth knowledge about the organization's business domain, mission, and strategic goals. By conducting user research and modeling the information space's target users, you've achieved a deep understanding of this audience and their needs. Once you've thoroughly analyzed the information space's existing and planned content, created a controlled vocabulary, and classified the content, you've provided a sound basis for making decisions regarding the information space's structure.

Now, you need to synthesize all of your learnings to devise a holistic IA strategy that clearly sets the direction for your IA program or project, ensuring that it meets the needs of both the organization and its users. Only by defining a clear, comprehensive IA strategy for an information space can you prepare yourself and your team to complete the transition from *Discovery* to *Design* as you progress toward implementation.

In this chapter, you'll learn about the following key topics:

- Understanding and aligning with business strategy
- Synthesizing UX research findings
- Understanding an organization's content
- Learning about implementation technology
- Looking at the big picture
- Other IA strategy concerns
- Envisioning, communicating, and validating your conceptual models
- Documenting and presenting your IA strategy

[2, 3]

Understanding and aligning with business strategy

"Strategy is the creation of a unique and valuable position, involving a different set of activities. ... Strategy requires you to make trade-offs in competing—to choose what not to do. ... Strategy involves creating fit among ... the ways a company's activities interact and reinforce one another."—Michael E. Porter [4]

"With a solid understanding of the goals of the project, you'll be able to create an IA that works for both the business and for people. You'll also be able to communicate and sell the draft IA more easily. When you can show how it will achieve the goals of the business, stakeholders will accept it more easily."—Donna Spencer

To succeed in the marketplace, each organization must have its own differentiated strategic mission, vision, and business goals. An organization's unique business requirements, goals for their digital information spaces and products, and existing and planned content should drive your IA strategy and help you determine the optimal structure and terminology for their digital information spaces and products. You must work with the organization to overcome any barriers that an IA project presents to your successfully defining an IA strategy for an information space or product.

Therefore, if an organization has not yet defined a clear strategic vision, collaborate with your business stakeholders and help them to accomplish that goal. If, during this process, you discover gaps in the organization's business strategy or identify any misconceptions in their strategic thinking, ask the difficult questions that are necessary to resolve such issues, seek allies on the business side of the organization, raise the organization's awareness of these issues, and work together to come up with a better strategy and plan. As necessary, conduct UX research to inform the organization's strategic decisions.

During *Discovery*, to develop business insights that inform both business strategy and IA strategy, you can conduct a competitive analysis to learn about other businesses that are competing in the same domain and understand both their business strategy and their IA strategy for competitive information spaces or products. Only by developing a strategic understanding of competitors can you discover how to differentiate your new or existing information space or product from those already in the marketplace.

You can also develop insights that inform your IA strategy by conducting evaluative UX research on an existing information space or product's information architecture during *Discovery*. For example, if an IA project's goal is to improve the existing information architecture, you can identify its deficiencies by conducting an expert review or heuristic evaluation. Fred Leise's eleven content-analysis heuristics, which I presented in Chapter 6, *Understanding and Structuring Content*, could be useful in evaluating an information space's content. Plus, by conducting competitive benchmarking or competitive benchmark testing, you can assess how well the current information architecture is performing against those of competitors, which can inform the definition of user requirements.

Interview your key business stakeholders to determine what core business problem the organization intends an IA project to solve, what their goals are for an information space or product, what the appropriate scope of the project should be—whether you're creating a new information space or product, making incremental improvements, or doing a complete redesign—and why they're undertaking the IA project now. Make sure that the scope of the IA project is appropriate to the business value it would ultimately deliver. Define your strategic goals for the IA project on the basis of your learnings from these stakeholder interviews.

If an organization is unable to clearly articulate their goals for an IA project or define the business outcomes they're seeking from it, you must drive that effort by working with them to propose clear goals and business outcomes for the project and get stakeholders' buy-in for them. Otherwise, when working on the project, you would struggle in trying to evaluate business requirements and user needs, determining your priorities, and making decisions, and you would lack any sound basis for assessing whether you're making progress toward achieving goals that would deliver business value.

Once an organization has clearly defined its strategic vision and business requirements for an information space or product or agreed to those you've proposed, you must develop an IA strategy that aligns with their business strategy and requirements. Achieving this alignment early during a project saves wasted time and effort throughout the project, enables you to avoid endless debates and misunderstandings that could result in your having to make changes unnecessarily, and thus, saves the organization money.

To devise an IA strategy that aligns with an organization's business strategy and optimally meets its unique business requirements, you must first gain a deep understanding of the business context in which an information space or product exists, including its competitive landscape, barriers to success, target audience, existing and planned content, and available technologies to support the implementation of your information architecture. Developing this understanding takes time, so you must rely on building strong relationships of trust and collaboration with your business stakeholders, including those among C-level leadership.

Because an organization's competitive environment continually changes, they may need to evolve their business strategy in response to competitive changes. To ensure that you and the organization maintain a shared vision for your IA project, you must be willing to adapt and realign your IA strategy to the organization's business strategy to accommodate such changes.

To verify that your IA strategy aligns with the organization's business strategy, you should communicate your strategic thinking to trusted business partners early on. Share the early drafts of your IA strategy documentation with your key stakeholders to validate that your understanding of the business strategy is correct. [6, 7, 8, 9, 10]

Facilitating strategy workshops

To learn from and share information with your key stakeholders across disciplines, facilitate a collaborative strategy workshop that takes place over a day or two. Limit the number of participants to between three and seven people to ensure that everyone you need to hear from has the opportunity to talk. Ideally, this should be a face-to-face workshop so you can form relationships of trust that let you ask the tough questions that can make the difference between your success and failure. Your discussion should be informal enough to allow you to pursue whatever interesting ideas arise.

Discuss the organization's business strategy, its implications for your IA strategy, how your IA strategy can align with that business strategy, and how your IA strategy could potentially have strategic impacts on the business by identifying and addressing previously neglected opportunities that could drive innovation. Consider the organization's areas of competitive strength and weakness, what differentiates their offerings from those of their competitors, and what competitive threats currently exist. Determine what opportunities and activities the organization should pursue to achieve success in their business domain and in seeking competitive differentiation.

One tool your team can use in quickly developing or evaluating an organization's business strategy is the *SWOT analysis*—an acronym for Strengths, Weaknesses, Opportunities, and Threats. You should ground your SWOT analysis in the organization's business data, your competitive analyses, and other UX research. Figure 8.1 depicts this model, in which the internal factors the organization can proactively address are its competitive strengths and weaknesses, while the external factors are the opportunities and threats that impact the organization positively or negatively.

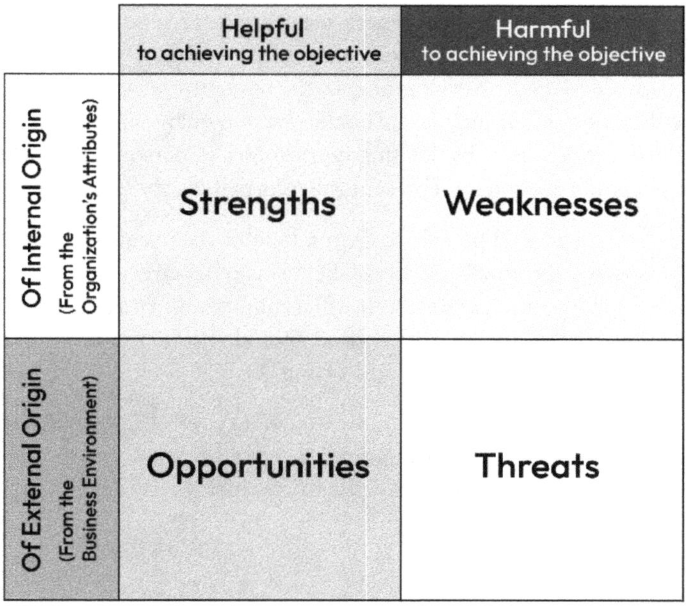

	Helpful to achieving the objective	Harmful to achieving the objective
Of Internal Origin (From the Organization's Attributes)	**Strengths**	**Weaknesses**
Of External Origin (From the Business Environment)	**Opportunities**	**Threats**

Figure 8.1—SWOT model based on a diagram from *Wikimedia Commons*

Information source: https://en.wikipedia.org/wiki/SWOT_analysis#/media/
File:SWOT_en.svg

Agree on the scope of the information space or product's requirements and the project's approximate timeline. Determine what roles the team responsible for planning, designing, and implementing the information architecture should comprise and what people should be part of that team. Only by working in close collaboration with a multidisciplinary team that is capable of helping you devise and successfully execute your IA strategy by implementing your IA design solutions can you ensure that your work provides competitive advantage and has strategic impact. [11, 12, 13]

Synthesizing UX research findings

"Personas ... are user archetypes.... Each persona represents a set of behavior patterns and goals."—Kim Goodwin [14]

Your North Star when planning and conducting UX research should always be gaining the knowledge and understanding you need to inform your IA strategy for an information space or product. Therefore, you should begin considering your IA strategy even before your team conducts UX research. You'll work collaboratively with your team throughout research and analysis, iteratively positing, testing, refining, and finally validating your assumptions and hypotheses regarding what your IA strategy should be. You'll find yourself in the midst of synthesizing UX research findings and devising your IA strategy long before you consciously start making the transition from research to design.

Synthesizing UX research findings from the generative user research your team conducted during *Discovery*—using such methods as user interviews, user observations, contextual inquiries, diary studies, or surveys—provides a deep understanding of the target users for an information space and their wants, needs, goals, and tasks. Triangulating qualitative and quantitative data from user research enables you to model these target users by creating user profiles or personas for inclusion in your IA strategy. Your personas should accurately represent the information space's actual target users.

On the basis of your personas, you'll be able to define reliable user requirements and, ultimately, design an information space or product that meets the needs of its target users. Your insights about users' needs could also enable you to create a well-differentiated information space that provides a sustainable competitive advantage in the marketplace. Include these user requirements in your IA strategy, as well as your proposed solutions for addressing them.

Your IA strategy documentation should also indicate what types of UX research your team has already conducted, describe your team's learnings and insights from any prior research, and specify any additional types of UX research your team plans to conduct during *Design* and the later stages of the project. [15, 16, 17]

Creating personas

Limit the number of personas that you create to only those that are essential—the personas that represent *all* the key target users your team identified through user research, whose unique behavior patterns, goals, and motivations are easily distinguishable from each other. The number of personas you need depends on the scope of the information space you're designing, as well as the number of roles its target users play. If many of a small information space's users share the same role, behaviors, goals, and motivations, you'll probably need to create just a few personas. In contrast, when you're designing a large, complex, enterprise information space whose users have many different roles, you might need to create 25 or more personas. In such cases, each persona's role, behaviors, goals, and motivations would likely limit the use of the persona for the information space to only specific usage scenarios.

To create a new set of personas that enable you to communicate specific users' needs to stakeholders and engender empathy for them, follow these steps:

1. **Determine users' roles.** If your target users' usage of an information space or product would depend primarily on their role, determine what those roles are. Verify that different people share the same role by considering what tasks are characteristic of that role. Then either recruit your user-research participants by role *or*, if the user research is already complete, organize the findings by role.

2. **Identify the characteristics of each role.** These characteristics include the key behaviors, goals, and motivations that are characteristic of each role, or persona. *If* certain demographics affect users' behaviors, you should consider those as well. You should compile a list of around twenty behavioral characteristics, then organize them in various ways—for example, by placing characteristics at the opposite ends of spectrums such as the frequency or duration of people's key tasks or differences in people's attitudes *or* by considering various categories of motivations or decision-making criteria behind particular behaviors such as the reasons people perform specific tasks.

3. **Map participants' behaviors to characteristics.** You can arrange the names of individual research participants along each applicable spectrum of behavioral characteristics relative to the behaviors of the other participants, as shown in Figure 8.2. To map participants to categories such as the motivations or reasons behind their behaviors, group the names of participants who share the same motivations or reasons under the appropriate categories, as shown in Figure 8.3. If *all* participants share a particular behavioral characteristic, it doesn't provide any differentiation between them so you can remove it from your list of behavioral characteristics.

4. **Identify potential behavioral patterns.** Begin by looking at all the behavioral characteristics together and noticing whether the names of multiple participants often appear together because they share similar behavioral characteristics and attitudes. Circle these groupings. If similar groupings of participants appear in more than a third of the characteristics, you might have identified a pattern comprising the behavioral characteristics of a particular role, or persona. Hypothesize possible explanations for the relationships between these characteristics—for example, a causal relationship, a conditional relationship, or a coincidental relationship between the patterns that you're observing. Once you're confident that you've identified a behavioral pattern, describe the characteristics of that pattern. Continue this process until you can't find any more patterns. In the same way, describe behavioral patterns for all of the roles, or personas, you identified earlier.

5. **Identify personas' goals.** Define the key goals that correspond to the specific behavioral patterns you've identified. Ultimately, by leveraging the findings from user research, you should be able to identify several goals for each role, or persona. Consider primarily two types of goals:

    -   *Accomplishment goals*, which are goals that an information space or product could help a persona fulfill

    -   *Experience goals*, which describe how a persona would want to feel when using an information space or product

    People also have basic human goals and life goals, which are not usually as relevant in the creation of personas. To clarify each persona's goals, ask *why* they need to accomplish each goal. Try to define two or three accomplishment goals and an experience goal for each persona. Express each goal clearly and concisely, using a persona's own words whenever possible, and beginning each definition with a verb.

6. **Differentiate the personas.** To ensure that the personas are distinctive and, thus, easier to remember and use, add more details to flesh out the personas and make them feel more real. Once you've identified the key differentiators for each persona, describe any other behaviors or characteristics that, although they did not play an essential part in defining the patterns that are core to a persona's identity, would be compatible with them. Describe the personas' current relevant activities, painpoints, contexts of use, capabilities, feelings, attitudes, interactions with other people, and demographics such as gender, age, and degree of familiarity with technology. Give each persona a distinctive, realistic name.

7. **Ensure that your set of personas is comprehensive.** Now that you've clearly differentiated your personas from one another, consider whether this set of personas adequately describes *all* the target users that your team has identified through user research. Also, consider whether you need to create *customer personas* or *served personas*. *Customer personas* represent the needs of the customers who purchase, but do *not* use, an information space or product. *Served personas* represent the people an organization's customer-service representatives serve when referring to an information space. To eliminate confusion among stakeholders who have become fixated on a type of user who is *not* generally representative of an information space's users and, thus, could lead design astray, you might need to define *negative personas* to describe these users.

8. **Prioritize the personas for each role.** Define the *primary persona* for each role, which represents the main target user for the role, and focus your initial design solution on meeting that user's needs. In choosing the primary persona for a role, consider the design and business impacts of focusing on each persona for that role. By meeting the primary user's needs, you'll also address the majority of the needs of the *secondary personas* for that role, which are similar, but have some minor differences. *Supplemental personas* do not have unique needs and exist primarily to satisfy the demands of stakeholders, so there is no need to consider them during design. Once you've created your design solution for the primary persona, you can incorporate minor changes that satisfy the needs of the secondary personas, as long as those changes don't detract from the user experience of the primary persona.

9. **Develop each individual persona.** When you're creating each of the personas in a set, be sure to focus *only* on behaviors and characteristics that are relevant to a persona's use of the particular information space, product, or section you're designing. A persona should comprise a name, a portrait, the persona's key differentiators—including key goals, behaviors, and attitudes—perhaps an illustrative quotation from your qualitative user research, key business objectives for the persona, personal information such as demographics or personal computing devices, and a one- or two-page, narrative description of the persona's behaviors. You might also include usage scenarios that describe a persona's interactions with the information space or product. Through effective storytelling, your personas and scenarios will come to life and elicit an emotional response from your teammates.

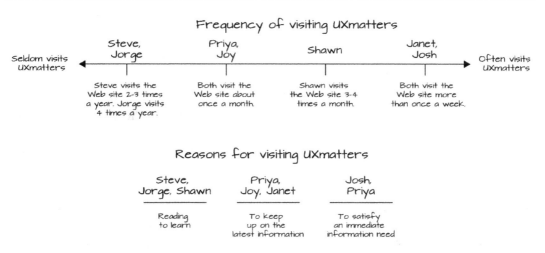

Figure 8.2—Mapping participants along a spectrum of behaviors

(Inspired by a diagram in Kim Goodwin's *Designing for the Digital Age: How to Create Human-Centered Products and Services*)

Figure 8.3—Mapping shared motivations behind participants' behaviors

(Inspired by a diagram in Kim Goodwin's *Designing for the Digital Age: How to Create Human-Centered Products and Services*)

Once you've created a set of personas for an information space or product, you'll be able to reuse them for its subsequent versions. Only when the target users of the information space or their user requirements change significantly would you need to revise or expand your set of personas. [18, 19]

Writing usage scenarios

"Scenarios put personas in motion. They are the stories of how a persona interacts with a Web site…. The persona is the character, and a scenario is a plot."—Steve Mulder, with Ziv Yaar [20]

"A scenario describes the future, not the present." —Kim Goodwin [21]

Usage scenarios tell stories about how people in particular roles would use an information space or product that does not yet exist. To fully realize the value of personas, you should also write usage scenarios that narrate their stories. Usage scenarios describe what certain target users would actually do to achieve their key goals when using a proposed information space.

Writing usage scenarios helps you to focus principally on envisioning an information space that would meet your personas' needs rather than letting business or technical constraints limit your thinking too early in the design process. By keeping your vision of the information space in the realm of the ideal and focusing on addressing unique user needs, you'll open yourself to opportunities for innovation.

Your usage scenarios must clearly communicate your vision for the information space to your team and other stakeholders. Your skillful storytelling can make reading the scenarios an engaging experience for them. Usage scenarios provide step-by-step narratives about how specific personas would employ the information space to accomplish particular tasks, but stay focused on their behaviors, *not* interactions with a specific user interface. These scenarios must be easy to read and comprehend and have a clear beginning and end.

Write a realistic, high-level scenario for each important user interaction that would take place within an information space—especially for interactions that are core to the main purpose for the existence of that information space or for tasks that no information space has previously supported. Developing scenarios can guide your conceptualization of the information space and ensure that your design solution covers all the interactions that are essential to your target users' achieving their goals and satisfying their information needs.

When writing scenarios, you'll focus primarily on describing the interactions of the primary persona representing each role. You should write at least one scenario for each primary persona and would likely need to write several scenarios, representing the persona's key interactions or tasks. Making the persona the main protagonist of your narratives, start with the persona's most important task, then write a scenario for each of that persona's key interactions with the information space, as follows:

1. **Establish the context.** Describe the broader context in which a persona's interactions occur. What specific event or need has prompted the persona to engage with the information space? In what situation did the persona realize that the need existed? Are any other people involved in the interaction?

2. **Describe the persona's goals and motivations.** What does the persona want to accomplish by engaging with the information space? Is there more than one goal the information space would enable the persona to achieve? What are the principal motivations behind each of the persona's goals?

3. **Analyze the persona's tasks.** What tasks would the information space enable the persona to accomplish? How do these tasks interrelate? What is the optimal workflow for each individual task? Should complex workflows comprise several interrelated tasks? In the abstract, what sequence of steps would the persona need to take to complete each task?

4. **Identify a happy-path scenario for each important task.** Describe a realistic, but somewhat idealized user experience for each of the persona's tasks, from beginning to end. How might the persona initiate the interaction? Outline each step the persona would take on the information space and the system's apparent responses to each action.

5. **Explore obstacles the persona might encounter.** Consider whether business or technical constraints exist that might necessitate making minor modifications to the scenario. What obstacles could prevent the persona's successfully completing the task or finding the desired information? How could the persona overcome them? How might specific challenges complicate the usage scenario? What constraints might exist? What is the persona's emotional reaction to the experience?

6. **Explore more complex variants of the task.** What additional possible outcomes might the persona hope to achieve in completing this task? What decisions would the persona need to make throughout the interaction and how might those decisions change the scenario's key pathway?

7. **Determine what success means.** What would constitute a successful outcome at the completion of the persona's task? How would the persona achieve that desired outcome? What steps would the persona take? What impact would the successful completion of the task have on the persona?

Writing usage scenarios helps you to evolve your vision for an information space or product from an abstract concept to specific requirements that define what your team should design and build. Throughout this process, your vision progressively becomes more concrete. You can also leverage your scenarios in devising test tasks for usability studies. [22, 23, 24, 25]

Understanding an organization's content

"Organizational goals, real-world resources, who your users are, what your users want, competitor activities, and many other factors all affect whether ... your Web content will be successful."—Kristina Halvorson [26]

Before defining an IA strategy for an information space or product, you must first analyze and understand the content that already exists—including its scope, the topics it covers, and the types of content—any gaps in that content, and what content the organization needs develop or acquire to fill those gaps.

If you're working in partnership with a content strategist, you'll be able to obtain much of the information you need about the organization's current and planned content from your colleague. Otherwise, you'll likely need to take responsibility for gathering that information yourself by interviewing your team's content managers, content creators, and other content owners. To accelerate your information-gathering process, you should consider facilitating a content-strategy meeting, bringing together five to seven key content owners from a variety of disciplines for a discussion about the information space's content.

In the section "Content-owner interviews," in Chapter 6, *Understanding and Structuring Content*, you'll find a long list of the questions to which you might need to get answers during your content-owner interviews or content-strategy meetings, including questions about planning an information space's content, assessing an organization's existing content, creating new content, and learning about the organization's content-development and content-management processes. In addition to content-owner interviews, Chapter 6 also describes a variety of other content-analysis methods in depth, including how to conduct content inventories, content audits, content mapping, and competitive content analyses. These methods enable you to gain a deep understanding of an information space's content that can inform the definition of an effective IA strategy and drive the design of a successful information architecture.

Once you fully understand the scope of the organization's existing content, the topics and types of content that an information space already comprises, the content's meaning and value to target users, its business value to the organization, what additional content your team needs to create or acquire, and how to create a coherent structure for the information space, you'll also need to determine how best to structure the information space's content on its individual pages and specific types of page templates.

Content modeling focuses on the design of detailed page structures by decomposing the pages' content into chunks and determining the order in which to sequence those chunks. The content objects on any given page can be static, dynamic, or a combination of the two. A content model's key elements comprise the information space's content types, content components, and access structures. You'll find more information about modeling content in the section "Bottom-up information architecture," later in this chapter.

If an organization has not yet created all the content an information space would require, help your team's content strategist to plan what content the content team should create. However, if there is no content strategist on your team, you might need to take on some of the responsibilities of that role as well to ensure that a viable content strategy exists and the content team can move forward with developing the necessary content for the information space. Plus, as you work on your IA strategy, you might discover gaps in the organization's content. Be sure to share this information with your content team.

As your organization identifies evolving content requirements for an information space, you must frequently realign your IA strategy with the organization's business strategy, and your information architecture must continually adapt to accommodate those needs. In fact, an organization's business strategy, content strategy, IA strategy, content development, and information architecture typically evolve in parallel with one another. Aligning the work of all these disciplines often introduces opportunities for improvement. [27, 28, 29]

Learning about implementation technology

When devising a strategy for the design of an information architecture, you must consider what technologies should provide the foundation for its implementation and management.

Designing an information architecture for an information space or product requires a basic understanding of the technologies underlying the Web or mobile devices and especially the specific technologies that an organization is using to implement their information space or product. These technologies present opportunities *and* impose constraints on the designs you can implement. So, before you devise your design strategy for an information space or product, you should first learn what technologies the organization is planning to use in implementing it—particularly what content-management system (CMS) the organization is already using or is planning to adopt and the capabilities it does and does *not* support.

Jim Kalbach recommends learning about the platforms on which you want an information space or product to run, as well as the backend and frontend technologies an organization is using to implement it. Ask the developers on your team to explain how the system works. When you're evaluating how these technologies might impact an information space or product, consider the following questions:

- On what platforms should it run? If on the Web, in what browsers and on desktop computers, mobile phones, tablets, or other types of digital devices? On what mobile operating systems: iOS, Android, or another mobile operating system?

- On what screen sizes should it display properly?

- In what ways can a user interact with it?

- How should you design the navigation system to function well on the various platforms you want to support? Will these different platforms require different navigation designs?

- How can you ensure the cross-platform compatibility of your designs?

- How would your design decisions impact its performance?

- Can you structure the database as necessary?

- Can you integrate content from different databases?

- Does the CMS manage hierarchical information, well-structured content from a database, or both?

- Does the CMS support content reuse?

- Does the CMS support personalization?

- What other features does the CMS offer that would help you to implement your preferred design?

- Would the CMS's lack of certain features prevent your achieving the design you want?

- Does the system support password protection?

- Does the system support user profiles?

- Do any specific features you want to include require knowing the user's identity?

- Does the system support customization?

- Do your target platforms support JavaScript and, thus, the interactivity you want to include in your design?

- Would your preferred design require any plug-ins?

- How would your design choices affect search-engine optimization (SEO)?

Depending on what you learn, you might find that you won't be able to implement certain functionality or design solutions using the existing technologies. On the other hand, what you discover about the system's capabilities might suggest features that you would not otherwise have thought to incorporate in your design strategy. [30, 31]

Considering existing versus new technologies

While, in an ideal universe, your organization's information technology (IT) or software-development team would develop or acquire whatever technologies would best support your design strategy, that is not always practicable—whether because of time, financial, or staffing constraints. So, if you're working on a new information space or product or a major revision of one, talk with your developers early on to let them know what capabilities your proposed design strategy would ideally require and learn about what technologies they're already planning to implement that could support it. They might be able to accommodate your design strategy to some extent, beyond what they had originally planned. Plus, you might be able to prevent their imposing unnecessary constraints on the design solutions whose implementation is feasible. They might even be planning to create new capabilities that you weren't previously aware of, but would want to integrate into your design strategy.

On most projects, you'll probably inherit prior technology choices that have already been implemented, which your design strategy must accommodate. So, when you're beginning your work on an IA strategy, you must learn about and assess the viability of any existing technologies. Meet with developers to discuss those technologies, as well as any that may currently be under development, then design within their constraints. Doing so ensures that your design strategy is feasible, and developers will be able to implement your information architecture as designed.

Even though your IA strategy might have to meet most of the constraints that the existing technologies impose, if their use would prevent the project from achieving key business goals or would inflict serious usability problems on users, you might be able to persuade your team either to acquire more capable technologies and integrate them into the system or to develop technologies that would better meet the project's needs.

Your IA strategy should describe the existing and new technologies that you plan to leverage in developing and managing the information architecture—for example, autoclassification tools; content-management systems, with capabilities that support the use of templates or content personalization or reuse; recommendation engines; or search systems. It should also identify opportunities for developing new tools and capabilities that would provide a competitive advantage. [32, 33]

Looking at the big picture

To devise an effective IA strategy for an information space or product, you must balance your goal of meeting your target users' needs with broader organizational goals such as driving business value, organizing new and existing content to maximize the benefits it provides to users and the overall organization, and leveraging the organization's preferred technologies in implementing the information architecture. You need to consider the big picture.

When you're developing an IA strategy, just exactly what the *big picture* comprehends depends on the scope of the IA work you're doing. An *IA project* whose focus is the design or redesign of a particular information space or product might have a relatively limited scope and duration—ranging from perhaps as little as six weeks to about eighteen months. In contrast, an *IA program* typically has a much broader scope, so its duration could be many months or even several years. Examples of IA programs include the following:

- Establishing an IA practice within an organization—as either a centralized function or a number of autonomous, dedicated teams

- Creating design standards and guidelines that apply to *all* of an organization's information spaces or products, for either one or more particular platforms or *all* platforms

- Creating a design system for use across *all* of an organization's information spaces or products, either on one or more particular platforms *or* across *all* platforms

- Designing a family of related information spaces or products

- Designing a cross-platform information space or product

Ideally, the information architects who establish the foundation of an organization's IA practice, including its design standards or design systems, or who lead major cross-platform or product-family design programs should be part of an in-house IA team. However, in less mature organizations that do not yet have a well-established IA practice, consultants often take responsibility for such work. In today's world of digital transformation, an information space can differentiate an organization from its competitors. Therefore, an organization can achieve competitive advantage by building an IA team in house and establishing a mature IA practice. [34, 35]

Other IA strategy concerns

At some point during a project's *Discovery* Phase, you'll have gained sufficient knowledge and understanding of an organization's business context, an information space or product's existing or potential users, and the organization's existing and planned content, as well as the big picture for the project to begin shifting your attention to the concerns of defining your IA strategy. You'll start discussing your ideas for structuring an information space and labeling its navigation systems. Through discussions and collaborative-ideation sessions with your team, you'll iteratively propose, test, refine, and validate your assumptions and hypotheses for your IA strategy.

To define an optimal IA strategy for an information space or product, you must address the common concerns of information architecture, including the following:

- Balancing top-down and bottom-up information architecture

- Factoring SEO into IA strategy

- Implementing and maintaining an information architecture [36]

Balancing top-down and bottom-up information architecture

When devising an IA strategy within a particular business context, there are two approaches you should take to information architecture, as follows:

- **Top-down information architecture**—This approach ensures the effectiveness of an information space's overall structure and navigation system.

- **Bottom-up information architecture**—This approach ensures the usefulness of an information space's content, as well as its supplementary navigation aids.

For an information space to optimally satisfy the user's information-seeking and browsing needs, you must balance top-down and bottom-up approaches to information architecture, each of which has its characteristic methods. Depending on the information architecture's current state, you may need to emphasize one IA approach *or* the other to bring them into balance. For example, if entrenched stakeholders have already taken ownership of the top-down information architecture and defined the primary information hierarchy, focusing on the bottom-up information architecture might be your only opportunity to add value. Your IA strategy should describe a holistic solution that balances top-down *and* bottom-up information architecture.

Top-down information architecture

The focus of a top-down approach to information architecture is on creating a home page, key sections' main pages, and landing pages that anticipate and serve visitors' immediate needs. When visitors land on an information space's home page, key sections' main pages, or one of its landing pages, they need to be able to figure out where they are and where to go to find the information they need. A top-down information architecture should help them find their way.

First, you should ascertain what organization schemes would be useful in structuring the information space's content, then determine which of these schemes should provide the basis for organizing the entire information space. Start sketching out the information space or product's information hierarchy. Clearly expressing that information hierarchy through primary, or *global*, and secondary, or *local*, navigation systems and the effective labeling of navigation links is key to enabling people to find the information they need.

Creating an effective top-down information architecture for an information space requires that an information architect ascertain, on the basis of prior user research, what specific questions visitors would most commonly have when they arrive on its home page, sections' main pages, or landing pages. The answers to these questions inform the design of the information space's global and local navigation systems. For example, when visitors go to an information space's home page, they might have the following questions:

- Where am I and what can I do here?
- What is the purpose of this information space?

- What tasks can I accomplish here?

- Does this information space provide the information I need?

- How is this information space organized?

- How can I find the information I need?

- Should I search for or navigate to the information I need?

- How do I search? What search terms should I use?

- How do I navigate? What do the labels of the navigation links mean?

- Can I create an account and log in to the information space?

- What is the organization's contact information?

- How can I get help?

- How can I contact a person at the organization?

- Does the information space provide a chatbot?

Visitors might have similar questions when they arrive on the main page for a specific section of an information space. For example, on Apple's Web site, customers can shop for specific categories of products on the following product-category pages:

- Mac

- iPad

- iPhone

- Watch

- AirPods

- TV & Home

- Only on Apple

- Accessories

For customers visiting one of these product-category pages, many of their questions would pertain to the products themselves, but they might also need answers to basic questions. Their questions could include the following:

- Does this page provide the information I need?

- How can I find the information I need?

- How do these products compare to one another?

- How much do these products cost?

- What accessories are available for these products?
- How can I get help?

Bottom-up information architecture

The bottom-up approach to information architecture focuses on defining the structure of an information space's pages in detail, or *content modeling*. Ensuring that the content objects, or *chunks* of content, on pages conform to a well-designed structure and sequence greatly eases visitors' information-seeking and browsing tasks.

When visitors arrive on a particular page deep within an information space—because they either clicked a link *or* conducted a Web search, then clicked a link in the search results—they must be able to figure out where they are, how to identify the information they need, and where they should go next. Effective contextual and supplementary navigation systems are essential elements of a successful bottom-up information architecture.

Working collaboratively with authors, content managers, and other content owners, identify all the types of content objects that currently reside on the information space, as well as any the content team needs to create to fill gaps in the existing content. Also, define descriptive, structural, and administrative metadata fields for indexers to use in describing the content objects. These metadata fields could apply globally, locally—that is, only to content objects within a particular section—or only to specific types of content objects.

To facilitate information seeking and browsing, you must create effective page structures that enable visitors to answer questions such as the following:

- Where am I and what can I do here?
- What is the purpose of this page?
- What information is on this page?
- What content should I look at next?

For example, on Allrecipes, each recipe page consistently has the same structure and comprises the following chunks of information:

- The title
- Rating and links to reviews and photos
- Lede
- Byline
- Photos of the dish
- Overview

- Ingredients

- Buy ingredients

- Directions

- Cook's note

- Nutrition facts

- Reviews

Such well-structured information makes this information space easy to use. [37]

Factoring SEO into IA strategy

"Search-engine optimization is the process of creating, editing, organizing, and delivering content—including metadata—to increase its potential relevance to specific keywords on Web and site search engines."—Kristina Halvorson [38]

"A sound information architecture for a Web site provides the essential foundation for search engine optimization."—Nate Davis [39]

Search-engine optimization (SEO) contributes to the findability of a Web site's content by both Web-search engines and a site's internal search engine. There are two important SEO ranking factors that impact a Web page's findability: the performance of the keywords in the page's title, section headings, and body text and the readability of the page's content. *Keywords* are the words that people typically use in seeking certain information. If your content's terminology is meaningful to your target users, they'll probably type those same words into search engines.

Of course, you must ensure that your content is findable without resorting to the use of any of the dark patterns that, in the past, gave SEO a bad reputation—such as keyword stuffing. Your main focus should be on the *findability* of the information on an information space or product. Therefore, your page titles, section headings, and content must be meaningful to people who are seeking information—and you should use semantic markup on your pages to make them comprehensible to search engines.

If the categories and other metadata that you apply to specific pages use terms that are meaningful to the people who are seeking the information on those pages, they're probably the same terms they would use in search engines.

In his *UXmatters* article, "Putting SEO in Its Place: An Information Architecture Strategy," Nate Davis refers to some different approaches for improving the findability of pages in Web-search results, as follows:

- **"SEO-friendly content"**—This content incorporates meaningful keywords in page titles, section headings, and the alt-text that describes images, videos, and audio recordings.

- **"SEO-friendly programming"**—This requires using semantic markup such as H1, H2, H3, H4, and alt-text for images and other content objects, as well as creating human-readable Web addresses that use consistent structures.

- **"SEO-friendly site architecture"**—This requires the creation of meaningful navigation and contextual links and, thus, relationships between the pages in an information space, as well as the consistent use of metadata on pages.

Thus, information architects can help ensure the findability of content by people and by Web-search and site-search engines. [40, 41, 42, 43]

Implementing and maintaining an information architecture

Excluding information spaces that have a very limited scope and change infrequently, most information spaces require some sort of content-management system (CMS) in which to create, publish, manage, and maintain their content *and* information architecture. The organization's decision regarding whether to implement or acquire a CMS depends on both the complexity of its information spaces or products and their broader business context. The following factors contribute to this complexity:

- The number of authors who contribute content to the information space

- The number of information sources from which the organization acquires content

- The numbers of existing and planned pages, page components, and reusable components

- The numbers of different types of components and content formats

- The extent of the changes the organization is planning to make to the content and the information architecture over time

- What degree of personalization they're implementing on an information space or product

- The numbers of information spaces and products an organization has created or is planning to create

- The number of redesigns they're planning

- The number of different platforms on which the organization is publishing content

- Whether the creation and publication of content is centralized or decentralized

- Whether the organization is adopting an existing CMS or developing its own

Depending on the scope of an IA project, an information architect or IA team might need to design one or several controlled vocabularies for implementation in the CMS. The content-management team, or whoever is responsible for content management, implements and maintains the controlled vocabulary by configuring the CMS. Implementing the controlled vocabulary in the CMS enables content authors and other indexers to categorize content objects by choosing preferred terms in the CMS and, thus, apply the proper metadata to content objects. In doing so, they control the flow of content into the CMS over time.

It is essential that those who are responsible for content management maintain the CMS's configuration properly over time—and thus, the information space's controlled vocabulary and information architecture. Plus, consistency in the use of a controlled vocabulary when tagging content objects with metadata tends to degrade over time, as different authors and indexers use tags in slightly different ways. So a content manager should periodically assess the tagging of contact objects and improve its consistency.

Your IA strategy should comprehend the tools your team uses, the roles of your team members and their purview, and what process they should follow. [44, 45, 46, 47]

Envisioning, communicating, and validating your conceptual models

Envisioning conceptual models for an information space or product *and* communicating your vision are two complementary aspects of the same process. Only by expressing your design concepts for an information architecture through some physical medium can you capture, refine, and expand on your ideas for the information space's organizational structure and labeling, navigation, and search systems. You should also validate your conceptual models, preferably by testing them with users.

Envisioning and communicating your conceptual models

There are many useful approaches to envisioning and communicating your conceptual models for an information space, including the following:

- **Usage scenarios**—When conveying your IA strategy to your stakeholders, you can create usage scenarios to express the ways in which you're addressing specific user needs through your strategy. Storytelling is an engaging means of conveying your design vision to stakeholders. Usage scenarios provide an excellent means of demonstrating how users with specific needs and characteristics might employ a future information space, as I explained in the sections "Creating personas" and "Writing usage scenarios," earlier in this chapter.

- **Storyboards**—Once you've started envisioning user interfaces at a later stage of the ideation process, drawing interactive storyboards, similar to the one shown in Figure 8.4, serves a similar purpose to writing usage scenarios. First, a *storyboard* should depict the most common pathway through a core user interaction on an information space, as a series of interlinked pages. Then add alternative pathways for any important conditions. Stakeholders may find storyboards more engaging and easier to understand than scenarios.

- **Metaphors**—The use of metaphors can help your stakeholders to see things in different ways. Since the introduction of the Apple Macintosh personal computer in 1984, designers have appreciated the usefulness of metaphors in designing digital systems. By devising possible metaphors for concepts, designers can leverage the familiar—people's existing knowledge of the world—to communicate new ideas in creative ways. In design, important types of metaphors include *organizational metaphors*, which enable users to transfer their knowledge of the organization of one system to another and, thus, are very important in information architecture; *functional metaphors*, which enable users to apply their knowledge of their real-world tasks to digital systems; and *visual metaphors*, which use people's familiarity with signage in the physical world to convey similar concepts through iconography and color. Although metaphor exploration can stimulate your creativity during ideation, you won't be able to apply most of the ideas the process generates.

- **Diagrams**—You can explain abstract concepts by creating diagrams rather than through words. The simplicity and clarity of conceptual diagrams often result in more powerful ways of conveying ideas to your stakeholders that have greater strategic impact. For example, *topic mapping* lets you explore what topics you should include within the scope of an information space's content and how they should interrelate. When drawing a topic map, start in the middle with the main topic, circle it, then cluster related topics around it and draw lines between them and the main topic. Continue this process with additional key topics, until you can't think of any more topics. Figure 8.5 shows an example of a topic map.

- **Site maps**—Creating a *site map* is an easy way of depicting an information space's structure and navigation paths in detail, in relative isolation from its user interface. During ideation and the early stages of design, you can create a site map either as a simple hierarchical list or a rough sketch; then as your design process progresses, a highly refined diagram. A site map can concisely communicate just specific parts of an information architecture or the full breadth and depth of an information space's structure, navigation paths, and the relationships between pages and other discrete content objects such as PDFs. Your site maps inform the design of an information space's labeling and navigation systems. We'll explore how to create site maps in detail in Chapter 10, *Designing and Mapping an Information Architecture*. (Some information architects refer to site maps as *blueprints*. However, having created blueprints for physical structures earlier in my life, using this term to describe site maps has always seemed like a misnomer to me. Wireframes are much more similar to architectural blueprints.)

- **Wireframes**—During design, an information architect creates wireframes that depict the logical structure of an information space's individual pages, laying out and grouping the discrete components that each page comprises—such as the information space's navigation and search systems. They render the structure of these pages only in outlines, without any visual-design details, but their visual hierarchy informs visual-interface design. They represent the labeling and content models for specific types of pages. Create wireframes for an information space or product's key pages such as its home page, sections' main pages, and important category pages; any page types that require templates such as content pages and search-results pages; and pages that are either unique or complex such as highly interactive pages. We'll consider wireframing in depth in Chapter 10, *Designing and Mapping an Information Architecture*. (Some information architects call wireframes *page schematics*, but I prefer the more accessible and now more common term *wireframes*.)

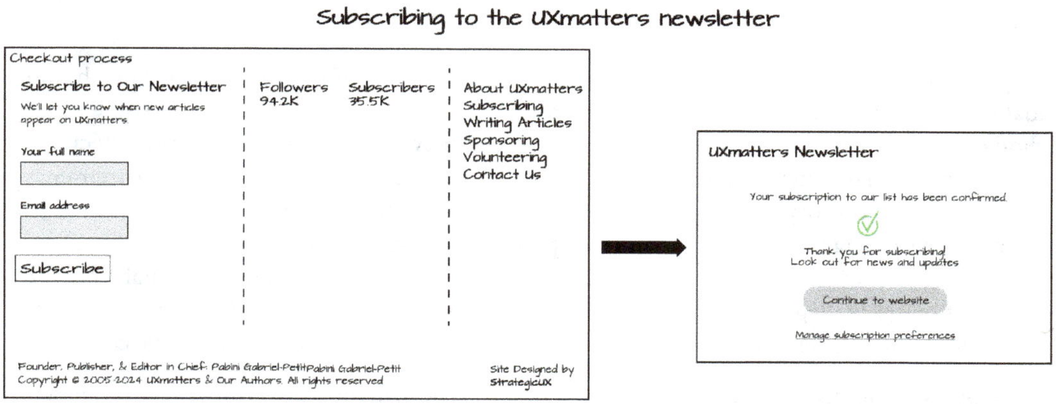

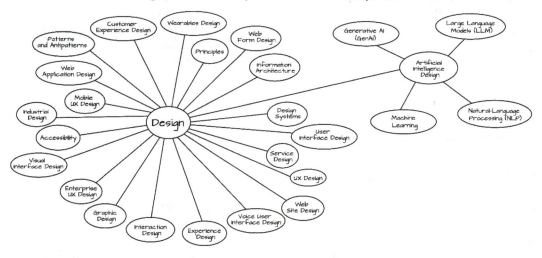

Figure 8.4—An excerpt from a *UXmatters* storyboard

Figure 8.5—An example of a topic map

(Inspired by a topic map in Christina Wodtke's *Information Architecture: Blueprints for the Web*)

Throughout ideation and, as you develop and refine your IA strategy for an information space or product, you'll probably create sketches, conceptual diagrams, site maps, and wireframes to capture your conceptual models and design ideas. IA strategy bridges user research and design, so you'll express your conceptual models and high-level design ideas for an information space's organizational structure, labeling scheme, navigation system, and search system through such visualizations. Once you've created your conceptual design deliverables for an information space, you should test your models and design solutions with users.

Validating your conceptual models

Once you've created your conceptual design deliverables for an information space, you should validate your conceptual models by conducting usability testing, expert reviews, or some other method of evaluative research. By testing sitemaps, you can evaluate the information architecture's overall organizational structure and validate both its categories and their labels. Testing wireframes or interactive Web prototypes lets you evaluate your navigation system within the context of an information space's pages, and you can learn how participants might use an information architecture. By conducting tests with a Web prototype, you can observe how participants would actually browse the information space. Regardless of how you're evaluating your conceptual models, ask participants to try to choose the appropriate navigation links to complete certain tasks or find specific content.

Even though your focus is on testing an information space's structure, categories, and navigation system, you cannot completely avoid the impacts of user-interface (UI) design choices on your test results. The organizational schemes you've chosen to use in structuring the content are themselves UI design choices. Plus, participants' ability to select categories successfully depends on how much context you provide. For example, when you're testing an information hierarchy, providing examples of some lower-level categories increases the information scent of the higher-level categories. It's also possible that your learnings about the information architecture might differ when you conduct testing within the more realistic context of a Web prototype.

Through your learnings from the evaluation of your IA strategy and conceptual models for an information architecture, you can begin to identify issues with your conceptual models, iteratively improve them, and ultimately, validate that you've devised a sound IA strategy. [48, 49, 50, 51]

Documenting and presenting your IA strategy

Ideally, you should clearly communicate your IA strategy by preparing both detailed strategy documentation and a presentation through which you can convey your IA strategy to your stakeholders. The purpose of this documentation is to codify your ideas and analysis and persuade your teammates and other stakeholders to buy into your IA strategy. The needs of your particular project and its scheduling constraints should dictate the degree to which you endeavor to create a highly detailed, polished document.

First, use your visualizations of abstract concepts and conceptual-design solutions to convey the big picture, then provide the details within the text of your document. Your *IA Strategy Document* could comprise the following sections:

- **Overview**—This section provides a summary of the essential ideas from your IA strategy, for busy executives who don't have time to read the entire document. It should identify the key issues your information architecture must address, provide your recommendations for solving them, explain how solving them would advance the organization's business strategy, and describe the expected business impacts.

- **Mission**—This section comprises your organization's mission statement for the information space or product, including its primary strategic goal.

- **Vision**—This section concisely communicates your vision for the information space or product, including your future goals for its information architecture.

- **UX Research**—This section describes the UX research your team has already conducted, as well as the research you're planning to conduct during *Design* .

- **Target Audiences**—In this section, you'll present the personas, or user profiles, that you've identified through your user research, which describe the roles they play within the information space's context of use, and describe key usage scenarios for each persona. These personas should accurately represent your actual target users. On the basis of these personas, you can define user requirements that enable you to design an information space that meets users' needs.

- **Learnings from Research**—This section describes the user requirements that you've identified through your UX research—especially your competitive analyses and generative user research—and your content analysis. Organize your learnings in subsections that focus on particular types of issues such as organization, labeling, global navigation, local navigation, search, and content. Create a table comprising the following columns:

  - **Source**—the specific means by which you've identified an issue

  - **Issue Identified**—a description of an issue that is strategically important—for example, a significant painpoint that research participants or actual users have encountered

  - **Analysis**—the conclusions that you've derived from your analysis

  - **IA Strategy**—the strategic implications of your learnings, including their impacts on the business and on users

  - **Requirements**—the high-level user requirements that you've defined to remedy the issue and meet users' needs

- **Recommendations**—This section defines your IA strategy for the information space in much greater detail, including explicit strategic recommendations for its information architecture. At this point, your focus on IA strategy begins to shift to design, as you transition to creating design solutions for strategically important issues. Organize your IA strategy recommendations into broader subsections that focus on design solutions for specific types of strategic issues—for example, issues pertaining to organization such as applicable structural patterns and organizational schemes, controlled vocabularies and labeling systems, global and local navigation systems, the search system, content requirements that address gaps or specify content reuse, content modeling for specific page types, and any customization or personalization capabilities. Then, within those broader subsections, devote a separate section to each individual design solution. You should include site maps and wireframes to illustrate specific design solutions, as necessary.

- **Content Management**—In this section, describe the various ways in which your IA strategy would affect content management, including its processes, workflows, roles, repositories, and systems; the impacts of your controlled vocabulary on style guidelines, metadata, and indexing; and the need to revise or create new templates that are based on your content models for specific types of pages.

- **IA Project Plan**—This section outlines your plan for realizing your IA strategy by designing, implementing, and testing an information space or product's information architecture. It should cover the specific tasks that are necessary to accomplish the goals you've defined in your IA strategy, who or what team is responsible for those tasks, and the deliverables they'll create; the project's scope, timeline, and budget; and what dependencies exist that could affect the timeline and budget. To both address near-term goals and critical issues *and* design a holistic solution that can transform a digital information space, you might want to develop short-term *and* long-term plans. [52]

Presenting your IA strategy

Ideally, you've involved your teammates and other stakeholders in UX research and analysis, strategy workshops and informal discussions about IA strategy, and design-ideation sessions so they already know about and understand much of what your *IA Strategy Document* explains and even feel a sense of ownership for the decisions it outlines. Many people have difficulty finding the time to review a long strategy document, but looking to see what impact they've had on the document can be a strong motivating force, as can their desire to ensure that your strategy won't have any untoward impacts on their work.

However, the only way you can ensure that your key stakeholders—especially busy executives—actually consume your IA strategy and give it the consideration it deserves is to deliver one or more strategy presentations to them. Presenting your IA strategy to your stakeholders gives you the opportunity to tailor your message to your audiences' needs, answer their questions, and participate in discussions that enable them to understand your recommendations—and, thus, persuade them to buy into and assimilate your strategy.

Accomplishing anything within an organization requires a unified team, so if you want to ensure that you can realize your strategic vision for an information architecture and get your team to implement your specific recommendations, you must get leadership's buy-in and backing. You must mobilize a team that is aligned behind, excited about, and capable of realizing your vision. [53]

Summary

In this chapter, we considered the various concerns of IA strategy—especially understanding and aligning with business strategy, synthesizing UX research findings, understanding the organization's content, learning about implementation technology, and looking at the big IA strategy picture. This chapter also covered how to envision, communicate, and validate your conceptual models, which are an essential part of your IA strategy, as well as documenting and presenting your IA strategy.

Next, we'll progress to *Part III: Designing Information Architectures for Digital Spaces*, starting with Chapter 9, *Labeling Information*, which describes how to design effective labeling systems, presents the characteristics of effective labels, and discusses how you can leverage search-log analytics in choosing effective labels. The chapter also covers labeling categories of information and explores some different types of labels. Plus, it describes how to construct optimal Web addresses and discusses some methods of UX research that are especially effective for testing labels with users.

References

To make it easy for readers to follow links to the references for this chapter, we've made them available on the Web: `https://github.com/PacktPublishing/Designing-Information-Architecture/tree/main/Chapter08`

Get This Book's PDF Version and Exclusive Extras

UNLOCK NOW

Scan the QR code (or go to `packtpub.com/unlock`). Search for this book by name, confirm the edition, and then follow the steps on the page.

Note: Keep your invoice handy. Purchases made directly from Packt don't require one.

Part III:
Designing Information Architectures for Digital Spaces

This part focuses on the design of information architectures for digital spaces, including the design of navigation and search systems, and comprises the following chapters:

- *Chapter 9, Labeling Information*
- *Chapter 10, Designing and Mapping an Information Architecture*
- *Chapter 11, Foundations of Navigation Design*
- *Chapter 12, Designing Navigation*
- *Chapter 13, Designing Search*

9

Labeling Information

"Content, users, and context affect all aspects of an information architecture, and this is particularly true with labels."— Peter Morville and Louis Rosenfeld [1]

Labels identify the various user-interface (UI) elements of a digital information space or product. When you're designing a digital space's information architecture, the most important labels that you need to create are those that describe its organizational structure—especially the labels for its global and local navigation systems, but also those that identify and provide structure to its pages.

To ensure the findability of the information that a digital space comprises, the labels for its navigation links must concisely and accurately describe their target destination and, thus, provide good information scent to users. For labels to be meaningful to users, they should employ the language that users would use in describing concepts that are familiar to them.

However, if you need to convey concepts that are unfamiliar to users and introduce new terms they might find confusing, you should both create clear, memorable labels *and* educate users on their meaning by providing clarifying explanations. Then, once you've defined new concepts, consistently use the same terminology for them throughout the information space to ensure that users are able to learn them. [2]

This chapter covers the following key topics:

- Designing effective labeling systems
- Attributes of effective labels
- Types of labels
- Enabling browsing and searching the Web
- Creating iconic versus textual labels
- Translating textual labels
- Discovering possible labels for categories
- Defining optimal labels for categories

- Designing labels for navigation systems and pages
- Testing labels with users

Designing effective labeling systems

Labels should be part of a well-thought-out labeling system that you apply consistently throughout an information space or product. The design of an effective labeling system is essential to communicating a digital space's information architecture. A labeling system comprises several different types of labels, each of which plays a specific role. The visual formatting of each type of label should clearly convey that label's role in relation to other types or levels of labels.

The labeling system for a digital information space must comprehend the full breadth and depth of its information architecture and clearly communicate the full scope of the space's content to its users. There should be no gaps *or* overlaps in either the information architecture *or* the labeling system. In addition to supporting a clear, comprehensive navigation system, the labeling system should convey the logical structure of specific pages or page types.

Labeling systems must continually evolve as an organization adds new content or removes obsolete content from a digital information space—especially if your team discovers that users are unable to find certain content. Conduct usability testing or analyze an information space's search logs or other analytics data to discover where your labeling system is failing users and your organization, then revise the labeling system accordingly. [3, 4]

Narrowing scope

Establishing a well-defined scope for an information space can significantly reduce the ambiguity of its labels and, thus, ensure a labeling system's effectiveness. By limiting the scope of an information space's business context, audience, and content, you can greatly increase the likelihood that users would interpret its language as you intended.

Business-context scope

Each business domain comprehends specific concepts and uses its own characteristic language—and, in many cases, jargon—to describe them. Within the scope of a particular business domain, each organization must have its own unique business strategy and, thus, goals for an information space or product that they're developing—including who would use it, under what circumstances, and for what purposes *and* what content it should comprise—as well as for its information architecture and labeling system.

Audience scope

It's essential that you identify the scope of an information space's audience—and determine whether it must accommodate multiple target audiences. When an information space has a specific, well-defined target audience that shares similar perspectives, the members of that audience are much more likely to interpret the language you use in your labels in the same way. However, the more the life experiences of your target audiences differ, the greater the likelihood that users might assume that certain words have different meanings. Therefore, if an information space has discrete target audiences—for example, the general public and people with expertise in a specific business domain or profession—you may need to devise separate labeling systems that address the needs of each key audience.

Content scope

By limiting the scope of an information space's content to only specific topics, you can make it easier for users to interpret the language you use for labels within the context of those topics.

If an information space's content must cover a broad range of topics for different audiences, create a well-defined structure comprising sections that focus on specific topics and target the needs of particular audiences. Then, you can target the content and labeling within each section to an audience that has the requisite knowledge to understand the language you're using. You might actually need to create a separate labeling system for each section of such an information space. For example, on an information space whose domain is medicine, one section might target the information needs of the general public, while the target audience for another section might be medical professionals. [5]

Attributes of effective labels

"A good label is so obvious it … doesn't make you pause (and it never makes you think.)"—Christina Wodtke [6]

Good labels should be highly descriptive and enable users to feel confident that they're on the right path to find the information they need. Ambiguous labeling causes confusion and frustration and can even prevent users from recognizing the content they need when they find it. Poor labeling choices could even destroy users' trust in your organization or offend them in some way, perhaps prompting them to flee your information space and choose that of a competitor instead. This section describes the attributes of effective labels. Many of these attributes reduce the ambiguity of labels. [7, 8]

Accessible

"The power of the Web is in its universality. Access by everyone regardless of disability is an essential aspect."—Tim Berners-Lee [9]

Good typography is an essential element of accessible labeling. Ensuring that an information space's labels are accessible to users with low vision, color-deficient vision, or other vision impairments such as farsightedness requires that you carefully consider font size and value contrast. Avoid using small font sizes. Be sure that there is sufficient value contrast between text and its background. For example, don't use light gray text on a white background, and avoid compounding that error by using small fonts.

Accessible wayfinding requires an information space to provide clear, consistent *navigation cues*—particularly clearly labeled, easy-to-use navigation links; have a logical structure and support exploration; and offer site search. It also requires consistent *orientation cues* such as page titles that clearly identify the user's current location within the information space; well-structured pages with clear subheadings; and ARIA landmarks, or *roles*, which define the sections of a page for assistive technologies.

Users with severe vision impairments must be able to transform onscreen text to enable them to read it. Satisfying this need means users with low vision must be able to increase the font size, without causing negative impacts on pages' layouts. The wide adoption of responsive Web design (RWD) has been helpful in accommodating this need. Users with low vision or color-deficient vision must be able to control the colors of text and backgrounds to ensure adequate value contrast.

Accommodating users who are blind requires support for screen readers and software that synthesizes speech, both of which have difficulty processing abbreviations and acronyms. Speech-synthesis software attempts to pronounce abbreviations and acronyms as words. To support the use of abbreviations and acronyms on Web sites, use abbreviation tags to spell out and display or pronounce the full words that they represent. Here is an example:

```
<abbr title="California">CA</abbr>
```

Modern operating systems offer many capabilities that support the needs of users who have vision, hearing, and motor impairments, enabling universal access to all users, regardless of their abilities. Plus, several accessibility standards exist for the Web, including the following:

- **Web Content Accessibility Guidelines (WCAG) 2.0**—This international standard—including WCAG 2.0, WCAG 2.1, and WCAG 2.2—describes how to make Web content, including labels, more accessible to people who have disabilities.

- **Web Content Accessibility Guidelines (WCAG) 3.0 Working Draft**—This incomplete draft of the WCAG documents updated Web accessibility standards that apply to Web content, apps, tools, publishing, and emerging technologies.

- **Web Accessibility Initiative–Accessible Rich Internet Applications (WAI-ARIA)**—This suite of standards provides greater accessibility to people who have disabilities—especially those who rely on screen readers or cannot use a mouse, trackpad, or touchscreen—by making dynamic Web content and rich Internet applications (RIAs), which have more advanced navigation systems and user-interface controls, accessible to them.

- **ARIA Authoring Practices Guide (APG)**—This guide describes how to create accessible Web applications by using ARIA accessibility semantics in implementing common Web-design patterns and widgets.

According to the WCAG 2.0, there are four essential principles at the foundation of Web accessibility, as follows:

- **Perceivable**—The presentation of content, navigation systems, and other UI elements must enable users to easily see them, listen to descriptions or narrations, navigate through keyboard or voice input, or read on a braille device. Supporting assistive technologies requires the coding of machine-readable content; providing alternative text for images and icons and descriptions of media such as charts, diagrams, illustrations, videos, and animations; providing transcripts or captions for audio content; developing semantic markup for titles, headings, and labels and style sheets that separate presentation from content; and following accessibility standards.

- **Operable**—People who are unable to use a mouse must be able to interact with navigation systems and other UI elements by using their keyboard, voice input, or assistive technologies. A well-structured information space ensures that users can navigate, find the content they're seeking, and orient themselves to their current location. Provide multiple means of finding content by both browsing and searching. Operability also requires clear labeling of navigation links, page titles, section headings, and controls; semantic markup; and sufficiently large controls. When automating the presentation of content, such as in a carousel, provide manual controls that let users override the automation to ensure that they have sufficient time to read or interact with the content. Avoid including content that could cause seizures or motion sickness—for example, flashing content, animations, or moving content—and provide controls to stop their motion.

- **Understandable**: An information space's content and user interface should be easy to understand. Ensure that all textual content is legible, clear, and consistent, especially labeling for navigation systems and interactive UI elements. Creating understandable content requires plain language, scannable sentences and paragraphs, defining unusual words; and semantic markup for *all* media, including images, animations, video, and audio. The labels, appearance, and behaviors of all UI elements should be consistent—for example, navigation elements that appear on multiple pages should reside in a consistent location on all pages. Maximize structure and minimize clutter to make content *and* user interactions easier to understand. Help users avoid errors and provide an undo capability that makes them easy to remedy when they occur. The user should always be in control; so, avoid causing unpredictable behaviors or making changes that could disorient users.

- **Robust**—Create labeling and content that are compatible with both current and future user agents, including Web browsers and assistive technologies, ensuring that they can reliably interpret and convey their meaning to *all* users. Always provide alternative content for multimedia content—including images, animations, video, and audio—and interactive UI elements.

[10, 11, 12, 13, 14, 15, 16]

Accurate

Accuracy is important in labels. Always use the correct terms for particular concepts—even if some people in your target audience refer to them using different terms—for example, obsolete terms or industry jargon that is not widely known. Using the wrong terms would make the use of an information space more difficult for the people who *do* know and use the correct terms. Through the consistent use of accurate terminology—perhaps including descriptions or definitions of specific concepts as well—you can help users learn and understand the terms the information space uses when referring to specific concepts and the correct terms to use when searching for content, enabling users to successfully find the information they need on the information space. [17]

Appropriate for the audience

Labels should use the terminology that is most appropriate for a specific target audience and the terms that the audience uses for particular concepts—as long as they're also the correct terms. If an information space includes specific content for particular target audiences, it may use different language and categories that are appropriate to each of its audiences.

For example, you should generally avoid using corporate or technical jargon in the labels on an information space for the general public. The trademarked names of popular products may be an exception to this rule. However, if you must use a trademark or introduce a term that visitors would be unlikely to know or understand, a description or a definition should accompany it. In contrast, if an information space's target audience consists of professionals who work in a particular industry or knowledge domain, use their professional terminology in its labels, including abbreviations and acronyms.

Employ a user-centered approach to determining what terminology to use. Conduct user research with members of an information space's actual target audience to learn what terminology and tone of voice they typically use, as well as each audience's reading level and perhaps their level of education in a particular knowledge domain.

Depending on the expectations of its target audience, an information space's tone of voice could be more informal or more business-like. Tone of voice could include the use of slang or more formal language, as well as the pronouns an information space uses when referring to the user. Tone of voice should reflect a brand's values. Therefore, using an inappropriate tone of voice could negatively impact a brand's credibility. [18, 19]

Clear

In deciding what labels to use, your primary goal should always be to achieve clarity. Your labels should clearly represent specific concepts, enabling an information space's users to understand and differentiate between its various concepts. On an information space for the general public, avoid the use of corporate or technical jargon, abbreviations, or acronyms with which an information space's users might be unfamiliar. Avoid choosing clever, cute, humorous, or deliberately cryptic labels in an attempt to pique users' curiosity, which can dissuade users from exploring an information space.

Plus, using witticisms, idioms, slang, abbreviations, and acronyms can be especially challenging for international audiences and non-native speakers of the language an information space uses.

The context in which any label exists can influence users' interpretation of its meaning and, thus, reduce its ambiguity. A label's context includes the other labels and text on a page, as well as any graphic elements, such as iconography, photos, animations, or illustrations, in its content. Using icons for labels or in concert with text labels can clarify the meaning of a link or other UI element for users who lack sufficient skill in an information space's language—for example, on an information space for children or international travelers.

Link labels should always clearly indicate what content users would find at their destination. In some cases, the necessity of creating longer and, thus, clearer labels might even require rethinking the design of your navigation system to accommodate them. An ambiguous or confusing link label could require a user to click the link to discover what information its destination might provide. But users are usually unwilling to click links whose labels they don't understand, so an unclear label could prevent them from finding the content they need, signing up for a service, or making a purchase. Plus, getting lost in hyperspace could be the consequence of a user's clicking the wrong link, which might cause the user to become frustrated and leave an information space, perhaps never to return. Poor link labeling could actually prevent an organization from realizing the expected return on its investment in designing, building, and marketing an information space. [20, 21]

Concise

The purpose of a label is to efficiently identify a UI element or, for a hyperlink, to briefly describe its target destination. The real goal of concision is to make labels easy to scan and comprehend. This is especially important for navigation links.

The labels in navigation systems must often communicate concepts using limited screen real estate, so should comprise only as many words as are absolutely necessary for clarity, but no fewer. All labels must clearly *and* concisely convey information, using terminology that is both meaningful and memorable to users. However, clarity is paramount. So, while you should endeavor to be concise, use as many words as are necessary to achieve clarity. For example, although the use of personal pronouns in labels can contribute to verbosity, you should use them if they're absolutely necessary for clarity. Use *Our* or *Us* in labels that must clearly refer to the organization; *Your* in labels that refer to the user. However, the overuse of personal pronouns, particularly *Our*, often results in repetition that obscures the more salient terms in labels, making the labels more difficult to distinguish from one another.

Although longer link labels, comprising seven to twelve words, could provide more information scent and, thus, result in higher rates of successful navigation, users' ability to navigate successfully diminishes when link labels are excessively long. Plus, underlined text is harder to read, especially in a long list of links where there is insufficient space between the links. For links within a page's body content, limit the link label to a key phrase and use the link's surrounding text to help provide information about the link's destination. Alternatively, you could pair a brief link label with a short description of the link's destination. [22, 23, 24, 25]

Consistent

Consistency unifies the sets of labels that a successful labeling system comprises and makes them easier for users to learn. There are three levels of consistency that you must consider when designing labels:

1. **Consistency in the use of specific labels**—Consistency in the use of specific terms in labels is paramount. Ensure the consistent usage of preferred terms for particular concepts and consistent forms of these terms, including their spelling, punctuation—for example, hyphenation or possessives—and the use of uppercase or lowercase letters. Consider consistently applying the same set of labels to the subheadings of all the pages in a particular section of an information space that have the same structure, comprising subsections that consistently have the same distinct functions.

2. **Consistency within a labeling system**—Particular attributes are characteristic of specific types of labels within a labeling system. Maintaining consistency in the visual formatting, or presentation, of specific types of labels is essential for users to differentiate various levels and types of labels and makes a labeling system easier to learn and use. The need for visual consistency includes the use of any grouping or graphic elements, the placement of labels on a page and the whitespace around them, and the use of fonts and their sizes, styles, and colors. This visual consistency makes labels easier to scan. Maintaining consistency in the style of specific types of labels is important as well. Try to use a parallel grammatical syntax across all the labels within a group of labels—for example, labels that are all nouns, all verbs, all gerunds, or all questions—*or*, if necessary, create subgroups of labels that share a parallel syntax. Use nouns or noun phrases and title case when labeling navigation links. Use verbs or verb phrases and title case when labeling action buttons or links that serve a similar purpose—such as **Read More**. Also, use title case when labeling menus, menu items, groups, tabs, page titles, column headings, and icons. *Title case* requires capitalizing all words *except* articles, conjunctions, and prepositions that consist of four or fewer letters. For subheadings, consistently use either title case *or sentence case*—which requires capitalizing *only* the first word, proper nouns, and acronyms.

3. **Consistency across information systems**—Leverage users' familiarity with certain labels whose use is common across many information spaces or specific industry domains. Be sure to employ whichever of these common labels that you choose to use consistently across an entire information space. Using such common labels makes an information space easier to use because users can leverage their prior experience with and knowledge of other information spaces.

Examples of navigation-system labels that are in common use across many different types of information spaces include the following:

- **Home**
- **About Us, About**
- **Products**
- **Services**

- **Press, News, News & Events**
- **Publications**
- **Resources**
- **Insights**
- **Support, Help**
- **Careers, Join Us**
- **Contact Us, Contact**
- **Your Account**
- **Search**
- **Site Map**

Note that a few of these common labels include personal pronouns. When a label refers to the company or organization to which an information space or product belongs, use *Our* or *Us*, as appropriate. When a label refers to the user, use *Your*. However, you should generally limit the use of personal pronouns in labels, which can contribute to verbosity. Use these personal pronouns only when necessary for clarity.

A high-quality information space must achieve consistency in its labeling. One that looks sloppy and unprofessional could fail to engender confidence in its visitors. To ensure that your information space is consistent, create or adopt a style guide that addresses *all* issues of consistency. [26, 27, 28, 29, 30]

Differentiated

"Same name, same thing; different word, different thing."—Caroline Jarrett [31]

Labels should be clear and unambiguous, and each label must be easily distinguishable from the other labels. Use unique terminology in describing different concepts and, thus, when labeling links. Most labels are part of a set or group of labels. Therefore, when choosing terms for the labels, you must ensure that they clearly differentiate the individual labels in the group from one another. Plus, the meanings of labels should *not* overlap one another. Never use the same terms for different concepts.

Repetition of the same verbiage in the labels for a long list of links makes the links difficult to distinguish from one another—for example, starting every link label in a directory of how-to documents with *How to...*. Instead, a page title or section heading should include that text and the individual link labels should consistently be specific verbs *or* gerunds such as the following:

- *Upgrade Your Operating System*
- *Upgrading Your Operating System*

Optimal labeling generally requires that you create all the labels in a group at once while focusing on adequately differentiating them from one another. If you must later add labels to a group, consider the context of the existing labels and be sure the new labels are easy to differentiate from the others in the group. Avoid changing existing labels whenever possible because doing so can be disorienting to your users. However, there are cases in which you must rethink an entire group of labels to ensure that they are clearly distinguishable from one another. For example, if the scope of information that a label represents has increased significantly or diverged, it might be necessary to split an existing category into two, giving each category a unique label.

When users are seeking information by browsing an information space, they're likely to click the first navigation link that seems to fit their need—and, thus, provides some information scent—rather than carefully evaluating all the links on a page. So, it is particularly important to clearly differentiate an information space's navigation links from one another. [32, 33]

Inclusive

Information spaces whose audience is the general public require the use of language that a broad audience of users can understand. To create inclusive labels, you should refrain from the use of plays on words, idioms, slang, corporate or professional jargon, abbreviations, and acronyms, which can be challenging for non-native speakers of an information space's language and international audiences.

Consider users who might currently be in challenging situations or contexts—for example, users in stressful situations such as those operating noisy, complex, dangerous machinery; or, users on mobile devices in bright sunlight, who would have difficulty reading low-contrast text. Also, consider the needs of users who have particular limitations such as low-literacy users, those who have temporary impairments because of an injury, or those belonging to communities who have unique perspectives and expectations, including those who belong to marginalized or disadvantaged communities. [34, 35]

Legible

Labels are essential UI elements that enable people to use and navigate an information space or product successfully. Therefore, they must be legible to users who have various levels of visual acuity. To make labels and other text easy to read, you should refrain from styling link text and other textual labels in ways that would reduce their legibility. Observe the following guidelines to achieve optimal legibility for digital typography, particularly for labels:

- **Use legible, sans serif fonts for navigation links, UI labels, and any text in smaller font sizes.** On the screen, the fine lines of serifs render less clearly, making serif fonts more difficult and more tiring for users to read. Therefore, the simplicity and clarity of sans-serif fonts make them the best choice for good on-screen legibility. This is especially important for users who have low visual acuity.

- **Reserve the use of serif fonts for display text and larger font sizes.** The use of a serif font in moderation can add visual interest to an information space's pages and provide another way of indicating visual hierarchy.

- **Avoid the use of light font weights that don't render well online.** The fine lines of their letterforms can break up on a screen. Plus, their lack of adequate contrast with their background can make them hard to read.

- **Avoid the use of small font sizes.** Larger font sizes are more legible and easier to read for everyone. Font sizes that are perfectly legible on paper might *not* be at all legible on a mobile phone's screen. Plus, people who are older or who have low visual acuity often struggle when trying to read smaller font sizes—or might not be able to read them at all. If there is insufficient screen real estate for your labels, links, or subheadings, revise the text to make it more concise. Don't reduce their font size in an effort to cram more text onto a page.

- **Avoid using too many different font sizes.** The need to express the hierarchy of a labeling system typically dictates the number of font sizes a page must use, along with the need for a legible font size for the body text. Don't use more font sizes than are absolutely necessary to express the hierarchy of a labeling system and the other content on a page.

- **Use font formatting to distinguish different levels of subheadings.** You can use font formatting such as color, size, style—normal, italic, or oblique—and weight—normal, bold, bolder, lighter, or other weights—to express hierarchy. To distinguish the different levels of subheadings in a hierarchy, they should progress from larger to smaller font sizes. Combining font size with the use of color and bold text is the most effective way of formatting subheadings.

- **Avoid overly long line lengths.** Long lines of text are more tiring for people to read because they require more eye movements, or *saccades*, to center the text a person is currently reading within the eye's foveal area—that is, the area that has the greatest visual acuity—and read the entire line of text. The exact number of characters each line of text should ideally comprise depends on the text's font size, but people generally prefer that the line length of page titles and subheadings be relatively short. Use a responsive layout whose line lengths adjust to a device's viewport and orientation—landscape or portrait.

- **Use sufficient line height.** Adequate line height, or *leading*—the space between lines of text—makes it easier for people's eyes to track individual lines of text. Long lines of text that also have an insufficient line height make it too easy for people's eyes to inadvertently jump from one line of text to another or, when trying to go to the beginning of the next line of text, skip a line or go back a line.

- **Visually group subheadings with the section of text that follows them.** To make it clear that a subheading relates to the section of text that follows it, the whitespace preceding the subheading should be greater than that after it.

- **Do not center text.** For optimal readability, use left-aligned text for languages that read from left to right. Whether readers are reading, scanning, or skimming text, aligning the text provides a consistent location to which the eyes can return when moving from the end of one line of text to the beginning of the next line. This makes all text easier to read—including page titles, section headings, lists of navigation links, and the headers in tables.

- **Use colors that provide sufficient value contrast between foreground text and page backgrounds.** Using a white background with black text *or* a black background with white text provides maximal value contrast. However, most people find reading large blocks of white text on a dark background tiring, so the use of dark backgrounds is most useful for navigation bars and page title bars. When choosing colors for subheadings, you'll often use brand colors, but if these colors do not provide sufficient value contrast with their background, choose a darker shade of a brand color on a light background or a lighter tint of one on a dark background.

- **Avoid using patterned backgrounds behind text.** Patterned backgrounds make text less legible and, thus, hard to read. Plus, they're distracting. Always use solid background colors behind text—colors that make it easy for people to focus on the text.

- **Avoid the overuse of all capitals.** Using all caps requires around 30 percent more space than mixed case and slows people's reading speed by about 15 percent because the relative lack of variation in the shapes of uppercase letters makes the text harder to read. Many lowercase letters have either ascenders (b, d, h, k) or descenders (g, p, q, y), which makes their letter forms much easier to distinguish from one another. So, avoid using all caps for long page titles and subheadings—a convention from an earlier time when typewriters offered few formatting options. Instead, use a combination of the more legible formatting alternatives that are now available. However, the link labels in a navigation system are brief—typically just one word or a *very* short phrase—so you could use all caps for the highest level of links to distinguish them from the lower-level links, *or* use all caps for a primary navigation system's links and mixed case for any supplementary navigation system's links. You could use all caps for concise group labels—for example, in a navigation panel or on a megamenu—and mixed case for subordinate links. Just make sure that you use all caps *only* for very brief labels. The effect of using all caps often seems to be that the text is shouting at people. So, if you want to completely avoid the use of all caps, rely on other types of text formatting, including font color, size, style, and weight—typically, the use of bold type for subheadings—to distinguish different types and levels of labels. Plus, if you want people to actually read larger blocks of important information, don't format them in all caps.

- **Minimize the use of italics.** Italic and oblique font styles don't render as clearly on screens, so are harder to read than normal text. Therefore, you should generally limit your use of italics to the emphasis of a single word or a very brief phrase. Avoid using italics for entire paragraphs of text—for example, for quotations—because the text would be hard to read. Avoid using them for subheadings because they don't provide sufficient emphasis. The need to limit the number of fonts that a page must load to optimize its performance would likely prevent the use of italics for emphasis *within* page titles or subheadings. You can still follow the convention of using italics to indicate a book or chapter title within a page's body text.

- **Avoid using underlined text for anything but link text.** On the Web, the convention of underlining the links in a page's body text makes it likely that users would assume any underlined text is a link. Therefore, you should *not* underline text for emphasis.

- **Do not use cryptic symbols instead of text labels.** Especially for navigation links, clarity is key, so avoid the use of incomprehensible symbols or images in lieu of clear text labels. People should never have to guess the meaning of a link label.

[36, 37, 38]

Parallel

Whenever possible, use a parallel grammatical syntax or the same parts of speech across an entire group of labels—for example, all nouns, all verbs, all gerunds, or all questions. Or, if that is not possible, create subgroups of labels that have a parallel syntax. For example, in a case in which most navigation links are nouns, or things, but a few are verbs, or actions, you might group all the nouns, then create a separate group for all the verbs. However, although using parallel forms of words could accelerate users' understanding of the labels, it might not bring consistency to their meaning, which is much more important. [39]

Persuasive

Most labels simply describe content at some level of granularity—for example, the sections of an information space, their pages, or the subsections of pages. However, persuasive labels take a more active role and encourage users to take specific actions or look at specific content that would enable them to achieve their goals and, thus, fulfill their needs. These button or link labels are typically in the imperative mood—for example, **Read More** or **Learn More**. They often provide an opportunity for users to get answers to their questions.

One type of persuasive label is the *call to action (CTA)*, whose goal is to motivate users to take a certain action that would meet a goal of the organization that owns the information space. Marketers refer to achieving such goals as *conversions*. Examples of conversions include signing up for an account, subscribing to a newsletter, or purchasing a product. A call to action typically takes the form of a direct instruction or command—such as **Sign Up Now**, **Check Out Now**, or **Donate Now**—or expresses a strong request or invitation—such as **Join Us**, **Get Involved**, or **Take Our Survey**. As Jim Kalbach says, "Calls to action speak directly to users." [40]

Readable

"People read differently on the Web. ... Users roam from page to page collecting salient bits of information from a variety of sources. They need to be able quickly to ascertain the contents of a page, get the information they are seeking, and move on."—Patrick J. Lynch and Sarah Horton [41]

People often confuse *readability* with *legibility*. Earlier, this chapter covered, in some depth, *legibility*, the design factors that make it easier for readers to visually distinguish the text on a page, and *accessibility*, making text perceptible to all people regardless of their physical abilities. In contrast to legibility, the attribute of *readability* derives from a reader's ability to understand textual content, which depends on familiarity with the language it uses and its reading level. While readability is a much more significant consideration for longer blocks of text, it also plays a role in the comprehensibility of labeling.

Let's briefly consider what makes textual content more readable:

- **Making text scannable**—Make it easy for people to find and identify the content they need. This requires providing clear page titles and subheadings.

- **Using words the reader is likely to know**—Use the same terminology that your users use when speaking. Avoid the unnecessary use of jargon, acronyms, or slang in labels, especially in the labels for navigation links.

- **Using simple rather than difficult words**—Your goal is clear communication, so choose simple words for labels. The reader must be able to accurately interpret the information that labels convey, so use words whose meaning would be clear in a given context.

Specific

When designing a labeling system, you should consider the granularity, or specificity, of the labels *and* the concepts that they represent. As much as possible, labels should be equally broad or narrow at a given level in a hierarchy and become progressively more specific at lower levels in the hierarchy. Thus, the specificity of each label of a particular type or at a given level should be comparable.

The best labels describe very specific concepts and, thus, have a fairly narrow scope. Always use the most specific label that is applicable to a given concept. Avoid creating meaningless links or category labels to group unrelated content—such as **Miscellaneous**. Labels should always be meaningful to an information space's users. If necessary, you can qualify a vague, overly general label to make it more specific—for example, **Account Details** rather than **Details** or **Company Information** instead of **Information**. [42]

Task Focused

To facilitate users' rapid learning, both the concepts an information space presents and the terminology it uses to describe them should ideally be task focused and familiar to people who work in its particular domain. You can learn what terms users employ in describing their tasks through user interviews and observations. Whenever you must introduce a new term for a new concept, be sure the term you choose focuses on users' tasks rather than the technology an organization is using to store or implement the information. Users' learning should focus on their tasks, not on deciphering unfamiliar terminology, which would place undue demands on their short-term memory and, thus, decrease their comprehension. [43]

Types of labels

Most user interface (UI) elements require text labels. Some of these labeled elements are *interactive* and enable the user to browse an information space or product, including the following:

- Navigation systems and the hyperlinks they comprise
- Contextual hyperlinks or buttons within the content of pages that display additional information on the same topic or reference information on a related topic
- Controls that progressively disclose groupings of additional information on the same page

Other elements on pages are *static* and have their precedents in print media. These include the page titles and subheadings that inform the reader about the content that resides on a particular page.

In the following subsections, we'll focus on the types of labels that relate to an information space or product's information architecture, including its navigation systems.

Labels for navigation systems

"From a user's perspective, … navigation labels are an [information space's] content, functionality, and structure. If navigation has a narrative role…, labels are the words that tell the story."—Jim Kalbach [44]

An information space or product's navigation system comprises sets of hyperlinks. These navigation hyperlinks typically reside on navigation bars, various types of menus—such as drop-down menus, mega-menus, hamburger menus, vertical menus, or hierarchical tree menus—or on an information space's footer. Many information spaces also provide lists of hyperlinks that take the form of a site map, an A–Z index, or a directory of topics or product categories. For more information about the UI elements that navigation systems comprise, refer to Chapter 12, *Designing Navigation*.

Because users typically click or tap navigation hyperlinks to browse an information space and navigate between its pages, the labels for these links must very clearly communicate the concepts they represent. Labels for navigation hyperlinks derive much of their meaning from their context—the navigation hierarchy itself.

From any page on an information space, users must always be able to easily navigate to its home page and the main pages of all its sections. Therefore, the labels for links whose target destinations are such key pages are the most important labels on an information space. [45, 46]

Labels for indexing terms

Content-rich information spaces and products employ a content-management system (CMS) that lets indexers assign indexing terms such as categories, tags, or keywords, as well as other descriptive metadata, to specific pages of content and, in some cases, to reusable chunks of content.

Defining indexing terms is essential to the optimal functionality of an information space's navigation and search systems, which enable browsing and searching, respectively. Indexing terms define the categories of information that an information space comprises. Menus or lists comprising these terms—such as lists of topics—facilitate browsing by providing an alternative to the primary navigation system. In some cases, indexing terms appear only in an information space's metadata and their purpose is solely to support search—both site search and organic Web search.

Indexers can assign multiple indexing terms to a specific page or chunk of content. Each of these indexing terms should be meaningful to users and accurately describe and, thus, be representative of that content. These indexing terms could derive from some type of controlled vocabulary such as a thesaurus. For more information about assigning indexing terms to content, see Chapter 7, *Classifying Information*. [47]

Labels for categories of content or products

Information spaces that comprise large collections of topical content or types of products typically provide a hierarchical list, or *directory*, comprising categories and subcategories, to aid browsing. Each category or subcategory is a link that displays the landing page for that specific category and provides access to a collection of pages that belong in that category. These indexing terms must clearly represent the categories and subcategories of content. [48, 49]

Labels for filtering controls

On an information space that uses indexing terms to define the categories of information that it comprises, filtering controls can enable users to control the scope of the information that the space displays to them. The user can select or deselect categories of information that correspond to specific indexing terms. Filtering controls often take the form of checkboxes, option buttons, drop-down list boxes, images, or sliders that let the user select a range. Selecting more categories increases the scope of the information while selecting fewer categories reduces the scope. The indexing terms must clearly represent the various categories of information.

When there are many filters, the filtering controls often reside within collapsible groups, whose labels must clearly comprehend all the individual categories of information within a particular group. This is especially important when a group is collapsed by default. [50]

Labels for contextual hyperlinks or buttons

Contextual hyperlinks reside within a page's body content and their labels should accurately represent a specific link destination. Clicking a contextual hyperlink could display either another page *or* a specific subsection of the same or a different page. Authors often add contextual links to the content they create to give readers access to more information on the same topic *or* to reference information—perhaps source information for the content on the current page—on either the same or a different information space.

Links to additional information on the same topic that resides on a different page on the same information space might be either link text or a button, often with a generic label such as **Learn More**, **Read More**, or **View More**. Such labels suffice when these links or buttons exist within the context of a clearly titled page or labeled subsection of a page. This pattern often occurs on home pages or the main pages of sections whose subsections introduce more detailed content on other pages within a particular section of an information space.

In contrast, labels for links to reference information on the same or a different information space should ideally be the title of the page on which the information resides or a concise and accurate description of the linked content. When creating links that refer to the pages of an information space—particularly the pages that anchor its major sections—always use their actual page titles rather than the labels of the navigation links that display them.

A reference link's meaning often derives from both its label *and* its context, including the text within which it resides, the subheading immediately preceding it, the page's title, and what the user knows about the information space's domain. Thus, a reference link's surrounding text often forms part of the description of the linked content and provides a meaningful context for the link, obviating the need for an overly long link label.

Within a given context, strive for consistency in your use of terminology across the content and the link labels it contains, as well as with the terminology that the link target uses. To ensure the quality and consistency of an information space's contextual-link labels, be sure to provide authors with some type of controlled vocabulary and clear guidelines for labeling contextual links. [51]

Labels for groupings of additional information

Labels for controls that progressively disclose groupings of additional information on the same page must be clear and concise, and their terminology must be consistent with that of the content within each of the groupings that they reveal. Various UI elements enable progressive disclosure, including tabs, accordions, cards, overlays, and panels. For more information about grouping content on pages, as well as implementing progressive disclosure to display more information, refer to the section "Structuring information on pages," in Chapter 10, *Designing and Mapping Architecture an Information*.

Labels for structuring content

Page titles and subheadings are static labels that derive from print media. They inform an information space's readers about the subject matter of a page's content and provide context. Together, a page's title and its various levels of subheadings form a hierarchical labeling system.

Page titles

Ideally, every page on an information space should have a descriptive page title that enables users to verify that they've arrived at their intended destination after clicking a navigation link. This page title orients users to their current location on an information space. It should appear prominently, in a consistent location at the top of the main content area of a page. Plus, the page title should use terminology that is similar to that of the navigation link that the user clicked to go to the page.

To optimize the visibility of page titles, they should be left aligned and distinct from the surrounding text—either in a larger font, a different color, or in bold type. Page titles should be highly legible, which requires that there be high value contrast between a page title and its background. [52, 53]

Subheadings

Most pages also have subheadings that structure their content, thus facilitating the scanning of that content. Each subheading should accurately describe the content of the subsection that immediately follows it.

On longer pages whose content requires multiple levels of subheadings, these subheadings constitute a labeling system that should clearly define a page's hierarchy. The context that this labeling system provides helps convey the meaning of a page. To create a clear visual hierarchy, the consistent use of unique visual formatting should distinguish each level of subheading, with higher to lower levels typically progressing from larger to smaller font sizes. To further differentiate specific levels of subheadings, you can use font styles, colors, layout, and whitespace.

If an information space comprises many pages of the same type that consist of similar types of information, create a consistent, but flexible labeling system that accommodates all the necessary types of information. For example, for the many product pages of an ecommerce app or site, you might be able to define standard labels for particular types of information.

When you're devising patterns of labels, keep them as simple as possible. Create labeling hierarchies comprising discrete levels that enable you to avoid any duplication or repetition of labels. Never rigidly enforce a labeling hierarchy that would result in redundant labels. For example, one way to prevent redundancy in breadcrumb trails is to avoid repeating the last item in a breadcrumb trail by making it the page title instead. [54]

So, instead of doing this:

Pet Supplies › Cats › Furniture › Beds

Beds

Do this:

Pet Supplies › Cats › Furniture ›

Beds

Enabling browsing and searching the Web

For information spaces on the Web, it is important to create Web addresses and browser titles that enable effective user interactions.

Constructing optimal Web addresses

When navigating the Web, people use the addresses for Web pages—that is, their uniform resource locators (URLs)—in a variety of ways. They might type a URL that they know into their browser or guess the URL for a Web site's home page. If they've provided a URL that is close to a site's actual domain name, the Web site should redirect them to its home page. Be sure to create redirects for all the common variants of a Web site's home page URL that people might type.

Ideally, the pages on your Web site should persist forever to prevent your breaking inbound links from other sites. Therefore, you should avoid removing pages or moving pages around whenever possible. However, if you must remove a page from your site, be sure to create a redirect to some other relevant page; and, whenever you move a page, create a redirect to the page's new location. Consider using a generic URL for pages whose content changes often—for example, *Recent News* or *Current Job Opportunities*. Plus, instead of creating permanent Web addresses for short-term marketing campaigns, contests, or surveys, you could create temporary redirects to display such landing pages.

People read and try to understand Web addresses in attempting to determine a Web site's structure. Therefore, constructing optimal Web addresses requires that you create concise, human-readable URLs that, depending on the point at which the user reads them, should either communicate a link's target destination or confirm that the user has arrived at the desired destination. Ideally, a Web site's domain name—for example, uxmatters.com—should be the name of the company, organization, or person to whom the site belongs.

You should generally spell out a company's name in full rather than using an acronym or abbreviation unless everyone in your target audience would know the acronym or abbreviation by heart. However, if necessary to avoid a very long URL or a difficult-to-spell or pronounce URL, you could alternatively use an acronym or abbreviation. But you should avoid choosing cryptic URLs just for the sake of brevity. People would be unlikely to remember them. Instead, it's usually preferable to create a redirect that would get people to your site if they do type an acronym or abbreviation into their browser.

A *compound domain name* comprises multiple words. For optimal usability, adhere to common practice and run these words together. However, if absolutely necessary, you could separate these words with hyphens or, in the case of subdomains, with dots. However, because people would be likely to omit the hyphens when typing a URL into their browser, you should try to avoid using them in Web addresses whenever possible.

Because users must be able to type Web addresses correctly into their browser, keep them as short as possible, use common words that most people know how to spell correctly, and use only lowercase letters, numbers, and either hyphens *or* underscores. Because of the strong potential for users to confuse zero (*0*) and the capital letter *O*, avoid using these characters in Web addresses. If you *must* use either of these characters, create a redirect that uses the erroneous character in place of the correct one and redirects it to the correct Web address.

Human-readable Web addresses should clearly convey a Web site's directory structure, helping to orient people to their particular location on the site. Therefore, people must be able to read all directory names, which should comprise words that accurately describe the Web site's structure. Use the same terminology in the Web site's navigation system *and* when assigning categories to pages, ensuring that users recognize the categories and, thus, the directories in which specific pages belong.

More advanced users should be able to tweak a Web address, either by deleting the latter part of the address in trying to navigate directly to a page that would provide an overview of a section or by a shortening it to just the domain name to go directly to the site's home page. Choose a content-management system (CMS) that you can configure to generate human-readable Web addresses and be sure to support users' manipulation of URLs in this way. Creating human-readable Web addresses is also good for search-engine optimization (SEO). [55, 56]

Choosing effective browser titles

The metadata for a Web page should include a title that Web browsers can display on a tab displaying the page, use by default as the name for any bookmarks that users create for the page, and include at the top of a printout of the page. Both Web-search and site-search engines display a Web page's browser title as the link text that appears on search-results pages. Ensure that a page's browser title accurately represents the content on the page to drive traffic from its link on a search-results page to the link's destination page. [57]

Creating iconic versus textual labels

Although most labels comprise text, the use of iconic labels to represent specific topics or types of information can be effective in some cases—usually in combination with textual labels. Icons have a much more limited vocabulary than textual labels do, so they are usually part of a small set. Use iconic labels only when all of the icons in a set clearly map to *all* the concepts you need to represent. Icons can be aesthetically pleasing and add visual interest to an information space's user interface, but their use should never impair usability. For an information space that has frequent users, users might learn its visual language, making it easier for them to recognize specific icons—especially if the icons are color coded—and facilitating navigation.

The use of iconic labels in an information space's primary navigation typically occurs only when it's easy to represent the information space's concepts as physical objects, as Figure 9.1 shows, *not* on those that comprise abstract concepts that it would be impossible to convey visually. Still, it's generally advisable to use icons only in combination with textual labels.

Figure 9.1–Booking.com

However, iconic labels are particularly useful for the navigation systems of information spaces for which young children are the target audience—especially for those who cannot yet read. Another common use of iconography in labels is for visual systems that represent the specific types of information on a page—for example, for notes or tips—*or* the logical subsections of pages that have a common structure. [58]

Translating textual labels

Translating an information space or product's labels into another language often increases their length significantly—by up to fifty percent—so, page layouts must either accommodate the longer labels *or* be sufficiently flexible to allow more space for them. In some languages, compound words can create awkward line breaks between words or have no breaks at all, which can result in very long, unwieldy labels.

Try to avoid creating labels that require a specific grammatical syntax, which can differ significantly across various languages. The differences between languages could include factors such as sentence structure, the existence of specific parts of speech, masculine and feminine nouns, and capitalization rules. So, if you're creating labels that rely on having a particular grammatical syntax, you should first assess the differences between languages to which you'll need to translate the labels.

Plus, sometimes the literal translations of words can have completely different meanings from those that you intend. So, be careful when translating textual labels. Consider what translations are necessary from the beginning, then take all these factors into account before deciding what terminology to use in your labels. [59]

Discovering possible labels for categories

In most cases, you won't need to design an information space's labeling system completely from scratch. Much of the terminology for an information space's domain and many of its concepts and topics are probably already familiar to your team, your target audience, and perhaps even the general public. As you consider what labels you might use for specific types or categories of information, gather viable labeling alternatives for each type or category into a document or table.

Analyzing existing content

Most organizations have a wealth of content that they could potentially publish, even if they haven't yet built an information space or product for that purpose. If an information space or product already exists, the terminology that you should use in labeling its categories of content should come directly from its content—or, more broadly, the information space's business or knowledge domain. Analyze an information space's content, compiling a list of descriptive keywords for each important content object—whether a page or a reusable chunk of content. You can typically derive keywords from page titles and subheadings—or, for academic papers, from summaries or abstracts. Conducting such a content analysis is an especially important source of labels for a new information space or product.

To learn more about analyzing content, refer to the section "Content-analysis methods," in Chapter 6, *Understanding and Structuring Content*.

If you're analyzing the content of an extensive information space or another large collection of content, relying on automated data-extraction software could save you considerable time and gather a fairly comprehensive collection of meaningful terms that could be candidates for a controlled vocabulary. However, most of these software tools are quite expensive, and setting them up requires significant effort, so their cost would probably be prohibitive for a single small- or medium-sized information space. You'll also need to fine-tune their output to ensure its quality and usefulness. [60]

Evaluating any existing labeling system

If you're revising the labeling system for an existing information space, evaluate its labeling system with an open mind. Use whatever labels are already working well. Users generally prefer that things remain the same unless you're clearly making significant improvements.

However, if it becomes apparent that insufficient consideration was originally given to the systematic nature of the existing labels, you may need to completely rethink the labels to create a coherent labeling system. If you're expanding an existing information space's scope beyond that of its competitors, think about what changes you might need to make to ensure that their existing labels work well with the new labels you need to add. To assess an existing labeling system, conduct evaluative UX research such as usability testing.

Conducting competitive analyses

Whether you're designing a new labeling system or revising an existing one, conducting a competitive analysis is an excellent means of determining what terminology and labeling are characteristic of the particular industry domain to which an organization belongs. Over time, labeling conventions emerge across domains. Following them can help an information space meet users' expectations. If competitors have already designed effective labeling systems, they can provide successful models that you might follow. Consider the similarities and differences between competitors' labeling systems, which labels are and are *not* working well, and what patterns emerge from your analysis of their labels. Adopt or refine the best of competitors' ideas for labels—keeping in mind the need to design a coherent labeling system that serves your users' needs.

Conducting user research

Understanding the needs of an information space's prospective and actual target users and identifying the language they use for specific concepts is essential to designing an effective labeling system. To learn about an information space's target users, you can conduct user interviews and other methods of generative user research with participants who belong to its target audience. Conducting generative user research enables you to learn about your users' prior experiences and, thus, their expectations of an information space.

You can employ a variety of methods to identify patterns in your target users' use of specific terminology—for example, by inviting users to participate in open card sorts or free-listing exercises or through a Web site's search–log analytics—as Chapter 5, *UX Research Methods for Information Architecture*, describes. These are the most reliable ways of learning users' own words for things and, thus, what labels would best meet their needs.

Conducting an open card sort

In the section "Open card sorting," in Chapter 5, *UX Research Methods for Information Architecture*, you learned that, as the final step of conducting an open card sort, research participants label the groups of cards that they've created. The group labels that they've chosen can inform the terms you should use when labeling navigation links at the higher levels of an information space's hierarchy.

Leveraging domain-specific controlled vocabularies or thesauri

A controlled vocabulary or thesaurus might already exist for an organization's industry domain and could be very useful in choosing labels or indexing terms. Typically, library-science professionals or other experts in specific industry domains create publicly available controlled vocabularies or thesauri, so they should already have high degrees of consistency and representational accuracy. To be maximally useful, a controlled vocabulary or thesaurus should ideally target a particular audience, working in a specific domain. [61] To learn all about classifying information and developing your own controlled vocabulary or thesaurus, refer to Chapter 7, *Classifying Information*.

Interviewing subject-matter experts

A good way of learning what terminology an information space's labeling system should use—specifically, what terminology is characteristic of a particular business or knowledge domain—is to talk with subject-matter experts (SMEs) within an organization or, for an existing information space, with either technical-support professionals who frequently interact with users and can advocate for their needs or an information space's expert users. If an information space has more than one target audience, you may need to track what terminology is characteristic of each audience, build a controlled vocabulary for each audience, and design labeling systems for collections of content that exist to meet specific audiences' needs. [62] Read the section "Content-owner interviews," in Chapter 6, *Understanding and Structuring Content*, to learn about the role that subject-matter experts can play in helping you to understand users' terminology.

Getting authors to label the content they create

Work with content authors to find out what labels they would use to describe the content they create—particularly category labels. Or, to spread the effort of labeling content across various members of your content team, ask content authors and other content owners to assign category labels to their content when publishing it in a content-management system (CMS).

To ensure consistency in content authors' and other content owners' use of terminology, they should refer to a controlled vocabulary when labeling their content. The controlled vocabulary should provide references to broader and narrower terms for each preferred term to ensure that the labels they use are at the most appropriate level of specificity. For cases in which the subject matter of new content goes beyond the scope of your existing controlled vocabulary, you'll need a standard approach for letting content owners add new terms to the controlled vocabulary, as well as a process for approving, refining, or rejecting their changes. To better understand how to work with content authors and other content owners, refer to the section "Content-owner interviews," in Chapter 6, *Understanding and Structuring Content*.

Leveraging search-log analytics

Any information space or product that has a search engine can provide data that is useful in discovering what terms its audience uses for particular concepts. Refer to a log of the queries that users have entered when searching for information—possibly information that they've been unable to find by browsing. People's search terms are an excellent source of candidate labels for navigation systems.

When you're deriving candidate labels from a long list of terms in an information space's search-log analytics, be sure to cull duplicate terms from the list. Then try to identify any patterns in the language that people use in describing the information space's content. Over time, as people's use of terminology evolves, you might see changes in the terms they use for particular concepts when searching an information space. In this way, you can leverage search-log analytics in choosing effective labels. Refer to the section "Site search–log analytics," in Chapter 5, *UX Research Methods for Information Architecture*.

Comparing the usage of similar terms across the Web

Once you've determined some likely terminology candidates for a specific concept, search for those terms using Google or another Web-search engine, note the term for which users most frequently search, and use that term for that concept in your labels and your content. [63, 64]

Defining optimal labels for categories

"For information architecture, labels are central. They're the ingredients we use for categories and classification schemes. … Users seek them out to pick up information about where they can go, how the environment is nested…. The purposeful definition of every component and connection within a … system or organization requires the use of labels."—Andrew Hinton [65]

Whatever the sources of the terms you might use in labeling the categories an information space or product comprises, crafting a coherent vocabulary from disparate information sources requires significant effort. Collect all of your candidate terms into an alphabetical list. Where there are multiple candidate terms for a single concept, determine which is the optimal, or *preferred*, term. In some cases, these candidate terms might be completely different words, while in others, they may be just slightly different variants of the same words. Reconcile any inconsistencies in the forms these preferred terms take, including their spelling, punctuation, and use of lowercase versus uppercase or possessive. Through this process, you'll establish style conventions for these terms.

Identify the *variant* terms for each preferred term, indicating that the authors and indexers who are creating and categorizing content should avoid using them. Also, identify useful terms that have either *broader* or *narrower* meanings than the preferred terms or that describe related concepts and add them to your list of terms. For detailed information about various types of controlled vocabularies that can be useful in classifying information and how to develop a controlled vocabulary, refer to Chapter 7, *Classifying Information*.

To complete this process, identify any gaps that currently exist in the vocabulary for the information space. Does it include a term for each important concept that the information space's content covers or should cover in the near future? Are there additional topics the information space should cover? If you're quite sure that you'll need to add certain topics to the information space, identify those topics, choose appropriate terms for them, add them to your list of terms, and create categories for them in your content-management system (CMS). However, you should be conservative in doing this. Don't bloat your system with categories that you may never use or use too infrequently, causing each topic or the scope of the entire information space to lose its focus.

Once you've decided what terminology you want to use in labeling categories, assign these categories to the information space's content. These keywords or tags describe and identify the content on each page of an information space and are also useful to users when they are searching for information.

Over time, people's use of terminology evolves. If you discover that you're not using or are rarely using certain categories, consider either of the following:

- Eliminating the categories and assigning other existing categories to any content to which you had assigned these categories

- Relabeling the categories to make them more consistent with the terminology that people are currently using for specific concepts

Also, identify and resolve any inconsistencies that exist in the use of specific categories to describe particular content on an information space, as well as inconsistencies in the use of specific terminology in category labels or the information space's content. [66]

Designing labels for navigation systems and pages

To provide good information scent, the labels for an information space's navigation systems and pages should ideally use the terminology that your target audience would expect to see when looking for particular information by browsing its navigation system—the language that would be meaningful to them. The difficulty is that different people use various words for the same things. But you must try to choose words that most users would recognize as referring to the information they're seeking.

Effective link labels are essential to the design of a useful navigation system. They should concisely and accurately describe the content at their target destination and convey what content would appear when the user clicks a link. Never make the user click a link label to discover where it leads.

When you're designing and documenting the labels for an information space's navigation system and link destinations, it could be helpful to create a table comprising the following columns:

- **Link label**—This column could include subcategories for primary navigation, secondary navigation, supplementary navigation, if any, body-content navigation, and footer navigation

- **Destination label**—Depending on whether a link is an *interpage link*—that is, a link to a different page—or an *intrapage link*—a link to a subsection of the same page—these labels might be either page titles or section headings, respectively

- **Destination label type**—Note the type of each destination label—*page title* or *section heading*

Such a table provides a concise means of structuring this information, which makes it easier to get an overview of the entire labeling system and identify any inconsistencies. [67]

Testing labels with users

You should always test an information space's labeling system with research participants who are representative of its target users, both during iterative design and after launch. When testing link labels, one approach is asking participants to describe the information at the links' destinations.

In Chapter 5, *UX Research Methods for Information Architecture*, you learned about some UX research methods that are especially effective for testing labeling with users. These methods of UX research include the following:

- **Card sorts**—The section "Card sorting," in Chapter 5, covers three methods of card sorting in depth: open card sorting, modified-Delphi card sorting, and closed card sorting. Card sorting provides a means of learning about users' understanding of structure, categories, and labels.

- **Free-listing labels**—The section "Free-listing," in Chapter 5, describes a way of exploring participants' common understanding of an information space's industry or knowledge domain to inform structure, categorization, *and* labeling. [68]

Summary

This chapter explored the design of labeling systems, taking an in-depth look at the various attributes of effective labels. It also covered the design of different types of labels, including labels for navigation systems, contextual hyperlinks and buttons, and progressive-disclosure controls. It briefly considered the use of iconic versus textual labels and the translation of textual labels. The chapter also looked at the discovery and definition of optimal labels for categories, navigation systems, and pages. Finally, it communicated the importance of testing labels with users.

Next, in Chapter 10, *Designing and Mapping an Information Architecture*, you'll learn how to apply everything you've learned throughout this book to the design of a coherent information architecture.

References

To make it easy for readers to follow links to the references for this chapter, we've made them available on the Web: `https://github.com/PacktPublishing/Designing-Information-Architecture/tree/main/Chapter09`

Get This Book's PDF Version and Exclusive Extras

UNLOCK NOW

Scan the QR code (or go to `packtpub.com/unlock`). Search for this book by name, confirm the edition, and then follow the steps on the page.

Note: Keep your invoice handy. Purchases made directly from Packt don't require one.

10

Designing and Mapping an Information Architecture

"It's difficult to write about design because the work in this phase is so strongly defined by context and influenced by tacit knowledge. ... The design decisions you make and the deliverables you produce will be informed by the total sum of your experience."— Peter Morville and Louis Rosenfeld [1]

The magic of great design happens only through the synthesis of your own foundational knowledge of information architecture (IA) and user experience (UX) design, your learnings from generative and evaluative UX research, and your prior experience on other design projects. The need to synthesize all this knowledge is the reason it's so important to hire information architects and UX designers who have deep experience and, thus, expertise in User Experience.

As Chapter 8, *Defining an Information-Architecture Strategy*, describes, your IA strategy for a digital information space or product derives from the business and user requirements you synthesized from your UX research findings, as well as your analysis and classification of the information space's content. This chapter discusses the *core* of IA practice: the design and mapping of an information architecture.

As your IA project's focus shifts to design, you'll build on your IA strategy for the information space, and the creation of detailed design deliverables will take on an increasingly important role. Producing these design deliverables helps you to crystallize your thinking, enables you to clearly communicate your design concepts to your teammates and stakeholders, makes your design solutions more tangible, and, thus, allows your team to identify design issues and provide actionable feedback on how to remedy them.

To some extent, the specific steps that the process of designing an information architecture comprises depend on the scope and complexity of a project—for example, whether you're designing a new information space or completely revamping an existing information space versus just expanding an information space's scope slightly or reworking only specific parts of an information architecture. The time and budget that are available for an IA project also impact what approach you can take, including how much user research and usability testing your team can do and what design activities you can pursue. Your need to make particular design decisions for an IA project also depends on whether and to what extent you can rely on decisions your team made during previous design cycles and how well those decisions serve the needs of your current project.

Nevertheless, as this chapter describes, the process of designing an information architecture typically comprises the following key steps:

1. Evaluating any existing organizational structure

2. Choosing an organizational model

3. Categorizing and labeling an information space's content

4. Diagramming an information architecture

Bear in mind that a well-defined information architecture is typically easier to maintain and, thus, requires fewer major redesigns. Plus, it better meets the needs of both users and the business. [2]

Evaluating any existing information architecture

If your goal is to redesign an existing information space or product—whether you're doing a complete redesign or just rethinking the information architecture for a few sections—or if you're expanding the scope of an existing information architecture, you must first evaluate and build your understanding of its current organizational model, structure, and labeling and assess the resulting findability of its content. There are several UX research methods that are especially useful in evaluating an existing information space's organizational structure and findability during *Discovery*, as follows:

- **Closed card sorting**—To determine whether an information space's existing information architecture conforms to the target users' mental model, conduct a closed card sort. Conducting a closed card sort can help you to evaluate and, hopefully, validate the usability of the information architecture's existing categories; understand whether the current organizational structure, groups of content objects, and their labels make sense to users; identify where gaps exist in the information architecture for which you need to create new categories and subcategories; and understand which categories your team should to assign to new content.

- **Reverse card sorting**—Employing this method of evaluative UX research lets you quantitatively rate *or* validate an existing information space's findability, establish a baseline for findability, and gain useful insights for refining or redesigning the top levels of an information architecture and navigation system. Consider conducting a reverse card–sorting activity *before* expanding the scope of an information space's content, modifying its information hierarchy, or making other structural changes.

- **Card-based classification evaluation**—The focus of this method of evaluative UX research is on assessing the findability of an information space's *taxonomy*, or information hierarchy, in isolation from its user interface. Because your research participants complete realistic, information-seeking task scenarios during a card-based classification–evaluation activity, you can gain a deep understanding of their perceptions of an entire information architecture's classifications and labeling, as well as insights about where they expected to find specific types of information.

- **Tree testing**—This method of UX research also focuses on assessing an existing information architecture's findability. By doing tree testing, you can evaluate the effectiveness of an existing information space's hierarchy and labeling, assess how well participants are able to complete realistic, information-seeking task scenarios, and identify any weaknesses in the information hierarchy. By conducting iterative cycles of tree testing and design, you can significantly improve an information architecture. Tree testing also provides insights regarding where participants expected to find certain information.

- **Usage-data analytics**—You can learn about actual users' behaviors when using an existing information space through the analysis of Web analytics, clickstream or path analysis, and search-log analytics, which lets you better understand their needs and how well the information space is meeting them. Using analytics data such as page views, conversions, bounce and entrance rates, and search-query volume is particularly helpful in identifying categories of information that users are having difficulty finding—as well as other painpoints—and, thus, informs the improvement of an information architecture. Plus, using analytics to measure key performance indicators (KPIs) such as reach, engagement, and sentiment lets you measure the effectiveness and impact of the information space's content.

You can conduct any of these methods of UX research, then synthesize insights from your findings to determine how to improve an existing information architecture. [3, 4, 5, 6, 7, 8] For in-depth discussions of how to conduct each of these methods of UX research, refer to Chapter 5, *UX Research Methods for Information Architecture*.

Choosing an organizational model

"The organizational model is the most concrete expression of how your...content fits together."—Bob Baxley [9]

At the very beginning of the *Design* phase for a new information space or the redesign of an entire information architecture, you must determine what organizational model would best serve the needs of the target audience by enabling users to easily find the information they need and, thus, help you to classify the information space's existing and planned content most effectively to meet those needs. Which organizational model is right for a particular information space depends primarily on the types of content it comprises.

To define this organizational model, you need to choose the primary structural pattern and organization scheme for the information space. Making these key decisions establishes the context within which you'll provide specific content to the target audience, determines what classifications provide the basis of the information space's structure, helps people understand the information space, and sets their expectations for its content. [10] Refer to Chapter 4, *Structural Patterns and Organization Schemes*, for an in-depth discussion of the various structural patterns and organization schemes that you might leverage in devising the organizational model for an information space.

Choosing a structural pattern

By choosing an optimal structural pattern for an information space, you can enable users to recognize, understand, and build their mental model of the space's organizational structure. *Structural patterns* define various types of structural relationships that can exist between groups of related content objects. You might employ the following structural patterns alone or use them in combination with one another to form hybrid structures:

- Hierarchy

- Relational database

- Hypertext

- Linear sequence

- Hub and spoke

- Matrix [11, 12]

The two most generally useful and, thus, common structural patterns are the hierarchy and the database, which information spaces frequently use in combination, including my Web magazine, *UXmatters*. If you're designing either a hierarchy or a combination hierarchy and database, you should begin by creating the information architecture's top-level categories, or topics, in which content objects reside, then add their subordinate levels. However, if the database in which an information space's content resides provides its primary structure, you must first define the content's attributes, or metadata. [13]

Choosing an organization scheme

"There's a simple reason why people find ambiguous organization schemes so useful: we don't always know what we're looking for."— Peter Morville and Louis Rosenfeld [14]

Choosing the right organization scheme for an information space facilitates your gathering its content into meaningful groupings that share specific attributes. This can help users to understand the information space's purpose. In their book *Information Architecture for the World Wide Web*, Peter Morville and Lou Rosenfeld define the following types of organization schemes that you can use to organize an information space's content:

- **Exact schemes**—These *objective*, mutually exclusive organization schemes require that users know exactly what they're looking for. They include alphabetical, numerical, chronological, geographical, spatial, and media-based schemes, as well as schemes that reflect an organization's actual structure.

- **Ambiguous schemes**—These *subjective* organization schemes are typically more difficult to design, maintain, and use, but support serendipitous information seeking so are generally more useful than exact schemes. They include topical, categorical, audience-specific, task-oriented, sequential, and metaphorical schemes.

- **Hybrid schemes**—Such organization schemes combine several complementary schemes that let users discover specific types of information in different ways. [15]

Categorizing and labeling an information space's content

"Categorizing Web content … is a process of drawing boundaries, of deciding what to include and what to exclude. It's a process fraught with subjectivity, bias, politics, errors, and cultural and historical memory. It's anything but simple."—Lisa Maria Martin [16]

Whether you're creating a new information space or redesigning an existing information architecture, you must determine what categories and subcategories the information space should comprise on the basis of your understanding of its content and the relationships that exist between its content objects. Early during your *Design* phase, you should define these categories and subcategories, determine what category labels would make the most sense to target users, and begin applying them to the content objects that the information space comprises.

The primary goals of classifying an information space's content effectively are meeting both the organization's business objectives for the information space and the needs of its target users; supporting the findability of its content by making it easy for users to recognize the content they need; creating an easy-to-browse navigation system; and improving any search system's retrieval of relevant content.

The ways in which you classify and, thus, label content objects influence users' perceptions and understanding of the information their labels impart and enable them to more easily, efficiently, and effectively satisfy their information-seeking needs, whether by browsing or searching. These basic decisions provide a solid foundation for the design of the information space's overall information architecture and determine its structure.

When you're designing an information hierarchy or a combination hierarchy and database, the process of categorizing and labeling the information space's content typically comprises the following steps:

1. **Conduct an open card sort.** Use this UX research method to understand how the information space's actual users would categorize and group the information space's content objects. This method is especially helpful in enabling you to design an information architecture that matches users' mental model of the information space. Conducting an open card sort can also help you to better understand what categories and subcategories the information space should comprise, as well as what relationships exist between them, and what category labels would make the most sense to target users. Ask participants *why* they've made particular sorting decisions during an open card sort to help you to understand their reasoning, and be sure to note cases in which participants want to place cards in more than one category. An open card sort can also help you to identify the information space's dominant structural pattern and organization scheme. For more details about open card sorting, refer to the section "Open card sorting," in Chapter 5, *UX Research Methods for Information Architecture*.

2. **Choose the information space's top-level categories.** These key categories must be *discrete* categories that are easy for users to distinguish from one another. To convey the full range of content the information space comprises and, thereby, orient users to its purpose, these categories must define the overall structure and breadth of the information space, the number of sections it comprises, and the links of the primary navigation system. Both these categories and their order must match the users' mental model of the information space's subject matter, which typically reflects the structure of similar or competitive information spaces. Employ users' vocabulary in labeling these categories to ensure that they'll understand the labels. Although the collective scope of these categories must be comprehensive, users' ability to visually scan the links in the information space's primary navigation system limits their number—perhaps to between about five to ten links.

3. **Define the subcategories that each top-level category comprises.** Once you have a deep understanding of the information space's content, you can define the subcategories for each of its top-level categories, thereby creating the space's taxonomy. Either create Post-it notes representing all of its collections of content—that is, its categories and subcategories—or list them. As you organize them into preliminary groupings, coherent collections of content inevitably begin to coalesce around the top-level categories, then their subcategories. Get feedback on these preliminary subcategories from your multidisciplinary team and from users, then refine them accordingly. Try to create mutually exclusive subcategories whenever possible, but the ambiguity of certain subcategories could require the creation of a polyhierarchy, as described in the section "Taxonomies," in Chapter 7, *Classifying Information*. Subcategories dictate the structure of the content in each of the information space's subsections. Therefore, the subcategories and their order must match the users' mental model of the subsections' content. The lower levels of the information space's navigation system comprise links that correspond to these subcategories, so must have clear labels that make sense to users. Consistency in your use of terminology in these labels is key and requires the use of some type of controlled vocabulary. You must balance the breadth and depth of each subsection's information architecture, while also limiting the number of levels in the overall hierarchy. While for relatively small information spaces, three or fewer levels might suffice; for larger, more complex information spaces, greater depth is often necessary, so you must ensure that the subcategories' labels provide good information scent. Users are happy to dig deeper as long as it is clear that they are getting closer to the information they need.

Note—Completing steps 2 and 3 is typically a highly iterative process, with decisions you make regarding one level impacting those relating to the other level. Plus, in some cases, such decisions might need to be audience specific. You'll likely encounter many trade-offs in making categorization decisions, so you probably won't get everything right on your first try.

1. **Create a coherent organizational system.** The hierarchy that you've defined by completing steps 2 and 3 dictates the information space's overall structure and navigation system. For a database structure, you also need to define the necessary metadata with which to tag each content object. To make good structural decisions, you must have a thorough understanding of the information space's content. Once you've decomposed the information space into its

component sections, consider how you could best logically structure each subsection. Identify any homogeneous collections of information that might require unique organizational models. *If* you're designing a complex organizational system for a heterogeneous information space, you might need to employ different structural patterns and organization schemes to provide structure for particular classes of content objects *or* subsections. The use of a variety of structural patterns and organization schemes for specific contexts can provide users with different ways of accessing the same information.

2. **Establish criteria for matching content to categories.** These criteria define the *organizing principles* that inform your team's decisions about categorization and, thus, the information space's structure. (You can learn more about defining these organizing principles later in this chapter.) When determining your criteria for matching content objects with categories and subcategories, consider your users' needs, goals, and expectations; the business's strategic goals; and both the existing and planned content. Doing this is an iterative process. Document your taxonomy in a spreadsheet or table, defining each of its categories and subcategories and adding notes regarding their usage. Educate content creators about the information space's structure, as well as how to apply the subcategories and metadata you've defined to the content objects that they create.

3. **Match content objects to the defined categories.** Consistently applying your organizing principles, assign one or more subcategories to each of the information space's content objects. Alternatively, for a database structure, you must define the necessary metadata, then tag each content object with the appropriate metadata. To organize content objects into narrow, homogeneous, logical groupings, you can either manually index them *or* tag them with metadata. Content creators must have a clear understanding of the meaning and purpose of each of the categories and subcategories into which they might place content when classifying content objects they're creating. As content creators add new content objects to the information space's content-management system (CMS), tagging them with categories and subcategories during the publication process, they might need to make difficult decisions about where they should place certain content. Thus, you might need to elaborate on the criteria for specific subcategories to clarify their purpose. However, if content creators are unable to place new content into an existing category *and* subcategory, you must define new subcategories to fill gaps in the information architecture.

4. **Iteratively test and refine your taxonomy.** Because so many trade-offs are necessary in iteratively making categorization decisions, you must continually refine your taxonomy to ensure that it meets users' needs. To validate your taxonomy, you might conduct usability testing or A/B or multivariate testing, or use data analytics to verify that users are getting where they need to go. For detailed information about methods of evaluating information hierarchies, you can refer to Chapter 5, *UX Research Methods for Information Architecture*. Even after you launch the new information space, you might need to make some additional refinements to its structure once you've observed how users are actually navigating the space. Plus, as the organization adds new content, an information architecture's categories and subcategories must evolve to some degree. However, if possible, you should avoid changing the top-level categories over time. Doing so would be very disorienting to users. Nevertheless, if you must significantly expand the scope of the information space, you might need to add a few new categories.

5. **Specify intrapage, internal, and external hyperlinks.** Beyond the hyperlinks that correspond to the categories and subcategories of the information space's navigation system, consider what other hyperlinks would form useful connections between its pages and other content objects and those of other information spaces. You can create intra-page hyperlinks to make it easy for users to jump directly to the subsections of long pages or specific content objects such as images or videos. You can create internal hyperlinks that let users go directly to related pages or to subsections of other pages or specific content objects on them. Finally, you can create external hyperlinks to related information on other information spaces such as subsites that are the properties of the same organization or information spaces belonging to other organizations in the same domain.

6. **Maintain the information space's taxonomy and metadata.** Over time, as the information space's content evolves, you should work with your organization's content creators, eliciting and implementing their feedback regarding adding new subcategories and metadata, placing new subcategories within the information architecture, removing any subcategories or metadata that have become obsolete, and improving the information space's overall structure. Ideally, you should create and document a process that would let content creators define their own new subcategories or metadata as necessary to fill gaps in the information architecture. Ensure that whatever means of defining new categories or metadata you provide to content creators meets their needs and those of the broader organization. Unless your entire team makes the effort necessary to keep your taxonomy and metadata up to date, the information architecture would degrade over time. [17, 18, 19, 20, 21, 22]

Later in this chapter, the section "Diagramming an information architecture" provides detailed information about methods of determining and organizing an information space's categories. Refer to Chapter 7, *Classifying Information*, for in-depth information about categorizing an information space's content, defining meaningful metadata to describe its content objects, and creating whatever form of controlled vocabulary would best meet an organization's needs. You can also enable an information space's search engine to deliver highly relevant search results by defining metadata and creating a controlled vocabulary. Chapter 9, *Labeling Information*, describes how to create optimal labels for content objects and navigation links, using the users' vocabulary.

Taking a top-down or bottom-up approach to information architecture

"In creating the structure, we identify the specific aspects of that information that will be foremost in the users' minds."—Jesse James Garrett [23]

In the previous section, I outlined a sequence of steps to take in categorizing an information space's content, assuming a top-down approach. However, when defining an information architecture, you can take either—or preferably both—of the following approaches to categorization:

*   **Top-down approach**—When taking a top-down approach, you'll derive the categories from your understanding of the business's strategic objectives for the information space and the needs of its target users. You'll first define the broad, high-level categories that provide the overall

structure of the information space, then create a hierarchy comprising the subcategories that make up each of these high-level categories, and categorize the information space's content accordingly. However, taking this approach could potentially cause you to miss important details regarding the content.

- **Bottom-up approach**—When taking a bottom-up approach, you'll conduct an analysis of the organization's existing *and* planned content, define the low-level subcategories based on your analysis, and group these subcategories into high-level categories to create the overall structure of the information space. However, taking this approach could result in an information architecture that is not sufficiently flexible to meet the future needs of the business as the information space grows.

To create an optimal information architecture that can accommodate changes over time, you should endeavor to balance these two approaches. Ultimately, at some point in the future, as the scope of the information space expands and your strategic objectives evolve, you'll probably need to rethink its organizational structure to ensure that it continues to meet users' needs. [24]

Choosing your organizing principles

These top-down and bottom-up approaches help determine the *organizing principles* that drive your primary decisions about categorization and structure—that is, the criteria for grouping certain content objects together or decomposing an information space's content into disparate groups. Depending on the types of content that reside at the information space's various levels, you may need to apply different organizing principles at different levels. For example, the organizing principles for the highest levels of the information space might include the following:

- Strategic business objectives for the information space
- Users' needs for information
- Audiences for the content and possibly their geographies as well

In contrast, the organizing principles for the information space's lower levels of content often require an inherently conceptual, content-specific structure, as follows:

- Concepts and subordinate concepts relating to specific broad topics of information such as business, science, technology, news, sports, or entertainment
- Broad *facets*, or attributes, that are characteristic of a specific collection of content
- Chronology—often based on recency rather than historicity
- Detailed business requirements
- Detailed user requirements [25]

Diagramming an information architecture

"Web sites should be visualized to facilitate their planning [and] analysis.… … Diagrams…pack many layers of information into the flatland of graphic 2D presentation. … Web sites are inherently multidimensional "—Paul Kahn and Krzysztof Lenk [26]

The creation of diagrams plays an important role in the design and documentation of a Web site's information architecture or that of any other information space or product. Creating effective maps or other visual representations of an information architecture enables the information architect to conceive of and visualize its overall structure *and* capture the details of the information architecture in deliverables that efficiently and effectively communicate that structure to stakeholders and teammates.

Fully communicating the complexity of an information architecture could require that you create multiple visual representations of the information space, each looking at it from a slightly different perspective. The essential elements of IA diagrams are *content objects*—including pages and other individual content objects such as documents, PDFs, images, and presentations, *and* components of pages—and the connections, or *links*, between them. Most information architects create the following types of diagrams:

- **Structure maps**—These diagrams could represent the structure of either the entire information space or just part of one. They can show the key relationships that exist between the pages, other content objects, and content components of pages that an information architecture comprises. Initially, a structure map might show only the information space's high-level structure—that is, just the top one or two levels of a hierarchy or database structure—then, as the IA project progresses, expand and evolve to show the complete information architecture. However, structure maps should *not* include every indirect link on every page of the information space. Providing too much detail could obscure the diagram's essential information. For detailed information about creating structure maps, see the section "Creating structure maps," later in this chapter.

- **User-journey maps**—By creating, then optimizing a user-journey map for each of a specific user persona's key information-seeking tasks, you can ensure that users need not take any unnecessary steps in reaching the information they want and that the path they should take through the information space is clear. You can depict an optimal user journey for a specific information-seeking task by highlighting specific pages within a structure map. For more information about creating user-journey maps, see the section "Creating user-journey maps," later in this chapter.

- **Sketches**—You can sketch the information space's navigation systems and the content models for specific pages, either on your own or in collaboration with your team at a whiteboard. For a detailed discussion of creating content models, refer to the section "Content modeling," in Chapter 6, *Understanding and Structuring Content*. For more information about sketching, see the section "Sketching pages," in Chapter 12, *Designing Navigation*.

- **Wireframes**—Low-fidelity wireframes typically provide minimalistic depictions of the information space's navigation and search systems; the content models and layouts for key pages—particularly the home page and other key pages; the *templates* for particular types of pages and page components; references to design specifications, implementation details, content requirements, and existing content; and annotations that describe conceptual relationships between page components or the reasoning behind specific design decisions. Although creating low-fidelity wireframes could be solely the purview of an information architect, creating wireframes at any level of fidelity should ideally be a collaborative effort involving multiple disciplines, including graphic design and other design specialties, product management, and software development. For additional details on wireframing, see the section "Creating wireframes," in Chapter 12, *Designing Navigation.*

Be sure that the level of effort you put into the refinement and elegance of your diagrams is commensurate with the value they provide. When creating diagrams to win new clients or present your work to C-level leaders or other key stakeholders who fund your projects, you might want to invest the extra effort that is necessary to create beautiful, highly refined diagrams. However, to keep up with today's agile-development teams, you must quickly create, then iteratively modify your design deliverables to incorporate your teammates' feedback. [27, 28, 29, 30]

Creating diagrams that meet your audience's needs

The optimal diagrams for you to create to visually represent the structure of an information space depend on your organization's needs. Before creating your IA diagrams, you should identify the needs of the specific audiences for the diagrams—whether clients, business stakeholders, other UX professionals, or a software-development team—to ensure that you don't base your decisions regarding their creation on any erroneous assumptions.

Creating IA diagrams that meet the expectations of clients or business stakeholders often demands a greater degree of visual finesse. A simple, high-level diagram's purpose might be to gain financial support for creating a complete information architecture, while that of a more detailed diagram might be to get approval of the work you've done. Stakeholders often appreciate seeing a structure map that provides just an overview of an information architecture rather than one that depicts every detail.

In contrast, if you're working for a Lean or agile organization, your audience might value rapidity more than refinement and, thus, need you to quickly produce simple whiteboard sketches. You can create minimalistic, practical IA diagrams for the downstream use of your UX design colleagues and developers. Because these diagrams change frequently, they don't warrant your investing as much time in visual refinement as the diagrams you create for the people who sponsor your IA work.

Present your IA diagrams rather than just posting them online—especially when you're new to a team and haven't shared similar diagrams with them before. Only in this way can you ensure that people interpret the diagrams as you intended and have the opportunity to ask the questions that would aid their understanding. [31, 32]

Creating templates for your IA diagrams and documentation

Whenever you need to provide formal documentation for an IA project, taking the time to create a template for your documentation first can save both you and the people who use your documentation a lot of time and effort later on. To increase your teammates' efficiency and effectiveness in using your documentation, the following information should appear in a header on each page:

- **Author**—If multiple people have worked on the documentation, include all of their names, listing the primary author first.

- **Email address**—Provide the email address of the primary author, who should moderate the IA team's responses to all questions and feedback from the broader, multidisciplinary team.

- **Project**—Indicate the name of the project to which the documentation applies to make it easy for people to refer to the documentation and to ensure its structure remains intact.

- **Section heading**—Include the headings of the top-level sections—for example, "Structure Maps: Subcategory: [Name]"—to help people orient themselves within the documentation and to enable them to refer to specific information in the documentation.

- **Page number**—Indicate the page number to make it easy for people to refer to specific information in the documentation and to ensure that its structure remains intact.

- **Version number**—Indicate the documentation's version number to ensure that people refer to the correct version of the documentation.

- **Date**—Indicate the documentation's date to ensure that people refer to the correct version of the documentation.

- **Approver**—Provide the name of the person who is responsible for approving the documentation for implementation.

Be sure to consider how your teammates would be likely to consume your documentation. Would they view it online or print it out? If the latter, can your organization's printers generally provide color or only black-and-white printouts? What form factors can the printers handle? Be sure to format your documentation in a legible font, using a font size that is easy to read. [33]

Documenting your design rationale in annotations

If you're creating a structure map or a user-journey map, always capture the rationale for your design decisions in your annotations. If you're creating detailed design documentation, you can explain the rationale for your design decisions in greater detail. In particular, be sure to document your rationale for your overall information architecture. Also, create a record of any alternative ideas that you considered in solving particular design problems, then rejected, and why you made the design decisions that you did—whether structural, categorization, or labeling decisions. It's essential that you document your design decisions and share any plans for future content that factored into structural decisions so your team can apply them to future iterations of the information architecture and in the development of new content.

To facilitate the maintenance of the information space and preserve the integrity of your information architecture, provide detailed documentation that describes the purpose of each section and what key types of content it should contain. Indicate what content and navigation elements specific page types should include. Explain the structural decisions that you've made—especially those that resulted in your removing certain sections or content. Offer guidance on how content creators should make categorization decisions. List all of the categories, or *preferred terms*, the information architecture includes in alphabetical order. Provide a detailed description of each category, as well as some examples. Indicate any related terms—such as broader or narrower terms—that those who make categorization decisions might find useful, as well as any synonyms and other variants that they should avoid using.

Always be prepared to present the reasoning behind your design decisions to both your stakeholders and teammates. Only in this way can you persuade them that you've given your decisions the consideration they deserve and have come to the right conclusions. By effectively communicating and documenting your design decisions, you can ensure that your information architecture gets implemented as designed and prevent your teammates from perpetually revisiting the same design decisions. [34, 35, 36]

Outlining information architectures

Creating visual representations of an information architecture is labor-intensive and, thus, time-consuming work. How labor intensive and time consuming depends on the complexity of the information architecture, as well as the elegance of the diagrams you're creating.

In contrast, if you're creating a simple hierarchy, using a word processor's outlining feature or a spreadsheet application to create a hierarchical list of links is the easiest, quickest way of designing and revising an information architecture. Therefore, information architects who are working within an agile or Lean environment and need to work quickly often document their information architectures in the form of lists, outlines, or spreadsheets. However, if the application you're using requires you to manually create a numbering system, wait to add the numbers until you're nearly finished organizing everything to prevent your wasting a lot of time and effort renumbering things.

An outline's indented levels provide an effective representation of a hierarchy, no matter how many levels it comprises. So, if you need to represent a deep hierarchy, an outline can be more effective than a structure map. Be sure to create a well-designed template and styles for your information-architecture outlines to ensure that they are as consistent and aesthetically pleasing as possible. To fit more levels on a page and prevent there being an excess of whitespace at the left of your outlines, avoid making each level of indentation too deep. Use plenty of whitespace between the items in an outline to make their text easier to read. In each item, render the link text in bold. If any description or annotation follows the link text, set off the link text with an em dash or colon.

For collections of pages of particular types or structured content such as the leader profiles, articles, or news items that an organization's content creators generate continually, rather than adding a specific title, provide a concise description of the page type within brackets, indicating a variable—for example, [Profile Name] or [Article Title].

Creating and editing a spreadsheet to represent an information hierarchy is also quite easy. If you've already conducted a content inventory and captured your data in a spreadsheet, you can use that spreadsheet as the basis for your information-architecture documentation. Be sure to document any changes that you make to the information space's structure. For a large, enterprise information space, a spreadsheet is much easier to maintain than a graphic representation of a structure map. The width of the spreadsheet can better accommodate all the indented levels of a deep hierarchy than an outline can. However, even though you could add columns for other types of information such as links to other internal pages, a spreadsheet doesn't accommodate adding descriptions or annotations relating to specific list items as well as an outline does. Plus, even though spreadsheet applications provide some additional formatting options, applying this formatting can be fairly laborious.

When preparing to diagram an information architecture, you'll typically first create sketches or use Post-it notes to figure out what the structure should be, then create your diagrams. However, creating text-based IA documentation need not be a two-stage process because it's so easy to edit or even restructure at a high level—often requiring just a simple cut and paste. Also, there is no need to worry about fitting the visual elements of a diagram on a page, which can require a lot of tweaking. [37, 38, 39, 40, 41]

Creating structure maps

"Mapping a Web site requires several steps. First, the kind of information to be represented must be defined. … Cataloging the content of the site according to these information types comes next. Third, the information must be organized into a visual pattern."—Paul Kahn and Krzysztof Lenk [42]

Note—Rather than referring to maps of information spaces as *site maps*, I've used the more generic term *structure maps* to communicate clearly that the information in this book applies to *all* digital information spaces, *not* just Web sites. Although some refer to structure maps as *blueprints*, my having studied the architecture of buildings causes me to reject this use of the term. Architectural blueprints bear a much stronger resemblance to the wireframes that map the structures of individual pages.

A *structure map* can provide an aesthetically pleasing, visual representation of the overall structure, or information architecture, for an entire information space *or* specific sections of an information space. This map depicts the relationships between content objects such as pages and content components. While mapping an information space is a very useful, fairly easy, and flexible approach to designing an information architecture, refining the layout of all the map's visual elements to perfectly align them can be labor intensive. Plus, revising the map can be a very laborious, time-consuming process—whether during iterative design when incorporating your research findings or your team's feedback or when revising the map over time, as the information space's content evolves.

When devising your IA strategy for an information space and initially defining its high-level, conceptual structure, you might have created a structure map to represent that structure. Such a structure map could provide an overview of the information space's entire top-level navigation system, including supplementary header and footer navigation links. Figure 10.1 provides an example of such a high-level structure map. Alternatively, a structure map could comprise just the top two levels of an information space's primary navigation system. In either case, the structure map should consist of the actual labels for the navigation system's links. Figure 10.2 shows a structure map that comprises two levels of navigation links. A high-level structure map could alternatively represent the information space's key pages and comprise their actual titles, as shown in Figure 10.3. Whether a structure map comprises navigation links or page titles, using the actual link labels or page titles helps establish a shared vocabulary for the information space.

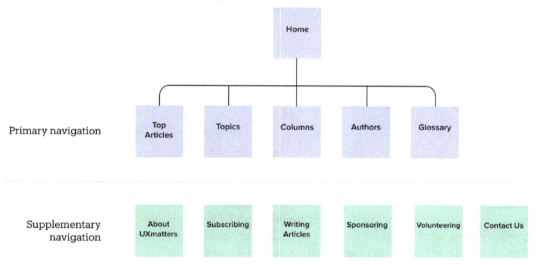

Figure 10.1—A high-level structure map of the *UXmatters* Web site

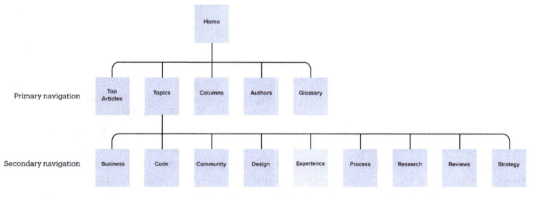

Figure 10.2—A structure map of the *UXmatters* Web site, showing two levels of navigation

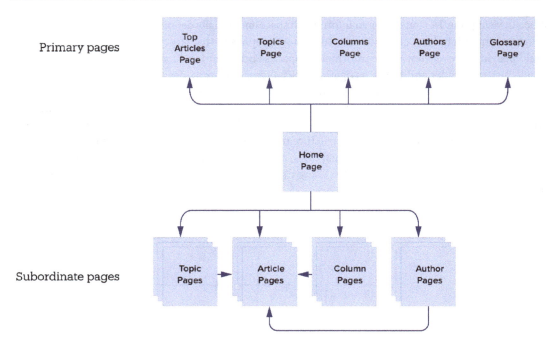

Figure 10.3—A structure map showing key pages of the *UXmatters* Web site

Depicting an information space in some type of high-level structure map can be very helpful in describing your IA strategy to your stakeholders and teammates. Your presentation of a structure map often raises questions that challenge your assumptions and design decisions, prompting interesting discussions about other possible information-architecture solutions and enabling you to resolve any design issues early in the *Design* phase when doing so is less costly.

A structure map is generally most useful either for representing just the high-level structure of a larger information space or for detailing the entire structure of an information space that has a fairly limited scope.

When preparing to create a more detailed structure map during *Design*, you should initially either create whiteboard sketches or use Post-it notes to figure out the information space's structure. If you're using Post-it notes to represent navigation links, write each link label on a Post-it; or if your Post-its represent pages, write each page title on a Post-it. Using Post-it notes offers the advantage of their being very easy to rearrange. If you're including both link labels and page titles in the same structure map, consider using different colors of Post-its for navigation links versus pages. *Or*, if your map comprises both simple pages and pages comprising structured content, use different colors of Post-its for the pages themselves and their individual components. *Or*, if an information space's pages or components of pages provide interactive functionality to support tasks other than information-seeking, consider using different colors of Post-its for that functionality.

Begin laying out the information space's structure by arranging the Post-it notes on a whiteboard or another large, flat surface. If you're mapping the information space's pages, you'll typically start by placing the information space's home page and the main pages for its key sections at the top. Then arrange columns of subordinate pages beneath each of the main pages, using indentation to indicate their level. Most structure maps also represent some groups of particular types of structured pages—rather than individual instances of pages—such as profiles, news items, and jobs on corporate Web sites, product pages on ecommerce sites, and article pages on *UXmatters* or other Web magazines.

In contrast, if you're mapping the information space's navigation system, you'll typically start by laying out the information space's primary navigation links across the top. Then arrange columns containing each section's subcategories beneath their corresponding primary navigation link, using indentation to show subordinate levels of subcategories.

Throughout the highly iterative process of mapping an information space, its structure will continually expand and evolve, and you'll progressively add greater detail. Capture a photo of each iteration of the structure as you work. Then, if you decide to undo some of your changes, you can easily backtrack to an earlier version. Continue making any necessary revisions until the structure stabilizes. Then, if the connections are not obvious, you can optionally connect the Post-it notes by drawing arrows or using Post-it arrows to show the connections between them. Take a photo of the final layout of Post-its, which could be all the documentation you need.

Once you've made most of your structural decisions, consider creating a more refined, graphic representation of your structure map for presentation or inclusion in any formal IA documentation. If you're designing an information architecture comprising only a small number of pages, you might be able to depict its entire structure in a single map. However, a single visual representation of an information space cannot comprehend every structural detail. For example, you should *not* generally include indirect, contextual links between specific pages. Therefore, if you're mapping a large, enterprise information space, you'll probably need to take a modular approach to mapping it—especially when a team of information architects is structuring different sections of the space.

Always keep your structure maps as simple as possible, focusing mainly on the information architecture's key structural elements. Nevertheless, you should provide a comprehensive view of the overall information space and represent all elements of the navigation system, all necessary pages and page components, and both existing and planned content. When creating a detailed structure map, employ the actual page titles or link labels.

As your IA project progresses, you might decide to create additional structure maps that focus on specific sections of the information space. Ultimately, you could map the entire information space. A clear structure map, comprising boxes and connectors or arrows, provides a very effective means of visualizing an information architecture—one that your stakeholders and teammates will find easy to understand. [43, 44, 45, 46, 47, 48, 49, 50]

Representing and providing a legend for the elements in your structure maps

To ensure that your teammates can easily understand these structure maps, you should limit the number of different types of visual elements that you use to represent content objects and the connections between them and include a legend that depicts their meaning. Structure maps commonly include the following types of elements:

Individual content objects:

- **Home page**—A page symbol represents the page that is the primary entry point for the information space or product.

- **Section's main pages**—Page symbols represent the information space's secondary entry points.

- **Subordinate pages**—Page symbols represent the individual pages that are subordinate to the higher-level pages.

- **Other individual content objects**—These symbols include PDFs, documents in other formats, spreadsheets, or images.

Groups of pages:

- **Collections of similar pages**—A stack of overlapping pages represents a collection of pages of a particular type or multiple instances of some type of structured content such as profiles, articles, or news items that an organization's content creators generate continually. Rather than adding a specific page title, provide a concise description of the page type within brackets, indicating a variable—for example, [Profile Name] or [Article Title].

- **Groupings of related pages**—Boxes contain groups comprising pages that are closely related to one another.

Components of pages:

- **Content components**—These are boxed content objects that appear on a page and are the components that make up the pages of the information space.

- **Integrated media**—These boxed content objects represent images, videos, audio clips, and presentations.

- **Integrated functionality**—These boxed content objects include search forms, filtering functionality, sign-up forms, sign-in forms, and other types of Web forms.

- **Groups of similar components**—These boxed groups comprise components of the same type—for example, people's profiles or job descriptions.

- **Groups of related components**—These are groups comprising closely related components—for example, on an ecommerce site, a product's description, pricing, product details, and reviews. Depending on the scope of such related components, all of them might appear together on a single page of a specific type, whose sections contain particular types of content and use a consistent set of labels *or* they might comprise a well-defined set of pages.

Connectors:

- **Line connectors**—These connectors depict the information space's structure and the direct, hierarchical relationships, or parent-child relationships, between pages and other content objects.
- **Arrow connectors**—These connectors depict the information space's workflow and, thus, the relationships between pages and components of pages.
- **Links**—These connectors also take the form of arrows and can depict either one-way or bidirectional links between pages and components of pages that the user can traverse.

Figure 10.4 shows an example of a legend for a structure map.

Figure 10.4—A legend

Defining a minimal set of clear visual elements for your structure maps ensures that your team can accurately interpret them and helps you to avoid overloading your maps with unnecessary information. Use the same visual elements consistently across all your maps, regardless of their scope—although you might use a more limited set of elements for maps with a limited scope. [51, 52, 53]

Organizing your structure maps and other deliverables

If you're designing a complex information space—especially a database-driven information space—you'll ultimately need to assign a unique identifier (ID) to each of its content objects, according to its level in the information hierarchy. This eliminates any potential confusion regarding the content objects to which you and your teammates are referring. The page IDs could derive from those that you assigned to pages when conducting any content inventory. For example, you could assign the ID 0.0 to an information space's home page; 1.0, 2.0, 3.0, 4.0, and so on to the main pages for the sections; 1.1, 1.2, 1.3, 1.4, and so on to the first section's subordinate pages; and 1.1.1, 1.1.2, 1.1.3, 1.1.4, and so on to the next level of pages; then extend this page-numbering scheme across all levels of the information space's hierarchy. Content objects in the same section and at the same level in its hierarchy have sibling relationships to one another. As Figure 10.5 shows, using this numbering scheme clearly indicates the various levels of the information space's hierarchy and helps organize what could become a large, modular set of maps.

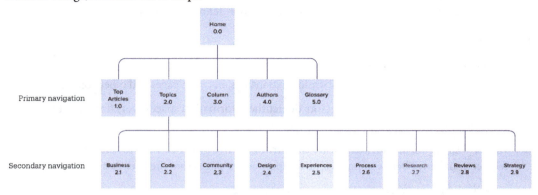

Figure 10.5—A structure map of *UXmatters* with page identifiers

Similarly, to identify the information space's content components, you could number each of the components on a particular page as follows: C1, C2, C3, C4, C5, and so on, as in Figure 10.6. Then, to provide a unique ID for each of the information space's components, you could combine the ID for a page on which a component resides with the component's ID—for example, P2.4–C4 or P5.2–C2.

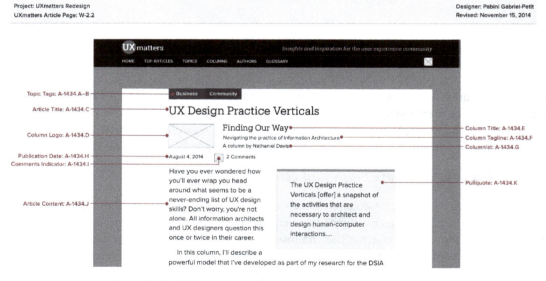

Figure 10.6—A *UXmatters* article page with component identifiers

However, if an information space employs some reusable components, you would need to create component IDs that are completely independent of any page IDs and are universally unique across the entire information space—for example, the IDs for components might be C1 through C214.

When creating structure maps, you can leverage the same scheme of identifiers that you used in conducting a content inventory for an existing information space. Identify any gaps in the content, then create new content objects with IDs that fit into the existing scheme. [54, 55, 56]

Annotating structure maps

Annotations to a structure map might indicate the following:

- Research findings regarding particular parts of the information architecture
- Feedback your teammates have provided on particular parts of the information architecture
- The metadata for a specific page
- Key internal and external contextual links on a specific page
- Whether a specific page comprises structured content
- What components a structured page includes
- What content needs to be created for specific pages
- Whether the content on specific pages requires revision
- What templates apply to particular pages or types of pages
- Any parts of an information space that constitute the equivalent of subsites
- Any parts of an information space that are available only to users who are logged in
- Any interactive functionality relating to user tasks other than information seeking
- What parts of an information space are to be implemented during specific phases of a project
- Any parts of an information space that are beyond the scope of the current project [57, 58]

Creating user-journey maps

A *user-journey map* provides a visual representation of the key steps in the end-to-end journey that a particular user persona takes to achieve a specific information-seeking goal when using an information space. Each user persona has unique goals, tasks, and needs so experiences different user journeys, comprising specific sequences of steps. Therefore, they encounter different challenges. Journey mapping is a simple, flexible approach to designing and optimizing the information architecture to support all user personas' key information-seeking tasks.

Regardless of whether you're mapping an existing or a new user journey, the first step of the process is identifying *all* user personas' key information-seeking tasks and, thus, the user journeys you need to map.

Once you've identified what existing user journeys you should map, the process of mapping them comprises the following steps:

1. Evaluate the user journeys by conducting usability testing, doing an expert review, or analyzing the usage-data analytics for them.

2. Conduct user research to better understand users' key information-seeking tasks.

3. Analyze your UX research findings to identify users' painpoints and the design or system issues that are causing them—which might be happening many steps earlier in the user journey—as well as any gaps in the existing information architecture.

4. Determine how best to remedy the issues.

5. Optimize all existing user journeys as necessary.

6. Design and map any new user journeys that are necessary to support the user personas' key information-seeking tasks and fill gaps in the existing information architecture.

7. Iteratively test and refine the design solutions for the new user journeys.

When creating a new information architecture, once you've identified what user journeys you need to support all user personas' key information-seeking tasks, follow these steps:

1. Conduct user research to better understand users' key information-seeking tasks.

2. Create personas and usage scenarios that are based on your team's user research.

3. Conduct a competitive analysis to evaluate similar user journeys in any competitors' information spaces.

4. Analyze the UX research findings to identify what user journeys are necessary to support all user personas' key information-seeking tasks.

Then, for each user journey, do the following:

1. Identify any phases that the information-seeking task comprises and determine their order. Phases are typically necessary for user-journey maps that have a broad scope—such as multichannel user experiences, the entire lifecycles of relationships or projects, or tasks that take a long time to complete.

2. Determine the steps of the overall user journey and their order. For user journey–mapping projects that begin with this step of the mapping process, each journey map typically relates to a specific task scenario.

3. Determine whether multiple steps in the user journey should occur on the same page and their groupings.

4. Design and map the path for the entire user journey.

5. Iteratively test and refine your design solution for the user journey.

Creating user-journey maps for users' key information-seeking tasks can help you ensure that the paths users must take in seeking specific information match their mental models for where the information should be and prevents users from needing to take any unnecessary steps in reaching the information they want. [59, 60, 61, 62]

Identifying users' key information-seeking tasks and journeys

When analyzing an information space's user journeys, you must first identify users' key information-seeking tasks and what overall user journeys would enable users to complete them. To determine what user journeys would align most closely with business and user requirements, answer the following questions:

- What essential information and links must users easily be able to find to achieve the business goals for the information space?

- What information is of greatest interest to the users of the information space?

- What paths do users commonly take when seeking information on any existing information space?

Whether you're evaluating an existing information architecture or creating a new one, consider the following questions regarding each key information-seeking task:

- How many and what steps must the user take to complete the user journey?

- What content is necessary for the user to complete each step in the user journey?

- How could you optimize the path of the user journey?

- Might the user need to choose whether to move forward or backward along the path of the user journey to complete certain steps?

- Might displaying all the steps on a page in the user journey require progressive disclosure?

- Might displaying the content that the user must consume to successfully complete the user journey require searching or filtering?

- Might the user need to perform certain steps in the user journey iteratively?

- Could you reduce the number of steps the journey takes without negatively impacting more important user journeys?

- Would the user journey cause the user to ping-pong back and forth between pages? Could you prevent this by providing all necessary information on the same page?

- What steps might lead the user astray? Could you eliminate those steps or enable the user to skip them?

- Might some users encounter dead ends because they've taken a wrong turn in the user journey, causing them to backtrack? At what point did users choose the wrong path?

- Where have users abandoned the user journey without reaching their goal, become frustrated, and left the information space? Could you provide information at that point that would get users back on track?

- Does access to a particular user journey belong in the primary navigation, secondary navigation, supplementary navigation, or contextual navigation?

- Could you offer a shortcut to expert users that would provide an alternative, more efficient user journey? [63]

Creating user-journey maps

Working either independently or collaboratively with other members of your team, sketch or diagram a user-journey map for each user persona's key information-seeking tasks when using an information space. Give each user-journey map a title that describes a user's specific information-seeking goal.

When you're preparing to sketch a user-journey map for a specific journey, write the title of each step in the journey on a Post-it note, then lay out the notes in order on a whiteboard or another large, flat surface. Connect the Post-it notes by drawing arrows or using Post-it arrows to indicate the user journey's path. Capture photos of the user journeys you create. Before making significant changes to your user journeys, be sure to capture their current state. Then, if you regret making certain changes, you can backtrack and restore a previous version.

You can diagram individual user journeys using a variety of software applications. When diagramming a user-journey map, you could employ the following visual elements:

- **Title**—The title of a user-journey map should have the following format: **Goal:** [A description of a specific information need].

- **Steps**—A rectangle represents each specific step in a user journey. The label for each step should describe the precise action that the user needs to take at that point in the journey—for example, clicking a link that would allow the user to proceed to the next step in the journey or reading content that would prepare the user for the next step. One or more steps could occur on an individual page. The key goal in decomposing a user journey into steps is making sure that each step makes sense to the user and flows naturally from the previous step.

- **Groupings**—If multiple steps in a user journey occur on the same page, group the steps using a rounded rectangle with a gray background. You could optionally group certain steps using a rounded rectangle with a gray background. Such groupings can be nested. For example, you might group steps that occur within specific phases of a process or relate to the actions of specific user personas.

- **Starting step**—Use a rounded rectangle to represent the first step in a user-journey map.

- **Decision points**—A diamond shape represents a step in a user-journey map that is a decision point, at which the user must choose what action or path to take and, thus, at which the user journey branches into two or more alternative paths.

- **Path**—The overall path of a user journey generally flows from left to right. You can represent a user journey's path and the directions in which the user can proceed by using arrows to connect its steps. These arrows are usually single-headed, one-way arrows, but in some cases, can be doubled-headed, two-way arrows. Use curved, single-headed arrows to connect the steps that form an iterative loop. If you're creating a user-journey map that depicts multiple journeys, highlight the overall path for each specific user journey using a different, contrasting background color.

- **Final step**—Use a rounded rectangle to represent the last step in a user-journey map. If a user journey branches—for example, depending on a decision point or a user persona—it could have more than one final step.

Figure 10.7 shows an example of a simple user-journey map, depicting the current repeat-purchase process for existing users on Amazon's iPad app. A complex user-journey map might show multiple variants of a single user journey or represent the evolution of a user journey across the stages of the development process. For example, following UX research and analysis, a team might incorporate insights about experience stages and steps, interactions, users' goals and motivations, positive moments, painpoints, and opportunities for improvement.

If necessary, provide a key, or *legend*, showing the meanings of various visual elements in your user-journey maps.

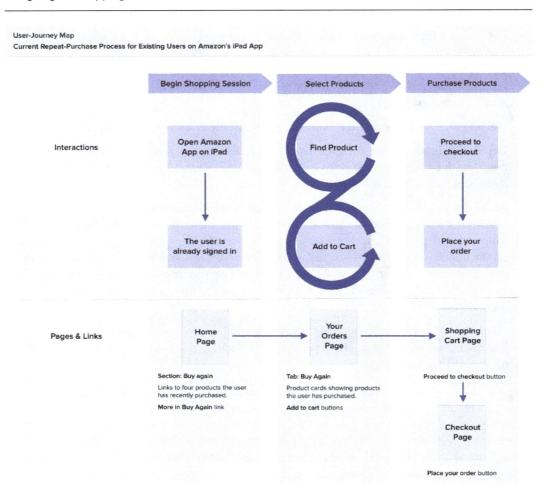

Figure 10.7—A simple user-journey map

If you've already created structure maps for an information space, they can provide the basis for your user-journey maps. One easy way of diagramming the optimal user journey for a specific information-seeking task is by highlighting specific pages within the context of a structure map, as shown in Figure 10.8. [64, 65, 66, 67, 68]

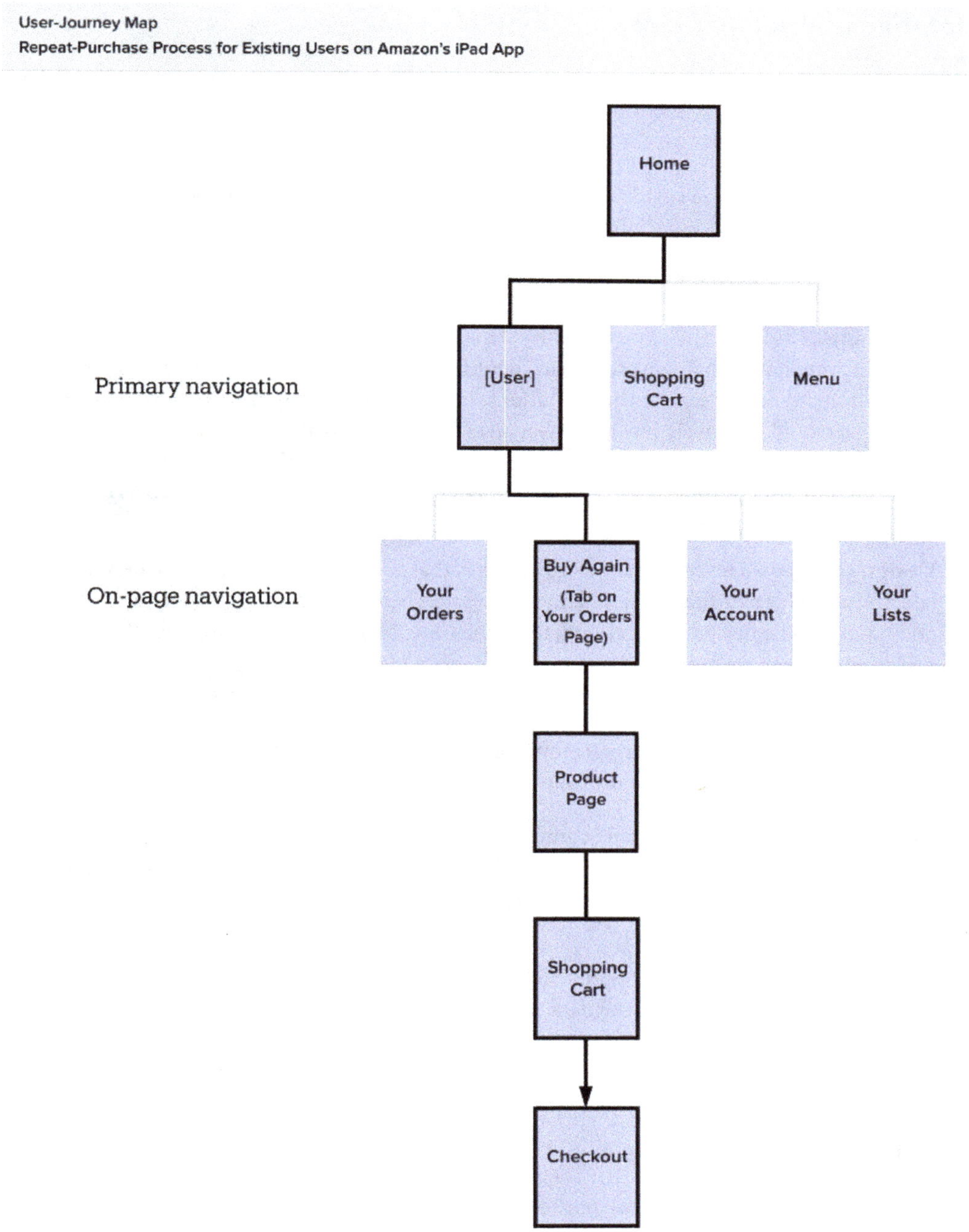

User-Journey Map
Repeat-Purchase Process for Existing Users on Amazon's iPad App

Figure 10.8—A user-journey map superimposed on a structure map

Annotating user-journey maps

Provide whatever annotations would be most salient to the audience for your user-journey maps. When you're creating a user-journey map to analyze an existing user journey, you should use annotations for the following:

- **Usage-data analytics**—Provide this data to indicate the usage levels for specific user journeys and elements on pages, especially those that comprise multiple steps. Your team should invest in improving the user journeys that get the most usage, especially those that also present significant painpoints to users or drive business revenues.

- **Painpoints**—Indicate and document any steps in user journeys that are painpoints for users, including any issues that your team has identified, as well as their possible solutions. Define clear requirements for each of these solutions. Invest in remedying the painpoints, particularly those that would negatively impact conversions. Or you could alternatively indicate what steps in user journeys elicit positive, negative, or neutral perceptions of the user experience. Redesign steps or entire user journeys that prompt strong negative emotions in users such as frustration, confusion, or indecision.

- **Design changes**—Describe the design changes that you're proposing and create sketches or wireframes to depict them. These annotations can help your team to understand the scope and nature of the design changes and, thus, the time, budget, and technologies that would be necessary to implement them; the number of pages and templates the project requires; and any new design standards and guidelines that would be necessary to ensure consistency.

To ensure that your maps of new user journeys convey all the information that your team members would find useful, you might provide the following annotations:

- **Essential elements**—Indicate any content components or links that are essential elements on particular pages in a user journey, without which the user would be unable to complete the user journey. Define clear requirements for each of these content components and create sketches or wireframes that depict them.

- **Proposed analytics**—Indicate points in the new user journeys at which you want to measure usage-data analytics to gather the data your team needs.

Consider color-coding particular types of information in your user-journey maps by using different colors of Post-it notes or markers. [69, 70]

Validating user-journey maps

Once you've conducted UX research and created well-informed user-journey maps for an information space, you should share the maps with your team, get their feedback, and incorporate that feedback into your maps. Conduct usability, A/B, or multivariate testing to validate that the user journeys you've created meet the needs of both the business and all the user personas.

You can conduct usability testing at any stage of *Design* or *Development*—by testing your sketches, wireframes, or a clickable prototype—as well as after launching a new or revised information space, then optimize its user journeys. Once you've launched an information space, you can conduct A/B or multivariate testing to evaluate the performance of two or more different design solutions for the same information-seeking task to determine which is most effective.

Regardless of what type of testing you decide to do, be sure to evaluate the following:

- User journeys' individual steps and their content and navigation elements
- Grouping and ordering of multiple content and navigation elements on the same page
- Decision points
- Painpoints
- User journeys' overall paths [71, 72]

Summary

In this chapter, you learned about the key steps of designing and mapping an information architecture. First, the chapter covered five UX research methods that you can employ in evaluating any existing information space's organizational structure and findability: closed card sorting, reverse card sorting, card-based classification evaluation, tree testing, and usage-data analytics. Then, it explained how to choose the right organizational model for an information space, which comprises its structural pattern and organizational scheme. Next, it provided an overview of the process of categorizing and labeling an information space's content, summarizing your learnings from the earlier chapters of the book. It considered taking a top-down versus a bottom-up approach to information architecture. The chapter concluded with an in-depth discussion of some methods of diagramming and documenting an information architecture, focusing primarily on creating structure maps and user-journey maps.

Next, Chapter 11, *Foundations of Navigation Design*, will discuss key concerns and objective of designing an information space's navigation system in depth. The chapter will also provide navigation design guidelines and describe several types of navigation.

References

To make it easy for readers to follow links to the references for this chapter, we've made them available on the Web: `https://github.com/PacktPublishing/Designing-Information-Architecture/tree/main/Chapter10`

11
Foundations of Navigation Design

"Navigation provides access to information, shows location in a site, shows [the] aboutness of a site, reflects brand, affects site credibility, and impacts the bottom line."—Jim Kalbach [1]

The navigation system for any digital information space comprises *hyperlinks* that connect its content in meaningful ways and, thus, enable users to browse the information space to find the content they need.

A well-designed navigation system orients users by conveying an information space's scope and the subject matter of its content, efficiently provides access to all of that content, and indicates the user's current location within the space. An effective navigation system enables users to build a mental model of an information space and, thus, prevents them from getting lost in hyperspace and experiencing the consequent frustration that could drive them to competitors.

The design of any successful navigation system balances the need to serve users' information-seeking needs with that to satisfy an organization's business goals, thereby determining the users' navigation experience.

In this chapter, I'll discuss the following topics relating to the design of navigation systems for information spaces:

- Key concerns of navigation design
- Key objectives of designing navigation systems
- Navigation design guidelines
- Types of navigation

Key concerns of navigation design

"The strategic foundation of a user interface is a dense mix of requirements and assumptions that may span a wide array of concerns—from business and user objectives to content, design-system libraries, and technical infrastructure."—Nathaniel Davis [2]

To ensure the effectiveness of an information space's navigation system, its designer must address the primary concerns at the foundation of navigation design. Let's explore these key concerns now.

Understanding and meeting the users' needs

The essential foundation of any user-centered, navigation-design project is your understanding of actual or representative users and their needs and behaviors. Gaining a deep understanding of how to meet users' needs requires UX research and analysis.

To learn about some UX research methods that you can employ when designing a navigation system, see Chapter 5, *UX Research Methods for Information Architecture*. Competitive analysis, open card sorting, and free-listing are generative methods of UX research that commonly inform navigation design. Some common methods of validating a navigation design include tree testing, A/B and multivariate testing, and data analytics.

Supporting various modes of information seeking

Information seeking is a nonlinear, exploratory process. Since people seek information for a variety of reasons—including monitoring information, exploratory seeking, known-item seeking, exhaustive research, and finding useful information again—and they explore and discover information in many different ways, it is essential that you design navigation systems that flexibly support their different approaches to finding information. See Chapter 2, *How People Seek Information*, which describes people's information-seeking behaviors in depth and informs the user-centered design of navigation systems.

Understanding and satisfying business goals and requirements

A navigation design project's unique business context and goals determine the business requirements for that navigation system. The business strategy for an information space informs its information-architecture and navigation-design strategies. When defining the business goals and requirements for an information space's navigation system, the members of your project team should refer to the section "Understanding and aligning with business strategy," in Chapter 8, *Defining an Information-Architecture Strategy*.

Creating an effective information architecture

Because an information space's structural navigation reflects the hierarchical structure of its information architecture, its navigation links should correspond to the categories of content the information architecture comprises. Structural navigation determines the main paths that users can take through the information space, and its groupings of links typically correspond to an information architecture's levels or types of links. While this entire book focuses on information architecture, to better understand the structures and organizational schemes that are characteristic of most information spaces, see Chapter 4, *Structural Patterns and Organization Schemes*. To learn about defining an information space's categories of content, read Chapter 7, *Classifying Information*. To better understand the core process of designing an information architecture, refer to Chapter 10, *Designing and Mapping an Information Architecture*.

Providing clear labels for navigation links

The clear labeling of navigation hyperlinks is key to ensuring users can feel confident that they are making progress toward their information-seeking goals. Because users quickly scan navigation options, then click the link that most closely matches their information need, the available options at any juncture in a navigation path must be easy to distinguish from one another. The terminology that you use in link labels should be consistent to avoid any confusion on the part of users. For in-depth information about creating effective labels for navigation hyperlinks, read Chapter 9, *Labeling Information*.

Employing optimal navigation elements

Your decisions regarding the navigation elements that form an information space's navigation system determine the scope of the navigation options that are visible—and, therefore, that are available to users at each level of navigation—as well as the interactivity that would provide the greatest flexibility in the presentation of links through progressive disclosure. To preserve the scalability of the navigation system and serve both the current and future needs of its users and the business, the navigation designer must have a deep understanding of the information space's existing *and* planned content. With this knowledge, the designer can choose the most effective navigation elements for the information space and provide the most efficient access to its content. For descriptions of common navigation elements and their uses, see the section "Navigation design patterns," in Chapter 12, *Designing Navigation*.

Designing effective layouts for navigation elements

The placement of navigation elements on an information space's pages determines users' ability to readily perceive them and easily understand the sequence of actions they must take to drill down through the information space's hierarchy. Consistency in the layout of the user-interface (UI) elements that the navigation system comprises across the pages of the information space is essential. For more information about how to lay out an information space's navigation elements effectively, refer to the sections, "Desktop navigation layouts" and "Mobile navigation layouts," in Chapter 12, *Designing Navigation*.

Key objectives for designing navigation systems

"Navigation plays a major role in shaping our experiences on the Web. It provides access to information in a way that enhances understanding, reflects brand, and lends [credibility to] a site."—Jim Kalbach [3]

To design a navigation system that meets the needs of both an information space's users and the business, the designer must achieve *all* of the objectives that I describe in the following sections.

Understand and serve the needs of an information space's users.

Determine who the information space's users are, then ensure that the design of the navigation system serves their diverse, exploratory information-seeking needs and supports their various approaches to browsing the information space. In determining your design priorities, you must first understand what is most important to these users: what types of information they're seeking, why they need it, what terms they use in describing that information, and how their information needs evolve as they browse the information space. [4]

You can answer all of these questions through your learnings from qualitative UX research. When designing a new information space and navigation system, you should conduct user interviews and observations or card-sorting exercises. When working to improve an existing navigation system, conduct usability testing or use a combination of A/B testing and data analytics to obtain the information you need to inform a user-centered design (UCD) process. Only by engaging in user-centered design can you develop a deep understanding of users and their needs and design a navigation system that fully meets those needs. The navigation system's structure should reflect users' understanding of the information space's content, and its functionality should meet their expectations of a browsing experience. [5]

Understand and convey the scope of an information space's content.

Before designing an information space's navigation system, you must understand the nature and structure of its content. The navigation system should clearly convey the scope of the information space's content and communicate both the breadth and depth of the subject matter that it comprehends. The scope of any information space's content typically grows over time, so it is important to design a scalable navigation system that can accommodate the existing and planned content, as well as content the organization might create in the future.

The navigation system must also let users quickly and accurately discern whether the information space's content can meet their information needs, as well as whether their current navigation path would lead them to the information they're seeking. The layout of navigation elements on an information space's pages plays a key role in facilitating users' recognition of the navigation paths that would enable them to find that information.

Balance an information space's structure and, thus, its structural navigation system.

An information space's structural navigation system derives from its information architecture, so you must first understand its structure and the categorization of its content. To achieve a balanced navigation system, it is essential that you first balance the breadth and depth of the information architecture. Ensure that each level of the information space's structural navigation comprises neither too many nor too few hyperlinks. Only by creating a well-balanced, clearly labeled navigation system can you ensure that the links the system comprises provide good information scent, enabling users to find the information they need.

Fulfill the business requirements for an information space.

The overall structure of an information space and, thus, its navigation system must reflect business priorities, which determine such factors as the following:

- The links, or categories, that the system comprises
- The order in which these links appear, as well as their groupings
- The layout of the navigation elements on pages
- The depth at which particular content resides

If an information space's visitors cannot find the information they're seeking, the company won't be able to achieve its desired business results. This is particularly the case for ecommerce companies. Customers cannot purchase products they cannot find. However, they'll happily continue browsing as long as they're finding products they want to purchase. Thus, the effectiveness of an information space's navigation system can have significant impacts on a business's profitability and ultimate success.

Investing in a high-quality navigation system for a company's intranet is essential to supporting employees' productivity. Too often, the navigation systems of corporate intranets lack a coherent design solution because different departments have failed to work together to create one. As a consequence, the time it takes employees to find the information they need negatively impacts their productivity, especially on very large intranets. Only by taking a holistic approach to navigation design and devising a solution that is consistent across its departments can a company deliver an intranet that works better for employees. [6]

Build the brand's credibility

The effective organization of the information space's content and, consequently, the design of its navigation system can help build the credibility of the business's brand and engender trust in the brand among potential customers and partners. Thus, a well-designed navigation system can help a business both reach its intended audience and persuade visitors to take actions that would enable the business to achieve its business goals. However, converting visitors to customers requires that they perceive the brand as credible. Whether a business wants visitors to consume specific information, sign up for a service, make a purchase, or take any other action that would benefit the business, persuading them to take the desired action requires building the brand's credibility.

A well-structured navigation system whose design reflects a business's brand identity can help build credibility, foster trust, and encourage customers' adoption of and loyalty to the brand. In contrast, a poorly designed navigation system that impairs customers' ability to discover the information or products they need can discourage customers from continuing to use an information space and drive them to those of competitors instead. [7]

Navigation design guidelines

"Usability guidelines endure because they depend on human behavior, which changes very slowly, if at all. What was difficult for users twenty years ago continues to be difficult today."—Jakob Nielsen [8]

Navigation design guidelines are a specific class of usability guidelines. Before we explore particular types of navigation and specific navigation design patterns, let's consider the following universally applicable guidelines for designing navigation systems. These general design guidelines apply to all types of navigation, as well as the navigation design patterns that you'll learn about in Chapter 12, *Designing Navigation*.

Design for maximal flexibility and freedom of movement.

In designing an information space's navigation system, make sure that the hyperlinks the system comprises adequately support both bidirectional, vertical movement within the information hierarchy *and* lateral movement across the information space without regard to the hierarchy. Fulfilling this guideline satisfies the design principle that requires providing all the appropriate navigation options at each wayfinding decision point, whether the user decides to explore further along the current path, choose a different path, or reverse direction and backtrack to an earlier decision point.

However, in designing navigation options, you must balance two goals: flexibility and simplicity. Although users should be able to go "anywhere from anywhere," [9] you should avoid cluttering pages with navigation elements and lateral links that are only marginally useful. The structure of an information hierarchy should always be readily apparent to users and any other navigation tools should clearly supplement or complement the structural navigation. You'll learn more about these different types of navigation later in this chapter.

Balance the information architecture's breadth and depth and, thus, the navigation hierarchy.

Balancing the breadth and depth of the information architecture from which an information space's structural navigation derives is essential. A well-balanced information architecture ensures that there are neither too many nor too few navigation links at any given level of the information space's structural navigation system.

A navigation hierarchy with greater breadth has more top-level categories, more categories at each subsequent level, and typically, fewer levels, while a deeper navigation hierarchy comprises fewer top-level categories, has fewer categories at each subsequent level, and typically has more levels. Creating a broader, shallower navigation hierarchy with very specific labels and discrete categories clearly conveys an information space's hierarchy and makes its content more findable. In contrast, a narrower, deeper navigation hierarchy with less specific categories generally makes content harder to find, which could frustrate users. However, having too many links at a given level can be visually overwhelming to users, and the limited screen real estate that is available for navigation might not accommodate all the links. Try to limit the number of navigation links at each level to only those that are necessary for the user to make particular wayfinding decisions. [10]

Having too few links at each level reduces information scent and, thus, requires greater effort on the part of users to drill down and discover the correct path to the information they're seeking. Including *all* the necessary links at each level of the information hierarchy—and *only* the necessary links—provides better information scent to users, avoids the addition of unnecessary levels to the structural navigation, and thus, prevents users from having to drill down through too many levels when wayfinding and potentially losing their way. [11]

Keep in mind that the overall navigation system must provide access to *all* the information on the information space. Determine what links or groups of links would be most useful to users within particular contexts—for example, at specific levels of the structural navigation or by using specific navigation elements. At each level of the structural navigation, include all of the links that would meet users' essential wayfinding needs, as well as those that support an organization's key business objectives.

Prioritize navigation links to an information space's key content.

The information space's navigation system imparts the business's brand values and priorities through the information architecture's categories, the layout and order of the navigation links, and the tone of the links' labels. Plus, the information space's structural navigation links convey the primary focus of its content, as well as the scope of its subject matter, enabling users to determine whether its content is relevant to them. [12]

When designing an information space's structural navigation, always give the most prominent placement to the navigation links that provide access to key information. On an information space whose language reads from left to right, this means placing the highest priority links in closest proximity to the upper-left corner of pages. When you're designing the layout of a page's content and its contextual navigation, place the highest-priority content nearest the top of the page and the corresponding links at the top of any list of contextual links. When designing a list of associative links, place the highest-priority links nearest the top of the list.

Logically group similar links to clarify a navigation system's structure.

When a navigation system comprises more links than users can readily comprehend at once, group navigation links that have a similar function or similar characteristics to simplify the navigation system *and* limit the number of visible navigation options by supporting progressive disclosure. Creating logical groupings of links defines the relationships that exist between particular links and distinguishes related links from unrelated links. These groupings of links should meet the needs of users who have different purposes for visiting the information space and are seeking different types of information.

Providing clear labels for these groups of links makes the meaning of the system's structure more apparent to users. Clear labels also convey good information scent, making it easy for users to recognize paths that would lead to the information they're seeking *or* to avoid groups that wouldn't be of interest to them.

Some characteristics that provide a meaningful basis for grouping similar links include the following:

- Navigation type
- Level of hierarchy
- Scope of the information space across which users need easy access to certain links
- Relatedness by category, topic, or function

According to the Gestalt visual-design principle of *proximity*, users perceive links that are near one another as being part of the same group, while links whose placement sets them apart either belong to a different group of links *or* stand alone. [13]

Make the links within groups of links easy for users to distinguish.

The labels of groups of links should help users to distinguish the links they contain from those in other groups. Clearly labeling the navigation links themselves, using unique, highly differentiated words or phrases, is key to making each link within a group of links easily distinguishable from the other links. To clearly differentiate the link labels within a group from one another, avoid repeating the same verbiage in different link labels and ensure that the meanings of link labels do *not* overlap one another. For detailed guidance on creating effective, highly differentiated labels for navigation links, refer to Chapter 9, *Labeling Information.*

Consistently lay out navigation elements across an entire information space or section.

To engender users' trust and give them confidence in their ability to successfully navigate the information space, the textual *and* visual wayfinding information that the navigation system provides must be clear and consistent across either the entire information space *or* a major section of the information space, as appropriate.

To enable users to leverage their learnings from using other information spaces' navigation systems, follow common design guidelines when you're designing the information space's navigation system.

Make sure that the visual hierarchy for the information space's navigation system clearly conveys how users can progress through each stage of a wayfinding journey by navigating the levels of the information space's structural navigation. To ensure that users can easily find specific content on particular pages, provide effective contextual navigation.

Users could potentially take a variety of paths to a particular wayfinding destination. Therefore, both the information space's overall user experience and its navigation system should be consistent across the entire space, regardless of what path a user takes. To maintain maximal cross-platform consistency, you must consider the space that is available for navigation elements. If an information space's navigation system differs too greatly across the desktop Web, mobile Web, and any mobile app, users who engage with the information space on these various platforms won't be able to transfer their knowledge of the navigation system from one platform to another. For more information about creating consistent layouts for an information space's navigation elements, see the sections "Desktop navigation layouts" and "Mobile navigation layouts," in Chapter 12, *Designing Navigation*.

Indicate the user's current location on an information space or in a section.

One key goal of the information space's navigation system is to orient users to their current location within either the entire information space or a specific section of the space. To facilitate users' ability to navigate the information space, the navigation system should orient users by clearly showing their current location within the space by doing either or both of the following:

- **Visually highlighting the user's current location on a navigation bar, menu, or other navigation element.** By highlighting whatever combination of links in the information space's structural navigation hierarchy corresponds to the users' current location, you can orient users to their position relative to all key wayfinding destinations on the information space—whether at the same level in the hierarchy or above or below their current location. This helps users to ascertain where they might go next. Figure 11.1 provides an example of highlighting the user's current location on the primary navigation bar on *UXmatters*.

- **Providing location breadcrumbs to indicate the user's current location within the information hierarchy.** [14] You can orient users to their current location on the information space by providing this complementary navigation element. For more information about location breadcrumbs, see the "Location breadcrumbs" section, later in this chapter.

Figure 11.1—Highlighted links showing the user's current location in the navigation hierarchy

In addition to providing orientation, users' awareness of their current location can convey other information such as the following:

- **Context**—Communicating the user's current location within the information space also provides context. The context within which a particular page exists can help users understand the page's content and better comprehend both the content on the current page and the page's relationships to other nearby pages. Both the information space's navigation system and the page's title give salient information about the user's current location and, thus, context. Having this context is especially important if the user has arrived on a page that is deeper in the information space's hierarchy by conducting a search, then clicking links, or by clicking an external link on another information space or a link in a list of Web-search results.

- **Depth**—Making users aware of the depth of their current location on the information space can also set their expectations regarding how general or detailed a page's content should be. Pages that reside at a greater depth in an information space's hierarchy tend to have a narrower scope than pages at higher levels. Plus, their content has greater specificity, which could indicate whether the user has plumbed the depths of the information that is available on a particular topic or could seek more detailed information by navigating to a deeper level of the hierarchy. [15]

Use complementary navigation elements to provide an overview of an entire information space or section.

Structural navigation provides a high-level overview of an entire information space, gives users their primary means of navigating the space, indicates the user's current location within the space, and thus, supports exploratory browsing, as described in the section "Structural navigation," later in this chapter.

In contrast, the primary purpose of complementary navigation tools such as site maps, directories, lists of topics, A-to-Z indexes, and step-by-step guides is to provide an overview of the entire information space or, in some cases, a major section of the space. Location breadcrumbs provide access to the higher levels of the information hierarchy, letting users choose a different branch of the hierarchy.

Although providing such complementary navigation aids is optional, they can help users to understand the scope and organization of the information space's content, what collections of content exist and what specific types of content they comprise, what content is readily available from the user's current location within the space, what wayfinding destinations are available from that location, and the possible routes to those destinations. [16] For more general information about such navigation aids, refer to the section "Complementary navigation tools," later in this chapter. For information about specific complementary navigation tools, refer to the subsection on each tool later in this chapter.

Types of navigation

"Various mechanisms come together ... to form a comprehensive navigation system, with each unit in the system playing a different role."—Jim Kalbach [17]

Users likely have various information-seeking needs when using a digital information space. Plus, they might take different approaches to finding the information they're seeking. To meet users' diverse needs, you must provide different types of navigation that fulfill specific purposes such as the following:

- Structural navigation

- Navigation pages

- Associative links

- Supplementary navigation

- Complementary navigation tools

- Web browsers' navigation capabilities

One goal of navigation design is to include whatever combination of these various types of navigation is necessary to create a navigation system that optimally serves the needs of the users *and* the business. Browsing an information space by using its navigation system is an engaging experience that lets users serendipitously discover content they might not have expected to find there. This can result in additional sales or other conversions and, thus, improve business outcomes. [18]

Some factors that influence which of these different types of navigation a designer chooses to employ include the following:

- **Purpose**—The various types of navigation have specific purposes, and they most effectively support particular information-seeking tasks or specific types of content.

- **Level of importance**—Specific types of navigation have different levels of importance, which can greatly influence the navigation system's design—for example, the placement of navigation elements on pages and the amount of emphasis their visual design conveys. For example, in decreasing order of importance, you might place navigation elements at the top, left side, right side, or bottom of pages.

- **Scope**—The extent of the content that a type of navigation covers—for example, an entire information space, a section of an information space, or an individual page on an information space.

- **Number of levels**—The *depth* of a structural navigation system—or the number of levels it comprises—determines the complexity of the navigation system's design, as well as the navigation elements whose use might be most appropriate.

- **Number of options**—The *breadth* of a structural navigation system—or the number of options at each level—is another important factor in determining what navigation elements to use.

- **Navigation design patterns**—The different types of navigation typically employ particular navigation design patterns and the navigation elements that best express their purpose. Although some behaviors of navigation elements are universal, others are specific to a particular navigation element. Some navigation elements are visually persistent, while others are not—for example, only an information space's global navigation might be visible by default.

Always consistently follow an information space's conventions for the use of specific navigation design patterns and components when applying them to particular types of navigation. This enables users to correctly interpret the purpose of each type of navigation and quickly identify and become oriented to its use when they land on different pages. [19]

Structural navigation

Depending on an information space's scope, its structural navigation could consist of only a single level of links. However, an information space's structural navigation typically represents a coherent, multilevel hierarchy. [20] Each level in a structural navigation system comprises a group of global or local hyperlinks at a single level in the information space's hierarchy. [21]

Except on very small information spaces that comprise only a few pages, global and any local navigation should be available in a prominent, consistent location on the home page and every content page on an information space, enabling users to easily traverse its information hierarchy in both directions—down, then back up. Figure 11.2 shows an example of structural navigation: the *Nielsen Norman Group* (*NN/g*) Web site's compact, two-level navigation system:

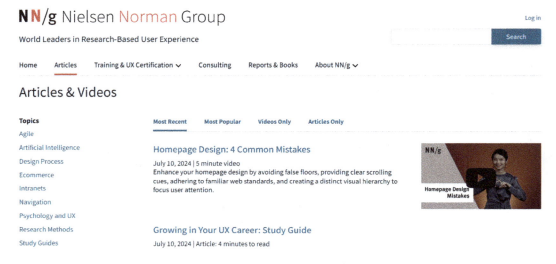

Figure 11.2—Structural navigation on the NN/g Web site

The placement of an information space's logo either left-aligned on the header above the structural navigation or to the left of the structural navigation provides context and, thus, orients users to the brand, as well as the subject matter and overall scope of the information space's content. In addition to offering essential navigation options to users, the structural navigation should always indicate the user's current location within the information space. For example, in Figure 11.2, **Articles** is highlighted on the *NN/g* Web site's primary navigation bar.

Since an information space's structural navigation typically occupies valuable screen real estate on almost every page of the information space, its visual design and layout should be very compact. In some cases, subordinate navigation elements might not need to be visually persistent. For example, on mobile devices, sometimes on tablets, and occasionally even on the desktop, many information spaces now use a hamburger menu to display an information space's global and local navigation in a navigation drawer. Figure 11.3 shows an example.

Figure 11.3—Apple.com displays a hamburger menu and navigation drawer on mobile devices

Although the links in most navigation systems consist of brief textual labels, they sometimes comprise icons or combine icons and labels. Textual labels offer the benefit of clarity, while icons are often somewhat ambiguous and do a poor job of communicating abstract ideas. Navigation links that combine a textual label and an icon usually occur at the global level of navigation. The icons are easier for users to distinguish from one another than textual labels and can be aesthetically pleasing. Plus, once users have become familiar with navigation icons, they might no longer need to read the textual labels. Therefore, by using navigation icons in structural navigation that resides at the left of pages, you can allow users to partially collapse the navigation, displaying only the icons in a compact rail, as the **National Oceanic and Atmospheric Administration (NOAA)** Web site does, as Figure 11.4 shows. [22]

Figure 11.4—A navigation rail comprising icons on the NOAA Web site

To ensure that navigation links are easy for users to click, a bounding box should demarcate the entire clickable or tappable region around a link—regardless of whether it is visible. This region may correspond, in some cases, to a selection rectangle with a different background color or other visual formatting that indicates whether a navigation link is currently selected.

In contrast to information spaces, many applications and ecommerce stores employ process funnels for their workflows. These process funnels restrict users' movement by *not* including any structural navigation elements on their pages, thus ensuring that users do not inadvertently exit the process funnel. Similarly, purely functional pages on Web sites, such as advanced-search pages and other Web forms, generally do *not* include structural navigation elements. [23]

Contextual navigation is a discrete type of structural navigation that provides page-specific navigation links and lets users navigate directly to particular subsections of a page. For more information about contextual navigation, see the section "Contextual navigation," later in this chapter.

Global navigation

"Global navigation system design forces difficult decisions that must be informed by user needs and by the organization's, goals, content, technology, and culture. One size does not fit all."—Peter Morville and Lou Rosenfeld [24]

An information space's global, or primary, navigation typically comprises links to the home page and the main pages of its major sections. These links orient users to the key topics that the information space's content comprises and provide universal, consistent access to the main pages across the entire information space. Global navigation links let users navigate directly to an information space's main pages or sections, so the placement and design of global navigation should enable users to easily identify its purpose. Since an information space's global navigation must also provide context, it should always indicate the section of the information space or the page that the user is currently viewing, as appropriate.

An information space's global navigation should ideally be highly visible. Therefore, most information spaces display their global navigation on a horizontal navigation bar that resides either immediately above or below the main content area on each page, as shown in Figures 11.5 and 11.6. Many global navigation bars incorporate an organization's logo on the left, which is typically a link to the home page, and provide access to an information space's search system.

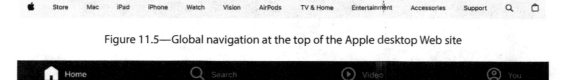

Figure 11.5—Global navigation at the top of the Apple desktop Web site

Figure 11.6—Global navigation at the bottom of the IMDb iPad app

However, on the smaller screens of mobile devices, most information spaces use some variant of the hamburger icon and menu to display both global and local navigation. This unfortunately makes the global navigation much less visible to users. As a consequence, users have greater difficulty discovering an information space's global navigation categories, and moving between them exacts a higher interaction cost. [25]

Tapping the hamburger icon displays a navigation drawer containing the information space's global navigation on *HBR*, as Figure 11.7 shows, or an accordion containing the information space's global and local navigation on *Fast Company*, as Figure 11.8 shows. The *Harvard Business Review (HBR)* mobile site combines its navigation and search capabilities.

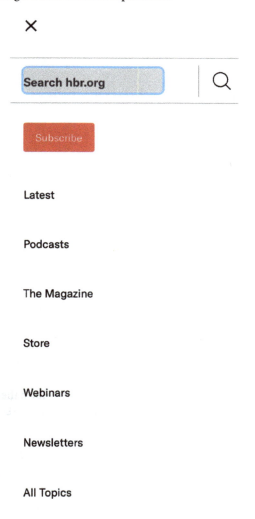

Figure 11.7—Global navigation on the *HBR* mobile Web site

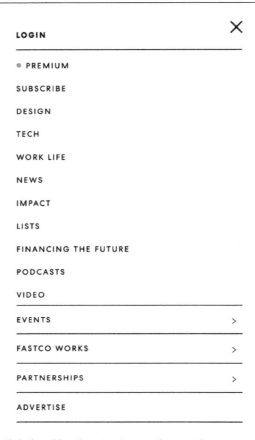

Figure 11.8—Global and local navigation on the *Fast Company* mobile Web site

Your team should always conduct user and stakeholder research to inform the design of an information space's global navigation, whose ability to meet users' needs and usability are essential to the information space's success. [26]

If the global navigation of a desktop Web site comprises too many links, accessibility can be a concern for keyboard and screen-reader users, especially those with motor disabilities. To enable these users to navigate directly to a page's main content, you can provide a skip-navigation link at the top of the page, before the global navigation and any utility navigation on the page's header. For keyboard users, provide a text link to which they can navigate by pressing the **Tab** key. *WebAIM* suggests using the following HTML code:

```
<body>
<a href="#maincontent">Skip to main content</a>
...
<main id="maincontent">
<h1>[Page Title]</h1>
<p>[First paragraph of text]</p>
```

Tabbing to the skip-navigation link moves the keyboard focus to the main content, and the user can navigate from that point.

For screen-reader users, activating a skip-navigation link can prevent the screen reader from reading every navigation link near the top of the page aloud before reading the page's content. Optionally, you can prevent the skip-navigation link from being visible by placing it off screen using CSS positioning, then use CSS transitions and the `a:focus` pseudo class to display the link at the top of the page when the user tabs to it. [27, 28]

Local navigation

"Local navigation indicates to users where they are and what other content is nearby in an information hierarchy."—Page Laubheimer [29]

Larger information spaces that comprise multiple levels require local, page-level navigation in addition to global navigation. An information space's local navigation should help users to orient themselves to their current context by indicating the user's current location within a particular section of the information space or category of content. Indicate the section, as well as the specific page the user is currently viewing, by highlighting the user's selections in the local navigation.

The links that the local navigation comprises are unique for each of an information space's specific sections. Plus, the nature of the content within a specific section could necessitate using navigation elements that are unique to that section. For example, a section comprising people's profiles might require an A-to-Z index.

Depending on the complexity and depth of the information space's navigation hierarchy, local navigation might provide links to all of a particular section's main pages *or* to specific pages of content at lower levels of the navigation hierarchy. Therefore, each section's local navigation could comprise one or more levels of navigation links to that specific section's pages. While, typically, these are secondary navigation links, in some cases, they could be progressively more local levels of links such as tertiary or quaternary links. However, to prevent the information space's local navigation from becoming overly complicated and, thus, overwhelming to users, avoid nesting levels of links too deeply.

Balancing the breadth and depth of an information architecture is key to obviating the need to create an excessive number of levels of local navigation. To reduce the number of levels that the local navigation comprises, one alternative would be to represent the lowest level of a section's hierarchy as a list of contextual links on that section's main page. Or you could use breadcrumbs to replace local navigation links at the lowest levels of the information hierarchy. However, doing so could impair users' ability to easily find content at the lower levels and cause related content to be less discoverable. Thus, relying on breadcrumbs alone would be inadvisable when important content resides at those levels.

Together, an information space's global and local navigation should form a highly coherent, structural navigation system that represents the space's entire, multilevel information hierarchy. This system should exhibit a clear visual hierarchy, with global navigation that is more prominent than local navigation, and enable users to easily distinguish the global and local navigation from one another. Otherwise, the user might mistake the local navigation for the global navigation and incorrectly believe that the scope of the information space's content is much more limited than it actually is.

For any information space that supports exploratory browsing, providing visible navigation for both global and local navigation would be beneficial. Accessing the information space's structural navigation would then incur little interaction cost and, thus, encourage users to visit multiple pages within a section. By clicking each global navigation link in turn, the user can display the local navigation links that are subordinate to that particular global navigation link. The user can then explore the content within each corresponding section by clicking other links at the same level or at lower levels of the local navigation hierarchy.

Ideally, you should conduct qualitative UX research and evaluate analytics data to determine users' specific needs for local navigation by answering the following questions:

- Do users frequently engage in exploratory browsing and, therefore, require clear orientation to their current location at all levels of an information space?

- Do users often browse multiple pages within a single subcategory during the same session?

- Do users need to refer to content that resides on multiple pages to make a decision?

- Do users typically need to visit the same pages to accomplish particular goals?

- Do users typically begin their journey by landing on pages that are deeper within an information space's hierarchy rather than on the home page?

[30, 31, 32]

Universal navigation

"A subsite is a home environment for a specific class of users … within a larger and more general site. … Subsites are a way of handling the complexity of large Web sites with thousands or even hundred of thousands of pages. By giving a more local structure to a corner of the information space, a subsite can help users feel welcome in the part of a site that is of most importance to them."—Jakob Nielsen [33]

Within large enterprises, various departments often independently create their own Web sites, each of which has its own home page and navigation system—comprising both global and local navigation—as well as a site-specific search system. Thus, a large, enterprise Web site could comprise several subsites. Although each subsite's homogeneous content might require some unique design standards or even its own navigation design, designers should endeavor to create as much consistency across all the subsites as possible. Most importantly, to integrate all these subsites into a coherent whole, you must create a *universal navigation* system that resides at the highest level of the enterprise's main site, connecting all the related subsites' pages to the universal home page for the entire enterprise site. [34, 35]

The purpose of such a universal navigation system is to provide easy access to the general, informational pages on the enterprise site, as well as *all* the individual subsites, as shown in Figure 11.9. To prevent the universal navigation from impairing the usability of the subsites' structural navigation, give greater visual prominence to the subsites' navigation system and make it easy to distinguish from the universal navigation. Since it is important to provide a clear visual hierarchy, a common location for the universal navigation is left aligned, at the very top of *all* the site's pages, above each subsite's header and structural and supplementary navigation. [36]

Figure 11.9—News subsite on Yahoo.com

Contextual navigation

While contextual navigation is another type of structural navigation, the information hierarchy that it displays reflects only the structure of an individual page rather than the information space's overall navigation hierarchy. Thus, the scope of contextual navigation differs greatly from that of the information space's global and local navigation. Contextual navigation's page-specific navigation links allow users to navigate directly to the page's subsections. Contextual navigation often resides in a sidebar near the top of a page. To distinguish the contextual navigation from any vertical, local navigation, it typically resides on the right side of the page, as in the **On this page** contextual-navigation block shown in Figure 11.10. [37]

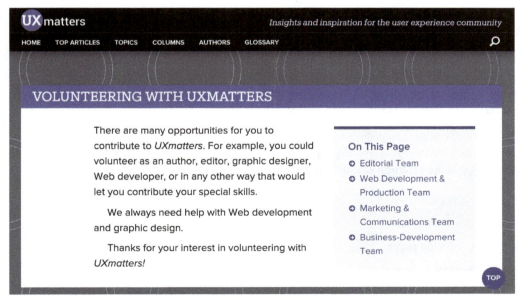

Figure 11.10—Contextual navigation in a sidebar on *UXmatters*

On Web pages, you can create named anchors that let users jump directly to specific subsections of the page by inserting an anchor `id` into each section heading—for example, `<h2 id="mission">Our Mission</h2>`. Then, links in the contextual navigation can navigate directly to the named anchors in the content—for example, `<li><a href="#mission">Our Mission</a></li>`. Thus, each link with an `href` value beginning with a number sign (#) refers to the content on the same page that has the same `id`. Clicking an anchor link scrolls the page to the corresponding section. [38]

Integrating global, local, universal, and contextual navigation

An information space's structural navigation system should provide consistent, coherent navigation across the entire information space. When integrating the various types of structural navigation on an information space's pages, be sure to avoid their collectively consuming too much screen real estate and overwhelming users with an excessive number of links. Together, the global and local navigation must visually convey a clear navigation hierarchy for the overall information space. For some large, enterprise information spaces, universal navigation constitutes the highest level of the structural navigation hierarchy. In contrast, contextual navigation displays the navigation hierarchy for only an individual page.

The unifying characteristics of global, local, universal, and contextual navigation are that they all convey hierarchy and support similar user interactions. Therefore, all the types of structural navigation should have the same behaviors and show state visually in similar ways. Although the scope of their hierarchies can differ greatly, consistency in their interactions and in certain aspects of their visual design can bring coherence to an information space's overall navigation-system design. Thus, an organization should establish and consistently follow navigation design guidelines. [39]

Navigation pages

"Most Web sites contain … two major types of pages: navigation pages and destination pages. Destination pages contain the actual information…. Navigation pages may include main pages, search pages, and pages that help you browse the site. The primary purpose of a site's navigation pages is to get you to the destination pages."—Peter Morville and Lou Rosenfeld [40]

The function of certain key pages on an information space is primarily navigational. The primary purpose of a navigation page is typically to provide direct pathways to a set of closely related, subordinate content pages. Some common examples of such navigation pages include the following:

- The home page
- Sections' main pages
- Product-listing pages
- Search-results pages

The home page

"Users rarely go to a homepage to admire it. They go there because they want to find their way to a destination elsewhere on the site."—Jakob Nielsen [41]

An information space's home page provides an overview of its content and orients the user to the space. Thus, a home page's content often comprises descriptions of the information space's main sections and provides associative links to them, as shown in Figure 11.11. These descriptions should elaborate on and, thus, clarify the meanings of the global navigation system's link labels, as well as the labels of the associative links that accompany the descriptions. [42] In *Information Architecture for the World Wide Web*, Morville and Rosenfeld refer to such descriptions as *scope notes*. These *scope notes* help users familiarize themselves with the content on an information space, introduce the terminology it uses to convey important concepts, and thus, reduce the ambiguity of the global navigation system's link labels. [43]

BUSINESS INSIDER

Newsletters   Log in   Subscribe

FINANCE

Grocery, robotics tycoon Rick Cohen hit by epic $9B wealth wipeout

POLITICS

Harris sprints to the middle ahead of 2024 home stretch

MEDIA

Disney created a PR nightmare with its response to a wrongful-death lawsuit

TRAVEL

Hotel employees share the 7 red flags to look for when checking into a hotel

CAREERS

20 high-paying and fast-growing jobs in the US

MILITARY & DEFENSE

Ukraine just unveiled a brand new drone military unit

As fast as some jobs are growing, others are losing steam

The fastest-growing (and fastest-shrinking) jobs in the United States.

Mark Cuban says Elon Musk's 'biggest power play' on X is letting users think they have free speech

It's getting harder to make big leaps at the frontier of AI. There will be huge winners and losers.

Figure 11.11—The *Business Insider* home page functions as a navigation page

Many home pages take the form of list pages or card-based pages. **Responsive Web design** (**RWD**) enables these two approaches to home-page layouts to work very well on both the desktop and mobile devices. Often, all items or at least featured items on both list pages and card-based pages include an image, consuming more screen real estate and increasing page-loading times. Therefore, the images should provide actual value rather than being merely decorative.

List pages

List pages comprise easy-to-scan, vertical listings of individual items such as articles, news items, products, or social-media posts and support exploratory browsing. The title of an article or news item typically links to the complete article. Figure 11.12 shows the home page of the *NN/g* Web site, which employs a compact list layout in which the articles' titles are link labels. In contrast, as Figure 11.13 shows, the listings on the *UXmatters* home page provide somewhat longer excerpts from the Web magazine's most recent articles, and clicking either an article's title or a **Read More** link displays the full article.

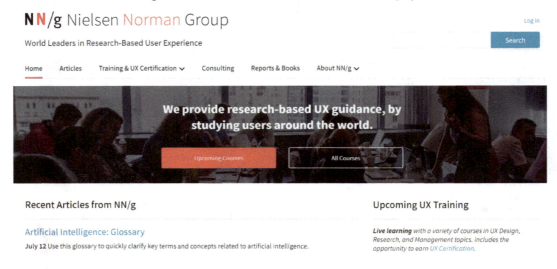

Figure 11.12 – Home page on the *NN/g* Web site

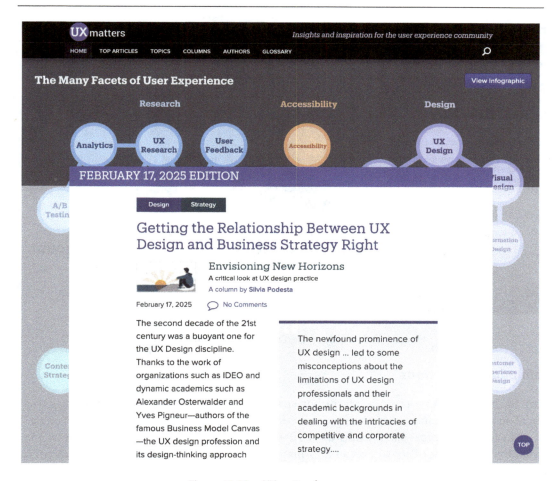

Figure 11.13—*UXmatters* home page

Card-based pages

Card-based pages display a collection of modular UI components called *cards* that support exploratory browsing. Each *card* is typically a rectangular container of a uniform width—and, in some cases, a uniform height as well—whose content comprises a standard set of closely related information elements. On a navigation page, each of the cards typically includes an image, a title, and concise summary information. The user can click a card to display more information about a topic. Figure 11.14 provides an example of a card-based user interface on the BBC.com home page. [44]

What Biden quitting means for Harris, the Democrats and Trump

President Biden has upended the 2024 White House race for the Democrats. Here is what it means for Kamala Harris, his party and Trump.

18 hrs ago | US & Canada

⊙LIVE Kamala Harris to speak for first time since Biden left race - as endorsements mount

The US vice-president is due to appear at the White House shortly for a pre-scheduled event - as key Democrats line up to back her candidacy.

Figure 11.14—Cards on the BBC Web site's home page

Main pages of sections

When the main page for each of an information space's sections functions as a navigation page, the content on these pages often comprises either brief excerpts from or descriptions of the content on the sections' subordinate pages and provides associative links to those pages, as shown in Figure 11.15. Similar to descriptions of content on an information space's home page, the content or category descriptions on a section's main pages clarify the meanings of the labels for the section's local navigation links, as well as any associative links within the main page's content. [45] A section's main page could also take the form of a list page or card-based page.

☰ Q

BUSINESS INSIDER

Newsletters   Log in   Subscribe

Strategy

CAREERS

A tech career coach says she listened to over 700 interviews — and the candidates who got big offers do these 3 things

Tech career coach Katie McIntyre said candidates should drop the perfectionist act in interviews and focus on these three things.

CAREERS

Introverts in the workplace should be their 'own best marketer,' Ancestry CEO Deb Liu says

Ancestry CEO Deb Liu — herself an introvert — says it's important for introverts at work to realize "you are your own best marketer."

STRATEGY

Corporations are accelerators of change and have to bring employees along the journey, says LiveRamp's people and culture executive

LiveRamp's chief people and culture officer shares how companies can bring out the best in their teams and prepare employees for the future of work.

Figure 11.15—Main page of the Strategy section on *Business Insider*

Product-listing pages

When looking for specific products to purchase on an ecommerce app or Web site, shoppers typically browse product-listing pages that take the form of list pages, card-based pages, or image grids. Product-listing pages must provide concise descriptions of the products that are available within specific product categories and associative links that let shoppers navigate to the product-details pages for specific products, as Figure 11.16 shows. Essential elements of such product descriptions include product images and product names. Placing a filter or a list of subcategory links at the top of a product-listing page lets shoppers reduce the number of product listings they must peruse. [46]

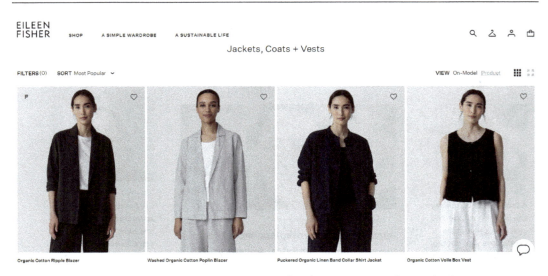

Figure 11.16—Jackets, Coats + Vests, a product-listing page on Eileen Fisher's site

Search-results pages

On any information space that has an internal search capability, search-results pages dynamically list the results for the user's search query, in the order of their relevance to that query, with the most relevant results first. [47] Each of the search results typically comprises a title that links to a page with relevant information, as well as a brief excerpt, or snippet, from that page, as Figure 11.17 shows. Depending on the typical needs of actual users and the types of data that an information space provides, search results might also include other information that would help users to differentiate between results and determine which might lead to information that would satisfy their needs.

The user can click links in the search results to seek relevant information by navigating to other pages. Therefore, search-results pages are navigation pages. For highly textual content such as that of a Web magazine, search results usually take the form of list pages that do *not* include images, as shown in Figure 11.17. However, in an ecommerce context, where an image of each product would help users to find what they need, search-results pages could be card-based pages, as shown in Figure 11.18. For in-depth information about designing search-results pages, refer to Chapter 13, *Designing Search*.

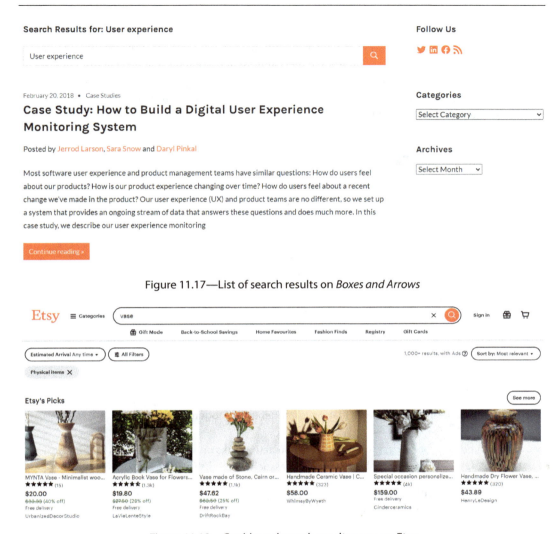

Figure 11.17—List of search results on *Boxes and Arrows*

Figure 11.18—Card-based search-results page on Etsy

Associative links

In contrast to the links that an information space's structural navigation comprises, *associative links* on words, phrases, or images *within* the content of a page connect concepts in a page's content to additional relevant information about the same or related concepts on another page on the same information space, in another section of the same page, or on a different information space. While associative links provide easy access to related information, they convey no information about an information space's structure or scope. In fact, associative links often cross structural boundaries. [48]

However, associative links—whether on specific text or alt text that describes an image link's destination—can define additional relationships between content objects on the same information space and, thus, support exploratory browsing. Therefore, associative links encourage visitors to read more of an information space's content or, on an ecommerce store, suggest other similar products that a shopper might be interested in purchasing. For example, Amazon suggests products that people frequently buy together, as shown in Figure 11.19.

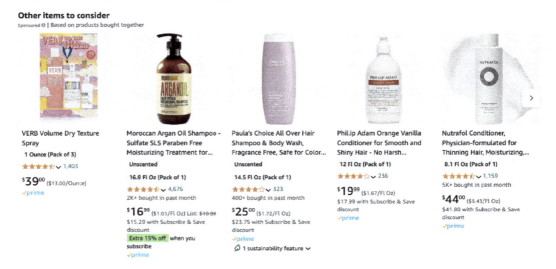

Figure 11.19—Products that users frequently buy together on Amazon

However, it can be easy for readers to miss inline, associative links on words or phrases within the text of a paragraph. Other more noticeable examples of associative links include the following:

- Links on article titles, titles of other pages, or product names that are subheadings for the subsections of a page or that appear at the beginning of a paragraph or list item

- Links that appear at the end of a section, paragraph, or list item on relatively generic phrases. Examples include **Read more**, which on a Web magazine might display the full text of an article; **See more**, which might display more articles on the same topic or more products of the same type; and **See also**, which, on an ecommerce store, might display related products.

Since associative links on such generic phrases provide no information scent, they are often coupled with another, more specific link such as a link on an article's title, a page title, or a product's name. For example, as shown in Figure 11.20, in excerpts from new articles on the *UXmatters* home page, there is a link on each article's title, as well as on the text **Read More**, both of which display the article.

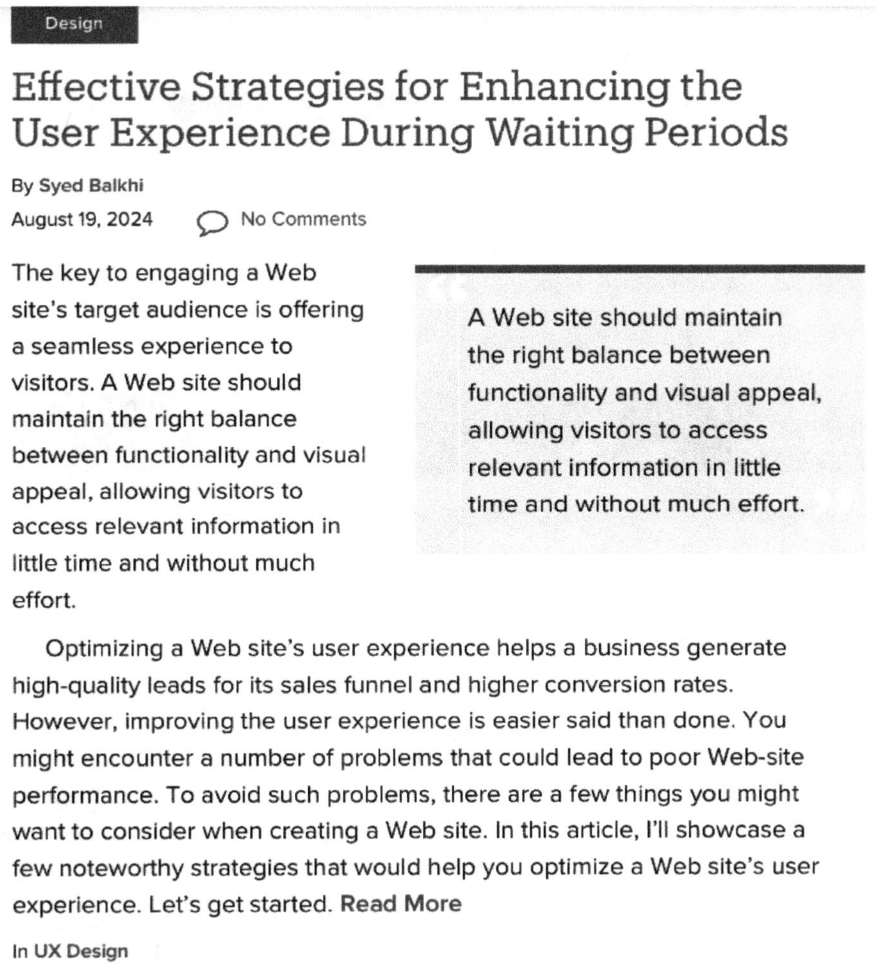

Design

Effective Strategies for Enhancing the User Experience During Waiting Periods

By Syed Balkhi

August 19, 2024   ○ No Comments

The key to engaging a Web site's target audience is offering a seamless experience to visitors. A Web site should maintain the right balance between functionality and visual appeal, allowing visitors to access relevant information in little time and without much effort.

> A Web site should maintain the right balance between functionality and visual appeal, allowing visitors to access relevant information in little time and without much effort.

Optimizing a Web site's user experience helps a business generate high-quality leads for its sales funnel and higher conversion rates. However, improving the user experience is easier said than done. You might encounter a number of problems that could lead to poor Web-site performance. To avoid such problems, there are a few things you might want to consider when creating a Web site. In this article, I'll showcase a few noteworthy strategies that would help you optimize a Web site's user experience. Let's get started. **Read More**

In **UX Design**

Figure 11.20—**Read More** associative link to an article on *UXmatters*

Another way of calling attention to important associative links on a page is to display them in specially formatted groups. For example:

- Display groups of associative links within specific areas of the page—such as in a sidebar or below a page's main content.

- Set off such groups of associative links from the page's main content by placing them in a box or separating them using horizontal hairlines.

- Give each such group a label such as **For more information**, **Learn more**, **More on this topic**, **Related information**, or **Related products**.

Figure 11.21 shows examples of such groups of associative links on *UXmatters*: **Other Articles by Pabini Gabriel-Petit** and **Other Articles on Interaction Design**.

Other Articles by Pabini Gabriel-Petit

- IA, Rosenfeld Media, and EUX: An Interview with Louis Rosenfeld
- An Interview with Alfonso de la Nuez, CEO of UserZoom
- An Interview with Mary Treseler, Co-chair of the New O'Reilly Design Conference
- User Experience, Entrepreneurship, and Redesigning Democracy: An Interview with Dirk Knemeyer

Other Articles on Interaction Design

- Empowering the User Experience Through Microinteractions: 7 Best Practices
- Spatial Computing: A New Paradigm of Interaction
- How Developing Use Cases Helps in Designing User Interactions
- Task Flows and the Process of Designing Interactions

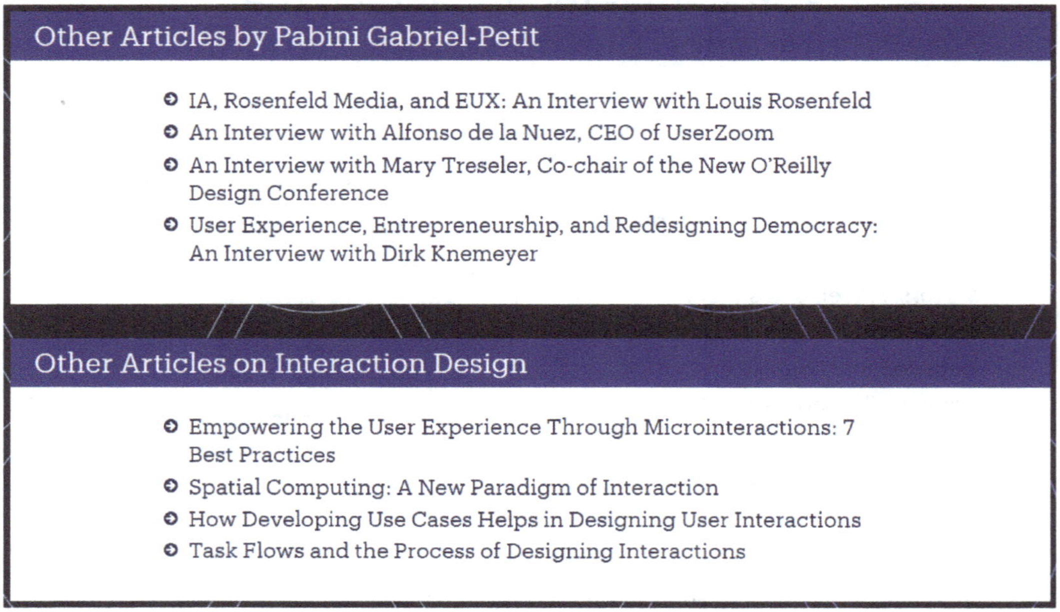

Figure 11.21—Groups of associative links on *UXmatters*

Although groups of associative links to other pages can complement the information space's structural navigation system, you should use them in moderation. You can also supplement an article's content by providing groups of external, associative links to content on other information spaces.

Link text should accurately convey a link's destination. Avoid using a meaningless link label such as **Click here**—even if the accompanying text describes the link's destination. Such labels would impair accessibility for users employing a screen reader, who might read only the link text. [49] Instead, provide link labels that accurately describe the links' destination and, thus, provide information scent.

It is no longer common practice to use an angled arrow that points to the upper right to indicate external links that would take the user to a different information space. However, it is helpful to add parenthetical text that indicates a link would display another type of document such as a PDF (Portable Document Format).

Supplementary navigation

"Utility navigation consists of secondary actions and tools [that] affect … satisfaction, user experience, and engagement. Put utilities where people expect and need them."—Susan Farrell [50]

An information space's *supplementary*, or *utility*, navigation provides links to useful tools that are *not* part of the main information hierarchy, but in most cases, should be available throughout the information space. [51] While the need for particular navigation links can vary across different types of information spaces, supplementary navigation might provide access to the following:

- **Secondary actions**—If an information space includes functionality that enables users to complete specific tasks that do *not* relate to information seeking, links to these tasks often appear in supplementary navigation. Such utility text links might display a concise form on the same page or, for a more detailed process, navigate to another page that lets the user complete a task. However, completing such tasks should require no additional navigation. Examples of links that support such tasks commonly include a **Sign Up** link that lets users sign up for an account; a stateful **Sign In/Sign Out** link that lets users either sign in to or out of their account; and a link to the user's account, which might be represented by a photo of the user, the user's name, or a label such as **Your Account**. As Figure 11.22 shows, Amazon's Web site integrates this functionality to reflect the current state of the user's account and, thus, minimizes the supplementary navigation. By default, before the user signs in, Amazon displays the rather verbose text **Hello, sign in** and an **Accounts & Lists** drop-down menu, at the top of which is a **Sign In** button, the text **New customer?** and a **Start here** link, allowing the customer to either sign in or sign up for a new account. The user can click **Sign In** to display the *Sign In* page or click **Start here** to display the *Create Account* page. Other links that commonly appear on the supplementary navigation include a **Contact Us** link, which could either display the organization's contact information or provide a contact form, or a **Subscribe** link that lets the user subscribe to an organization's newsletter.

- **Tools**—Although familiar icons often represent tools in supplementary navigation, some other tools might require text labels for clarity. The tools typically include search and occasionally include a language selector—as you can see in Figure 11.22—or a locale selector. On information spaces or in sections whose content comprises articles or other primarily textual content, supplementary navigation might include tools that act on the current page's content—such as tools that let the user download, share, save, or print its content or a font-size selector. On ecommerce apps and sites, a shopping-cart icon typically has prominent placement on the supplementary navigation.

- **Social-media icons**—Supplementary navigation often includes a grouping of easy-to-identify icons that represent the various social-media platforms. The text **Follow us on** might precede the icons that the social navigation comprises.

- **Help**—A Help system provides information that facilitates users' ability to use an information space. A question mark icon often represents online Help. [52]

Figure 11.22—Supplementary navigation on the Amazon Web site

Displaying an information space's supplementary navigation in a somewhat less prominent, but still highly visible position on each page ensures that users do not confuse its supplementary navigation with its structural navigation system. Thus, supplementary navigation typically resides on the header, at the upper-right corner of each page, or on the footer.

Header navigation

The standard placement for supplementary navigation is on an information space's header, right aligned, in the upper-right corner of each page, as Figure 11.23 shows. It is common to find secondary actions, tools, social-media icons, and a **Help** icon in the supplementary navigation. In some cases, a **Search** bar appears on the header instead, or a **Search** icon appears at the far right of any horizontal navigation bar or beside a hamburger menu. Content tools might appear on a toolbar immediately above the main content area. However, content tools such as a **Share** or **Download** link, button, or icon might appear at the beginning or end of an article instead. [53]

Figure 11.23—Supplementary navigation on *A Book Apart*

Footer navigation

In some cases, a Web site's footer navigation replicates certain links from the site's header navigation—such as links to information about the organization to which the site belongs, social-media links, or links that enable users to subscribe to a service, newsletter, or RSS (Really Simple Syndication) feed. Or such links might appear *only* on the footer, as on *UXmatters*, as shown in Figure 11.24.

Figure 11.24—Footer navigation on *UXmatters*

Complementary navigation tools

Digital information spaces can provide a variety of complementary navigation tools to give greater flexibility in how users can access information. Unlike the information space's structural navigation, which serves as the users' primary means of navigation and is an integral component of the home page and content pages, each of these complementary navigation aids serves a particular purpose and, thus, typically resides either on its own page or set of pages *or* in an area on the home page or a section's main page.

For information spaces with a deep information hierarchy, providing complementary navigation aids to offer shortcuts that lead directly to specific content at lower levels of the hierarchy is often necessary to make that content more visible to users. [54]

Complementary navigation tools such as site maps; directories, comprising lists of topics or products; and A-to-Z indexes might provide a visual overview of either an information space's main content or a specific subsection's content. Other types of complementary navigation tools include a variety of step-by-step guides such as wizards, tutorials, and tours, *and* location breadcrumbs. On very large, complex information spaces, the inclusion of some of these tools becomes particularly important in ensuring usability and findability. [55]

An information space should usually incorporate just one complementary navigation tool whose purpose is to provide an overview of its main content. However, if an information space must include more than one of these tools, group the links that display them together, and place these links in a consistent location on all content pages. You might also want to place links to complementary navigation tools in a more prominent location on Help pages and pages that offer accessibility features for users who have special needs. [56]

All of an information space's content should be accessible using a complementary navigation tool that provides an overview of its content. As your team adds new content, be sure to keep the links in any complementary navigation tools up to date.

You'll find detailed information about each of these complementary navigation tools in the following sections.

Site maps

"A site map is only as good as the site's information architecture."—Donna Tedesco, Amy Schade, Kara Pernice, and Jakob Nielsen [57]

A *site map* could provide a compact visual representation of any information space's overall structure, as well as an overview of its content. However, although a detailed site map could provide comprehensive access to an entire information space, site maps typically present only the top few levels of an information space's hierarchy, serving as a directory of its key content. A site map should typically take the form of a multicolumn list of links and, ideally, occupy only a single page to ensure maximal visibility of as many of its links as possible. A site map's links enable users to directly access an information space's main pages.

To provide additional structure to a site map, making it easier to scan and better serving the needs of different types of users, you could optionally add category headings to groups of related links, linking each category heading to a content category's page to provide easy access to general information about the category. You could organize the category headings using the information space's hierarchy *or* by task or audience. Category headings can make it easier for users to browse an information space's content using the site map. However, you must make sure that users can easily distinguish these headings from the category links.

For an information space whose content employs a strong categorization scheme, you could create a *hierarchical* site map that gives users an overview of all the content categories. Make sure that the site map clearly expresses the structure of the information hierarchy. However, you should avoid creating a nonstandard, overly complicated, interactive site map that would overemphasize the relationships between the information space's categories. Multicolumn site maps that comprise only textual links provide a more useful and usable overview of an information space's categories and the pages on which they reside load quickly. So, minimize the use of images in site maps, using only those that would visually convey sets of categories and subcategories with greater clarity—such as an ecommerce store's product categories—and would make it easier for users to identify the categories.

When deciding whether to create a site map for an information space, consider the space's scope. For a small information space whose information hierarchy comprises only one to three levels, a site map might be wholly unnecessary. However, a site map is a necessity for a complex information space that comprises many categories and levels.

Although the main purpose of a site map is to help users understand and remember the structure of an information space and, thus, facilitate its navigation, many users simply do *not* care about an information space's structure, and there are few who actually use site maps. Instead, most users rely on an information space's structural navigation system, along with the content on its home page, to discover whatever information they need. While many information spaces do *not* have a site map, a site map could nevertheless contribute to their usability. Users might turn to a site map out of frustration if they become lost or cannot find the content they need because an information space is too disorganized or cluttered. Plus, the internal links that any site map comprises could play an important role in search-engine optimization (SEO).

Users can typically access any site map via a **Site Map** link on an information space's footer or other supplementary navigation, making the link available on every page of the information space and, thus, easy for users to find, as Figure 11.25 shows. You should always place the **Site Map** link in a consistent, sufficiently visible, but not too prominent, location. Avoid burying the **Site Map** link in the midst of too many other links—especially among links that users are likely to ignore such as links that lead to legal disclaimers. Follow all visual-design guidelines that would ensure the legibility of both the **Site Map** link and the links in the site map itself, including providing sufficient foreground-background value contrast and using highly legible fonts at adequate font sizes.

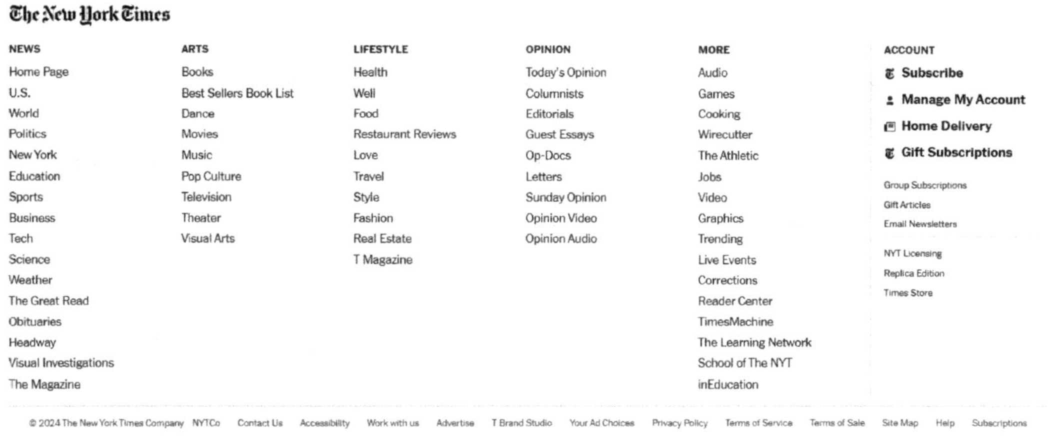

The New York Times

NEWS	ARTS	LIFESTYLE	OPINION	MORE	ACCOUNT
Home Page	Books	Health	Today's Opinion	Audio	ⓔ Subscribe
U.S.	Best Sellers Book List	Well	Columnists	Games	± Manage My Account
World	Dance	Food	Editorials	Cooking	⌸ Home Delivery
Politics	Movies	Restaurant Reviews	Guest Essays	Wirecutter	ⓖ Gift Subscriptions
New York	Music	Love	Op-Docs	The Athletic	
Education	Pop Culture	Travel	Letters	Jobs	Group Subscriptions
Sports	Television	Style	Sunday Opinion	Video	Gift Articles
Business	Theater	Fashion	Opinion Video	Graphics	Email Newsletters
Tech	Visual Arts	Real Estate	Opinion Audio	Trending	
Science		T Magazine		Live Events	NYT Licensing
Weather				Corrections	Replica Edition
The Great Read				Reader Center	Times Store
Obituaries				TimesMachine	
Headway				The Learning Network	
Visual Investigations				School of The NYT	
The Magazine				inEducation	

© 2024 The New York Times Company   NYTCo   Contact Us   Accessibility   Work with us   Advertise   T Brand Studio   Your Ad Choices   Privacy Policy   Terms of Service   Terms of Sale   Site Map   Help   Subscriptions

Figure 11.25—Link to *The New York Times* site map on the information space's footer

Apply the same text formatting to a site map's links as to the other associative links on the information space, indicating visited and unvisited links and, thus, ensuring that users don't inadvertently traverse the same links again and again. Interacting with a site map should be easy, requiring only that the user click the links that it comprises and, if necessary, scroll through the site map. Multicolumn site maps are more usable than single-column site maps because they require less scrolling. Scrolling up and down a site map could cause users to become lost or inadvertently skip over parts of the site map. Plus, users might not realize that they need to scroll to get to the bottom of the site map, so be sure to avoid adding any horizontal elements that users might falsely interpret as the end of the page.

A site map usually resides either on a separate page or on a set of pages, but a partial site map could appear within the body of a content page. For example, you might add a partial site map to a *Page not found* error message to provide related links and give users another way forward.

It's usually best to keep a site map as simple as possible. However, for an information space that comprises more essential content than the links on a single-page site map would be able to convey, consider providing an easy-to-use means of progressive disclosure to display different parts of the site map. Because of its long publication history, *The New York Times* uses progressive disclosure to provide full access to all of its content, as shown in Figure 11.26.

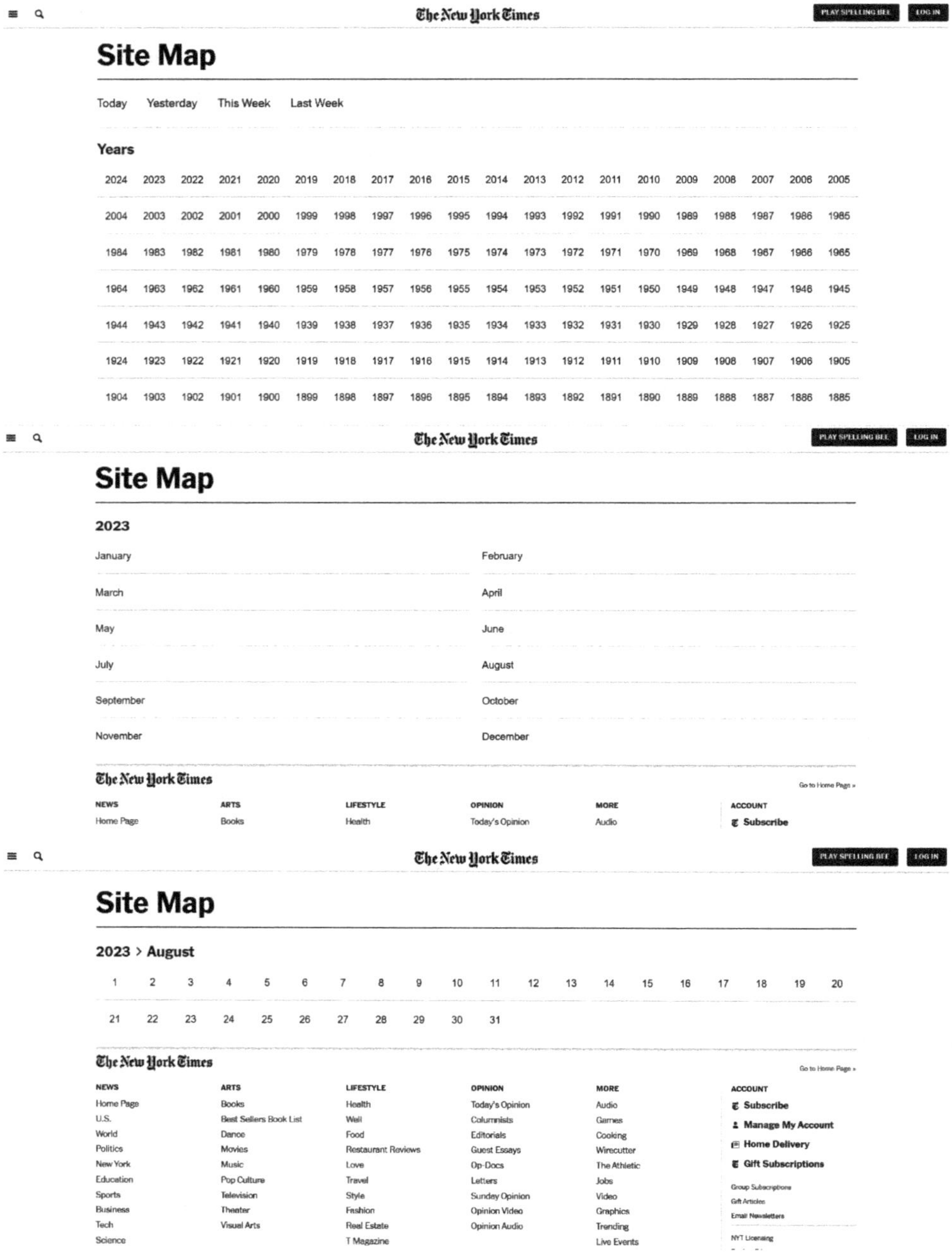

Figure 11.26—*The New York Times* site map

The depth of a site map could depend on the number of levels it can display clearly within the allocated space. Thus, in some cases, a site map might provide links to all the pages on an information space, while in others, it would not. For example, the site map for a large ecommerce site with thousands of products might extend only to the categories of products the site offers. However, do not eliminate useful links just to limit a site map's scope. Be sure to include all the links that would enable users to achieve their important goals. However, if providing a link to a parent category provides adequate coverage—as in the ecommerce example—or users would rarely visit the lower levels of the information hierarchy, consider eliminating lower-level links from the site map. [58, 59, 60, 61, 62]

Many Web sites employ two types of site maps, both of which can benefit users:

- **HTML site maps**—These Hypertext Markup Language (HTML) site maps visually represent a Web site's structure and give users an alternative means of navigating a site.

- **XML site maps**—These Extensible Markup Language (XML) site maps let search engines crawl and index an entire Web site and, thus, deliver high-quality search results to users. [63]

Directories, topic lists, and product lists

Most information spaces are topical, or subject oriented, and therefore, should provide some form of access to their content by topic. To facilitate users' exploratory browsing on an information space that comprises a large collection of either topical content or product categories with many top-level categories, provide a multicolumn list, or *directory*, giving users direct access to specific topics or product categories, respectively. Such a directory is particularly useful to users who don't know exactly what they're looking for.

By using a directory of topics, the user can access an information space's content categories and subcategories. A product directory provides access to categories and subcategories of products. In a directory, each category or subcategory is a link to the main page for a specific category or subcategory, respectively, and represents a collection of pages that belong in that category or subcategory. The labels of the links are nouns that describe the categories and subcategories. [64, 65, 66] For example, the **TOPICS** page on *UXmatters* provides access to the broad topics and more specific subtopics of the site's articles and columns, structuring them alphabetically, as Figure 11.27 shows.

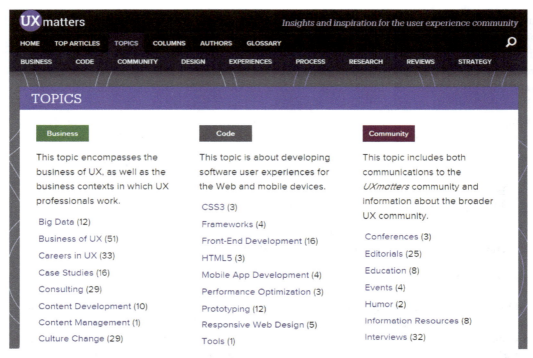

Figure 11.27—*Topics* page on *UXmatters*

You can organize a directory's topics or product categories either alphabetically, hierarchically, or sequentially, as appropriate to the content. [67]

A-to-Z indexes

One means of facilitating exploratory browsing on information spaces and, thus, providing direct access to large collections of content whose structure is not strictly hierarchical is the A-to-Z index. However, A-to-Z indexes are most useful when the user is seeking known items of information, using specific terminology. A digital A-to-Z index consists of alphabetical entries and is relatively flat, typically comprising only one or two levels. It comprises key words and phrases that provide links to specific content. However, links could also include topics that derive from the information space's content and describe particular concepts or themes *or* the titles of specific units of content such as articles or blog posts.

An index should help users with different levels of knowledge find the information they need, by including both popular and technical terms that describe the same concepts rather than just the terms the information space uses. For users who are already familiar with the terminology an information space employs, who are seeking known items of information, *or* who don't care about the information space's hierarchical structure, an A-to-Z index is a very effective complementary navigation tool. Links in the index take users directly to specific items of information they need. [68, 69, 70]

For an information space with a long A-to-Z index, you might divide the index into multiple pages, with one page for each letter of the alphabet or for a range of letters. Whether an A-to-Z index resides on a single page or multiple pages, you should provide an A-to-Z navigation bar to let users navigate directly to each letter in the index. While large, dynamic information spaces with many content pages aren't as amenable to indexing, an A-to-Z index might be the only complementary navigation tool that is necessary for a smaller information space.

A large, complex information space that comprises different types of information could use an A-to-Z index to structure just a specific section—for example, a section that consists of author or employee profiles or a glossary. The *Authors* section on *UXmatters* is an A-to-Z index, as shown in Figure 11.28. This section has an A-to-Z navigation bar. If there are any letters for which no content exists, the letters should not be links, but should instead appear dimmed.

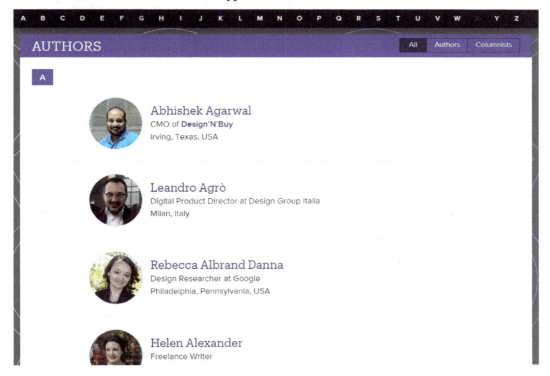

Figure 11.28—Authors section on *UXmatters*

You can evaluate the clarity of an A-to-Z index by conducting an expert review or usability study. However, the A-to-Z index is generally familiar to most users and, thus, easy to use. While indexes primarily support known-item seeking, their *see also* references can also aid in discovery. Similar to site maps, A-to-Z indexes play a role in search-engine optimization (SEO). [71]

Best practices for digital indexing

When creating a digital A-to-Z index, observe the following best practices:

- **Deciding how to create the index**—If you're creating an A-to-Z index for a small information space, you can probably do it manually, based on what you know about the information space's content and the users' terminology. For a large information space, use a content-management system (CMS) to create a controlled vocabulary, assign its terms to specific pieces of content, and generate the index automatically. Using a CMS ensures greater accuracy.

- **Determining the right level of granularity**—Levels of granularity range from links to the main sections of an information space to collections of closely related pages to individual pages to the subsections of pages to paragraphs to concepts to terms that represent these concepts. Which of these would be most relevant for a given information space? To make this determination, conduct user research to understand the audience, their needs, and the terminology they use—and, if the information space already exists, analyze its analytics and the information space's search logs.

- **Ensuring accuracy**—To create a useful, usable A-to-Z index, accuracy is essential. You can assess the quality of an index by checking the accuracy of a sampling of index entries across the index. Click the links to ascertain whether their destinations provide the expected information. Is the index's usage of terminology appropriate? Check a sampling of popular, technical, and ambiguous terms in the index against the content to which they refer. Are the spelling, capitalization, and punctuation of the index terms consistent with those in the content? Is the use of singular versus plural forms of terms consistent? Is the alphabetization of the index accurate?

- **Determining the indexing depth**—Before you begin working on an A-to-Z index, decide what depth would be appropriate for the information space's content and audience. A typical index includes three to five main entries for each screenful of text; a brief index, two main entries; and a detailed index, ten main entries. Limit the number of subentries for each main index entry to a maximum of twelve. To assess an index's depth and organization, identify the key topics on a few pages, then check whether the corresponding headings appear in the index. Does the index cover all key concepts?

- **Achieving completeness**—Whether you are creating an A-to-Z index for an entire information space or just one of its main sections, you must balance its level of exhaustiveness with its precision, faithfulness to the wording of its text, and usability. Any index must be complete and provide access to every major topic that the information space or one of its main sections covers. It should also provide *see* references to the preferred terms for particular concepts and include *see also* cross-references to additional relevant information. An index that is incomplete can frustrate users by making it difficult or even impossible for them to discover important information and could possibly lead them falsely to conclude that an information space does *not* provide the information that they're seeking. To create a usable index, leverage user research and analysis to determine which terms users would most likely seek—and, if an information space already exists, analytics—then use those terms in the index entries. A usable index must be highly coherent. Plus, it should use cross-references to link similar terms and related concepts together.

- **Choosing the components of an index entry**—An index might comprise both its main index entries, or *entries*, and second-level index entries, or *subentries*. A main entry's heading typically comprises a noun, a noun phrase—with an adjective modifying the noun—or a proper noun and is a link. Headings for the main entries could be key words or phrases that appear in the information space's actual text, topics that describe specific concepts or themes and derive from its content, *or* the titles of specific units of content. A subentry's heading should have a logical or grammatical relationship to the main entry's heading to which it is subordinate. Are the headings of index entries and subentries concise and specific, each referring to one easy-to-distinguish topic? If checking an index by clicking some key terms fails to find important content, but trying some broader terms is successful, add some more specific entries to the index. Look for index entries that have an excessive number of subentries and break some of them out into separate index entries. In some cases, it might be helpful to invert a noun phrase, placing the key word first. Plus, an index entry can optionally include a cross-reference.

- **Designing an A-to-Z index**—An A-to-Z index should have a title that employs the same styling as other page titles on an information space, as well as a letter heading for each letter of the alphabet for which the index includes entries. Use capital letters and a larger font for the letter headings. Ensure that there is sufficient whitespace between each letter heading and the entries under it, as well as between the last entry under one heading and the next letter heading. A-to-Z indexes usually comprise one to three columns of entries, each of which should be left aligned. Any A-to-Z index should be easy to scan, and its layout should clearly differentiate subentries from the main index entries. The main index entries should have run-in headings that are left aligned. Subentries are indented and alphabetized beneath the main index entries to which they are subordinate. Each index entry and subentry should begin on a separate line. If an index entry continues onto multiple lines, wrap and indent the additional lines of text. The indentation should be greater than that of the lowest level of subentry in the index.

- **Capitalizing index entries**—The first word of the heading for an index entry or subentry should generally have an initial lowercase letter, unless the word is capitalized in the information space's text, as for a proper noun or a title.

- **Alphabetizing headings**—The first word of the heading for an index entry or subentry always determines where the entry appears in an alphabetical listing. For example, inverting a main index entry's heading—by placing the key word for which users might be looking first—would change the index's alphabetical order. Before you begin working on an index, you must decide whether to alphabetize the index letter by letter *or* word by word. In *letter-by-letter alphabetization*, continue alphabetizing until the first parenthesis or comma in an index entry, then resume following that punctuation. The proper order of precedence when alphabetizing is one word, a word that is followed by a parenthesis, a word that is followed by a comma, a word that is followed by a number, and a word that is followed by letters. Ignore any spaces or other punctuation. In *word-by-word alphabetization*, continue alphabetizing only to the end of the first word—even if that word is an abbreviation or a hyphenated compound—then consider subsequent words only if other headings begin with the same word. The order of precedence is one word, a word that is followed by a parenthesis, a word that is followed by a comma, a word that is followed by a space, a word that is followed by a number, and a word that is followed by letters. Ignore spaces and other punctuation. For indexes that include symbols and numbers, the proper order for alphabetization is symbols, which could include punctuation and special characters; numbers, uppercase letters, and lowercase letters. Or you might instead spell out the name of a symbol—for example, *ampersand (&)*.

- **Employing term rotation**—*If* an index entry is a phrase and comprises more than one keyword, using *term rotation*—also called *permutation*—changes the order of the words by placing a specific key word at the beginning of the entry. By including each word order in a different index entry, you could enable users to find the same information at more than one location in the index. Include such permutations in an index only if users would be likely to look up more than one phrasing. Avoid the overuse of permutations. Conduct user research to determine where permutations are necessary.

- **Including cross-references**—A cross-reference directs the user from one index entry to another. Cross-references include *see* references and *see also* references. *See* references might refer the user, for example, from an informal term to a technical term, from a pseudonym to an actual name, or to a synonym, alternative spelling, or abbreviation. *See also* references refer the user to one or more index entries that provide additional relevant information. The headings of cross-references should link to the content to which they refer. Place the full index entry under the term to which users are most likely to refer. Use *see* references only in cases where the user might miss the full entry. Avoid overusing cross-references. Never create a *see* reference or *see also* reference that leads to another *see* reference. Occasionally, it's best to omit a *see* reference and instead duplicate an index entry under two alternative terms. Both *see* references and *see also* references should appear in italics and immediately follow below the heading for an index entry. In evaluating an index, assess whether *see* references guide users to the main index entries for specific topics and *see also* references lead to additional pertinent information. Consider whether duplicate index entries are appropriate. [72, 73, 74, 75]

Sequential navigation

Sequential navigation lets users move forward and backward through consecutive pages or sets of data or complete the steps of a process sequentially. Common forms of linear, sequential navigation include process funnels that support users' specific tasks; various step-by-step guides such as wizards, tutorials, and guided tours; and pagination controls for search results or other paginated data.

The UI elements that let users control sequential navigation typically comprise either a series of buttons with labels that indicate specific progressive actions *or* a pair of < **Previous** and **Next** > action buttons or links, each comprising a text label and an arrow. For languages that read from left to right, a left-pointing arrow precedes the **Previous** text label, indicating that clicking the button or link would let the user take a step backward; a right-pointing arrow might precede or follow the **Next** text label, indicating that the user can take a step forward. Thus, a control with a left-pointing arrow would navigate to the previous step, page, or dataset, while a control with a right-pointing arrow would navigate to the next one. In contrast, languages that read from right to left reverse the placement of the arrows. Whenever possible, include a text label as well as an arrow to avoid ambiguity.

Sequential navigation can be useful for navigating information spaces that comprise multiple chapters or sections the user should read in a particular order, advancing through structured datasets, completing a step-by-step process, *or* progressing through the pages of a survey or quiz. [76]

On the *UXmatters* home page, **Previous** and **Next** buttons let the user navigate to the home pages for prior monthly editions of the magazine—that is, to the edition for the previous or next month, respectively, as shown in Figure 11.29.

Figure 11.29—**Previous** and **Next** buttons on the *UXmatters* home page

Process funnels

Many information spaces incorporate *process funnels*, whose purpose is to guide users in successfully completing particular step-by-step tasks—such as the check-out process in an ecommerce application. When designing a process funnel, eliminate all distracting UI elements that could prevent users from reaching their goal, such as extraneous content or hyperlinks that could take users away from the task's key steps.

Process funnels should minimize the number of steps that are necessary for users to accomplish a goal. According to Van Duyne, Landay, and Hong in *The Design of Sites*, a process funnel must always consist of at least two steps, but never more than six steps. They recommend either splitting a process that requires more than six steps into two or more process funnels or combining multiple steps of a process into one. However, this is not always possible because the steps in a process funnel are often conditional, depending on the user's prior interactions, and the particular pages that a process funnel comprises might not be able to accommodate all the necessary information.

Clearly indicate the user's current location within the process funnel. To enable the user to easily progress to the next step in the process funnel, always provide a highly visible, clearly labeled action button in a prominent and consistent location on the page, above the fold. For a fairly simple process, the action button's label indicates the next stage of the process, as shown in Figure 11.30.

Figure 11.30—Amazon *Shopping Cart* page's **Proceed to Checkout** button

For more complex processes, you should show the current step on a progress bar, providing greater context regarding the progress the user has made toward the overall goal, as in Figure 11.31. Allow the user to easily skip any unnecessary steps in a process. Always track the user's current status on the progress bar—even if some actual steps were conditional and, thus, not entirely consistent with the steps that appear on the progress bar. [77]

Figure 11.31—Progress bar on Amazon's *Track Order* page

Wizards

A wizard comprises an ordered sequence of numbered steps and guides users—in many cases, new or infrequent users—through a complex process, an unfamiliar task, or a task that users perform infrequently. Wizards often simplify or automate parts of a task, thus, saving the user time and effort. Typically, a wizard consists of the following key elements:

- A *title bar* that displays the title for the overall wizard, which describes the process that the wizard supports

- A *left pane* that lists the wizard's numbered steps, indicating the user's current step. Although a wizard's steps usually occur in a specific sequence, some wizards support conditional processes in which, depending on the user's previous inputs, the wizard might skip entire steps or change the steps later in the process. Sequences of steps that branch or loop require different users to perform different steps to complete the wizard, so displaying a list of steps in the left pane can be challenging. You could choose to display *only* the wizard's main steps, with each step comprising multiple subpages. When the wizard skips steps, you could simply highlight the user's current step in the list of steps. You could dynamically modify the list of steps to conform to the wizard's actual process, displaying only the steps of the branch that the user is following. If none of these approaches is suitable for a wizard, it might be necessary to omit the left pane.

- A *right pane* that displays the wizard's pages of content, each with a subtitle that describes the page's content. In a complex wizard, the content that initially appears there often provides an overview of the process or other introductory information, enabling users to determine whether the wizard would meet their needs. For each step of the process, a *user-input page* appears in the right pane. Any wizard must include at least two user-input pages and can include as many pages as a process requires. In general, keep each user-input page simple rather than creating a few complex input pages. On each user-input page, inform the user about the current step's purpose; provide complete instructions and all the information the user needs to complete the step, as well as input fields to gather information from the user—if necessary, indicating its proper format; and describe the wizard's response to the user's input. Once the user fills in a step's input fields and clicks either **Next** or **Finish**, the wizard processes the input the user has provided and automatically either proceeds to the next step *or* completes the overall task, respectively. A wizard's final page typically confirms that the process is complete and provides a summary of its results.

- A *button bar* that appears at the bottom of all the wizard's pages. In a left-aligned group on the button bar, < **Previous** and **Next** > buttons—*or*, preferably, equivalent buttons with more specific labels that describe the previous and next steps—constitute the wizard's sequential navigation, which lets the user navigate backward and forward through the wizard's steps. Plus, a **Finish** button lets the user complete the task. A right-aligned group of buttons typically comprises **Cancel** and **Help** buttons. Any button that is currently unavailable should appear dimmed—as is the < **Previous** button on the first page of a wizard.

Experienced users often prefer quicker, more autonomous methods of performing tasks, so whenever you implement a wizard, be sure to provide an alternative way of performing the same task. Figure 11.32 shows an example of a wizard that lets potential buyers build their own MINI Cooper. [78, 79]

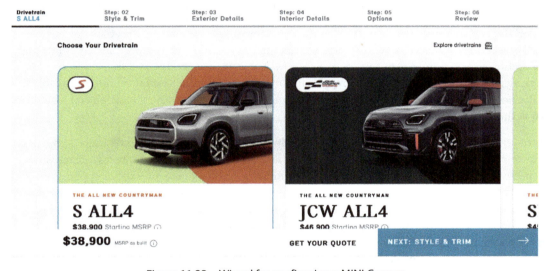

Figure 11.32—Wizard for configuring a MINI Cooper

Tutorials

An information space might offer one or more interactive, online tutorials, each providing training on a specific topic, procedure, or user task, enabling users to complete its lessons and learn at their own pace. Each page of a tutorial typically focuses on a single concept or one step of a procedure, so the content should be fairly brief. Traditionally, a tutorial's sequential pages comprise an introduction that makes the goal of the tutorial clear; a series of up to 15 tutorial pages that present progressively more difficult skills or concepts—including basic, intermediate, and advanced skills or concepts; a summary page that lets the user review the lesson; a quiz; and finally, a page that congratulates the user on completing the tutorial. Ideally, each tutorial page should link to separate pages that provide examples and practice exercises, enabling the user to better understand and apply the learnings from the lesson. In general, limit tutorials to seven to ten simple concepts or skills. Completing a tutorial should usually take the user only about 20 to 30 minutes.

Many tutorials—especially those that teach concepts rather than procedures—incorporate quizzes that test how well the user is absorbing the information. Such tutorials evaluate the user's responses to the quiz's questions, then personalize the sequence of the tutorial's subsequent steps accordingly—often by using artificial intelligence (AI)—enabling the user to review any information on which further instruction is necessary.

Tutorials provide sequential navigation, typically comprising **Next >** and **< Previous** buttons that let the user advance or go back and forth through its pages. A **Home** button sometimes lets the user return to the beginning of a tutorial. If appropriate, let the user end the tutorial at any point by clicking a **Close** or **Pause** button, then resume the tutorial by clicking a **Resume** button. Always place **< Previous, Next >**, **Home, Close, Pause**, and **Resume** buttons in consistent locations on a button bar at the bottom of all the tutorial's pages.

A traditional, stand-alone tutorial for either a hardware device or an application typically comprises activity-centered lessons and consists of instructional text and either illustrations or photos *or* screenshots, respectively. Such tutorials provide step-by-step guidance on how to use a device or application. In addition to teaching the concepts and procedures the user needs to learn, these tutorials often include task scenarios that allow the user to perform tasks, albeit using dummy data. While such a tutorial is *not* integrated with the actual user interface of the application that it documents, it can simulate the experience of using the application. In a progress indicator, display the tutorial's overall number of steps and the user's current step—for example, **Step [#] of [#] steps**.

In-app tutorials run on top of an application's user interface and provide guidance on how to use the application to complete the user's actual tasks within the context of the application. An in-app tutorial's activity-centered lessons guide the user through step-by-step workflows by providing instructions in dialog boxes, ToolTips, or other forms of overlays. Figure 11.33 provides an example of an in-app tutorial.

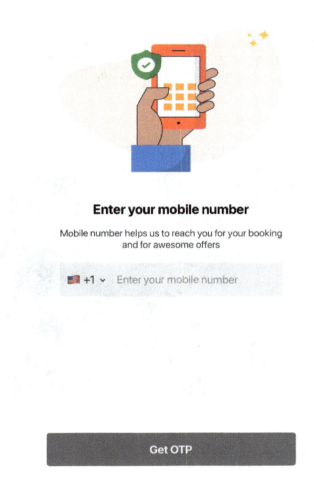

Figure 11.33—Eazydiner's in-app onboarding tutorial

Creating a series of storyboards can help you define a meaningful sequence of steps for a tutorial, define what necessary conditions must exist before the user begins the tutorial, and ensure that you include all of the task's steps in the proper sequence. Storyboarding should be an iterative process, so review your storyboards rigorously, then make revisions as necessary. Be sure to include all necessary details on each of the tutorial's pages. Also, because users are not yet familiar with the concepts or tasks they'll be learning about, you should be careful to avoid providing superficial explanations or omitting key information. To ensure that the tutorial's sequence of steps represents a logical workflow, test with paper prototypes.

A tutorial could alternatively show how something works by playing an animation or a video recording, as shown in Figure 11.34. However, such tutorials disrupt the user's workflow, might not enable users to perform a task more effectively, and might not be memorable. Plus, planning and producing multimedia content can be challenging and absolutely requires iterative storyboarding. Be sure to include all the necessary details in an animation or video, and keep its pace slow enough to ensure that users can absorb the material. Test the performance of any media files and limit their size to ensure they don't take too long to download. Although an animation could play automatically, navigating through an animation or video typically relies on standard controls, including **Play**, **Pause**, **Stop**, **Rewind**, and **Fast Forward** icons.

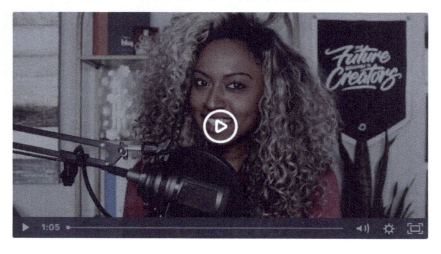

Figure 11.34—A video tutorial

Many users consider working through tutorials to be a waste of their time, so they frequently skip them. Therefore, you should carefully consider whether a tutorial is even necessary. For a simple information space or application that employs standard interaction models, a tutorial probably isn't necessary. However, a tutorial might be helpful in cases where users need to learn novel interactions such as in a virtual-reality (VR) or augmented-reality (AR) app.

Any tutorial must provide an engaging user experience. So, in addition to offering an activity-centered learning experience, you should consider including hypertext links to other tutorials or related information to encourage users' further exploration. [80, 81, 82, 83, 84, 85, 86, 87]

Guided tours

Similar to an in-app tutorial, a guided tour runs on top of an information space or application's actual user interface and guides the user by highlighting each element that the tour describes in turn and displaying ToolTips, speech bubbles, or other forms of overlays that contain brief descriptions of the elements. An arrow should indicate the element to which the current description refers.

Some information spaces present an *automated, guided tour* that introduces users to their content or user interface. These guided tours often take the form of a *product walkthrough*—making them useful for onboarding new users; providing an overview of new features to existing users; *or* marketing a product or service's features and, thus, enabling potential users to perceive its value. The onboarding tutorial for ConvertKit is shown in Figure 11.35. Although some guided tours start running automatically when a page loads, the user should ideally be able to start a tour by clicking a UI element such as a call-to-action button.

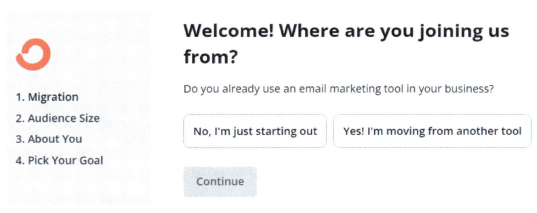

Figure 11.35—ConvertKit onboarding tutorial

Other information spaces might let the user choose what features, content, or UI elements to explore by clicking elements on particular pages to highlight them and display overlays that contain brief descriptions.

Whether taking an automated guided tour or navigating autonomously through a tour, the user should be able to exit the tour at any time by clicking a close box on the current overlay. Completing a guided tour should typically take the user only about ten minutes.

For a more extensive guided tour, a list of links to the left of all the tour's pages could serve as a table of contents, and the user could navigate through the tour by clicking links in that table of contents. When the user clicks a link in the table of contents, a particular page of the tour appears. To ensure consistency in the tour's behavior, *all* pages of the tour should be either automated *or* under the user's control.

The greater an information space's complexity, the more important it might be to provide a guided tour. However, many users don't feel the need for a guided tour and would want to skip all or parts of the tour. So you should ideally let users both initiate and control a guided tour. Also, few users would need a guided tour more than once. Therefore, an organization should make only a moderate investment in creating a guided tour. While any guided tour must effectively serve its purpose, you should carefully consider whether a guided tour is even necessary and avoid going overboard in trying to create a compelling, feature-rich tour. [88, 89, 90, 91]

Pagination controls for navigating paginated data

Another form of sequential navigation, *pagination controls* let users navigate pages of data in systems that limit the number of search results or other data points that appear at once on a page. Pagination controls typically consist of **Previous Page** and **Next Page** links, with an indicator in between them that displays the current page and the total number of pages of data—for example, **Page 3 of 18 pages**. Pagination controls typically appear at the bottom of each page of data to let the user view successive pages of data. Clicking the **Previous Page** or **Next Page** link lets the user page through previous or next pages of data, respectively.

Some information spaces with pagination controls also provide direct access to specific pages of data by displaying a horizontal list of page-number links that the user can click to navigate to a particular page of data and highlight the number of the current page. However, this design approach isn't all that useful because there's no way to accurately predict what might be on a particular page of data. Listing all the page numbers takes up too much space, designs often become overly complex when there are many pages of data, clicking the tiny page-number links is difficult, and trying to tap them on mobile devices is impossible. [92]

Today, with the prevalence of *lazy loading*—in which a system loads more search results or other data as the user scrolls down a page of data—pagination controls are now uncommon.

Location breadcrumbs

On information spaces with a deep navigation hierarchy, where users are more likely to lose the scent of information and become disoriented, location breadcrumbs can provide a complementary means of displaying and enabling users to navigate the hierarchy. Thus, location breadcrumbs could help users to understand and orient themselves to their current location within an information space, which is especially important if a user has arrived on a page by clicking an external link. However, many users ignore location breadcrumbs, and most users rarely click these links.

Location breadcrumbs comprise a sequential series of textual links that represent the optimal path to the user's current page on an information space. Each link in the series corresponds to a page at a particular level of the information hierarchy, from the home page—*or*, in some cases, the main page of a key section—to the user's current page, enabling the user to jump directly to a higher level of the information hierarchy. On an information space whose architecture is a polyhierarchy, location breadcrumbs should represent what is usually the best pathway to the current page.

Location breadcrumbs provide an easy-to-use, very compact wayfinding aid. On the desktop, they consist of just a single line of textual links that resides immediately below the primary navigation system and above the page title, as Figure 11.36 shows. Ideally, **_angle brackets (>)_** should separate the breadcrumb links. When the user's current page is the last item in the series, make sure it is *not* a link. Alternatively, the location breadcrumbs could appear immediately above the page title, and the layout of these elements could demonstrate the continuity between the breadcrumbs and the page title.

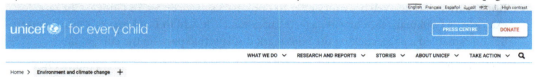

Figure 11.36—Location breadcrumbs on the UNICEF site

However, location breadcrumbs can be difficult to represent within the limited screen real estate of a mobile device. Usability requires that they *not* wrap onto multiple lines and that their links be easy to tap, so their font size must be large enough, and sufficient space must separate the links. For mobile devices *only*, you might need to minimize the screen real estate that location breadcrumbs require. Therefore, you could consider including links to just the pages that are one or two levels above the user's current page. [93, 94, 95, 96]

To make the function of location breadcrumbs absolutely clear, consider adding the label **You are here** preceding the breadcrumb links. Doing so also helps the users of screen readers by clarifying the context of the breadcrumb links. A page's code could optionally hide this label from sighted users. [97, 98]

For an information space whose navigation hierarchy has many levels, you could consider hiding the lowest levels of the local navigation and representing the user's current location in the hierarchy using location breadcrumbs. However, this approach is *not* ideal because it prevents the user from being able to see all the links at each level of the local navigation. [99]

Web browsers' navigation capabilities

When you're designing the navigation system for a Web site, be sure to consider the navigation capabilities of the user's Web browser, which typically include the following:

- **Open URL**—By typing a Web address, or URL, into a browser's Web address bar, the user can go directly to any page on a site. Since doing so bypasses the Web site's home page, it is important to provide adequate context on *all* the site's pages, including branding that clearly indicates the organization to which the site belongs and a structural navigation system that shows the site's information hierarchy, as well as the user's current location within it.

- **Back and Forward buttons**—This pair of buttons supports backtracking in both directions. Users commonly use the **Back** button when navigating the Web. Therefore, it is particularly important that Web sites support people's use of these buttons to ensure that things work as expected.

- **Tabs**—The user can open pages on multiple tabs at once, then navigate between the tabs.

- **History menu**—The options on this menu comprise pages the user has visited recently and appear in reverse chronological order. By clicking an option on this menu, the user can go directly to the corresponding page. To ensure the optimal usefulness of this menu for users, always give pages concise, meaningful browser titles.

- **Bookmarks**—Some Web browsers refer to bookmarks as *favorites*. In Google Chrome, for example, to bookmark the current page, the user can click **Add to Bookmarks**. Once the user has bookmarked a page, clicking the bookmark displays that page on the current tab.

- **Link previews**—In some Web browsers on desktop computers, when the user's pointer hovers over a link, the browser displays the destination URL for that link in the lower-left corner of the browser window. Since most URLs provide information about a site, its hierarchy, and the content at the link's destination, the user can get some help in deciding whether to click the link before clicking it.

Although Web sites should always support browsers' navigation features, they should *not* rely on them for functionality that a site's navigation system itself should provide. [100, 101]

Summary

In this chapter, you learned about some basics of designing navigation systems for information spaces, including the primary concerns at the foundation of effective navigation design and the key objectives of designing navigation systems to meet the needs of both the information space's users and the business. You also learned some universally applicable design guidelines that relate to all types of navigation, as well as to the navigation design patterns that you'll learn about in Chapter 12, *Designing Navigation*.

This chapter also provided in-depth information about employing the various types of navigation that an information space might incorporate—alone or in combination with others. The types of navigation whose design we covered included structural navigation, navigation pages, associative links; supplementary, or utility, navigation; and complementary navigation tools, as well as leveraging the navigation capabilities of Web browsers.

Next, in Chapter 12, *Designing Navigation*, you'll learn about the many navigation design patterns that are in common use and how to employ them effectively. You'll also learn about various means of visualizing navigation design and producing the design deliverables that are an integral part of designing navigation systems. Finally, we'll cover the basics of testing your navigation designs.

References

To make it easy for readers to follow links to the references for this chapter, we've made them available on the Web: `https://github.com/PacktPublishing/Designing-Information-Architecture/tree/main/Chapter11`

12
Designing Navigation

"Navigation systems should be designed with care to complement and reinforce the hierarchy by providing added context and flexibility."—Peter Morville and Lou Rosenfeld [1]

Designing an effective navigation system for a digital information space depends primarily on creating well-organized, well-labeled *navigation hyperlinks* that have strong information scent and, thus, result in good *findability*, enabling users to correctly identify links that would lead to the specific information they're seeking. The labels for navigation hyperlinks derive much meaning from their context—the navigation hierarchy itself.

When users are browsing an information space, they quickly scan groupings of navigation hyperlinks for labels that closely match the topics or types of content they're seeking. Clicking a navigation hyperlink displays either a different page on the information space or other information that resides on the same page. Browsing an information space can be a highly interactive and engaging experience that enables users to discover what other content is available on the information space.

Navigation design resides at the intersection of information architecture (IA), visual user-interface design, and interaction design (IxD), which respectively determine the structure, look, and feel of an information space's navigation system.

Therefore, information architects are typically responsible for designing a navigation system's structure and labeling, while visual-interface designers are responsible for creating the layout and look of the navigation elements, and interaction designers design the interactivity of these elements—especially when the user interface is more complex than simple links on a navigation bar. To achieve optimal design and wayfinding outcomes, this team of designers should engage in multidisciplinary collaboration. However, a UX designer might alternatively be responsible for *all* aspects of a navigation system's design.

This chapter covers the following topics relating to the design of navigation systems for information spaces:

- Navigation design patterns
- Visualizing navigation design
- Testing navigation design

Navigation design patterns

"Patterns derive from principles and heuristics that are based on human physiology and cognitive psychology."—Steven Hoober [2]

Navigation design patterns describe the various ways of organizing and representing groups of links and, in a few cases, individual links that make up an information space's navigation system. The links that are part of each navigation pattern share a common look and feel and have similar functions, the same behaviors, and depending on their state, the same appearance.

When designing the user-interface elements that an information space's navigation system comprises, you can employ a variety of design patterns or combinations of design patterns that typically play specific roles within a navigation system. Some patterns express and enable users to traverse an information space's navigation hierarchy, while others represent linear sequences of interactions. Still other patterns provide an overview of an information space's content and help users navigate directly to specific content. Page-specific navigation patterns let users jump to particular sections of a page. By clicking the text or images that represent associative links, users can go directly to related information on other pages on the same information space, in other sections of the current page, or on a different information space.

When designing a navigation system, you must choose whatever navigation design patterns optimally represent the structure and content of a particular information space and meet its users' needs. The typical roles of each of these navigation design patterns determine which patterns to employ in creating particular elements of a navigation system. Let's look at each of these patterns in detail. [3]

Fundamental navigation elements

The fundamental elements of any navigation system consist of various types of hyperlinks or action buttons.

Hyperlinks

Hyperlinks provide connections between the nodes, or pages, and chunks of information, or sections of pages, that any digital information space comprises. The forms that hyperlinks can take include the following:

- **Text links, or hypertext links**—These could be either *structural links*, which collectively form the structural navigation for an information space, including both its global and local navigation and the contextual navigation links for specific pages; or *associative links* on words or phrases in a page's content that users can traverse to view related information on the same information space—whether on another page or in another section of the same page—or a different information space. Advantages of text links include that they load more quickly than image links and are accessible to visually impaired users via the use of a screen reader.

- **Icons**—Iconographic links usually accompany *structural* hypertext links to facilitate their recognition, but common icons that most users would recognize can also stand alone, especially in contexts where constant usage would enable users to learn the meanings of the icons. Some common icons include **Home**, **Search**, **Share**, **Help**, **Play**, **Pause**, **Sound**, **Previous**, and **Next**. Icons sometimes also accompany *associative* hypertext links within a page's content—for example, a font icon might depict a right-pointing arrow that accompanies a **Read More** link. Formerly, information spaces often used icons that indicated clicking a link would display content on a different information space, but that is no longer common practice.

- **Organizational logos**—On an information space, an organization or company's logo should link to its home page. This logo typically resides in the upper-left corner of each page on the header or navigation bar, but some companies center their logo at the top of each page—particularly on mobile devices. According to research from *Nielsen Norman Group (NN/g)*, users are six times more likely to successfully navigate to the home page when the logo resides in its standard location in the upper-left corner than when it is centered. To ensure that users can easily find the home page, also provide a **Home** link in the global navigation. Alternatively, you can place a **Home** link in any location breadcrumbs. [4]

- **Images in carousels**—Large photographs or illustrations within carousels often provide associative navigation links to key pages on an information space. However, not all users will be able to correctly interpret an image as being associated with the content to which the image links. Therefore, such images should also incorporate some identifying text.

- **Hot spots**—Traditional hot spots are predefined areas of an *image* that provide contextual navigation links that let users view related information, either on another part of the same page or on a different page. For example, in a map of Europe, each country could be a hot spot that lets users view information relating to that country. However, one disadvantage of hot spots in images is that users might have difficulty distinguishing them and might not even realize that they can interact with parts of an image. In 3D *virtual-reality environments*, interactive hot spots are visible user-interface elements that enable point-to-point navigation. The user can click these hot spots, which often take the form of subtle white rings, to navigate a virtual analogue of a physical space such as a store, museum, or model home, as shown in Figure 12.1. Interacting with these hot spots provides an immersive experience that encourages exploration and discovery. In cases where the intent of hot spots is to draw users' attention and guide them—whether for purposes of onboarding users to an application or new feature by providing a tour or more in-depth training *or* to provide a call to action to get users to take a desired action such as registering for a Webinar or other event—hot spots are typically pulsing circles. Once the user clicks this type of hot spot to get more information or take an action, the hot spot disappears. [5, 6]

Figure 12.1—Example of hot spots in a virtual house tour

Across an information space, hypertext links of a particular type—for example, structural navigation links or associative links—should consistently indicate state, as shown in Figure 12.2. For example, hypertext link states might include the following:

- **Available**—This is a hypertext link's initial state and indicates that the user can click or tap the link to go to its destination. The text formatting for associative hypertext links in an information space's content should be different from that of its body text—by default, in blue (#0000EE) and, optionally, in bold, but many information spaces use a color from the organization's branding palette instead. Traditionally, associative hypertext links are underlined as well—by default, with a solid underline, but alternatively, a dotted underline. However, because the hypertext links in an information space's structural navigation system already have unique placement and text formatting, they need *not* be underlined. Plus, most information spaces do *not* underline featured links such as article titles or author bylines. Many information spaces do *not* underline associative links in their content because underlined text is harder to read. Nevertheless, underlining hypertext links or using some other unique text formatting makes links more accessible, especially to people with color-deficient vision.

- **Hover**—This state exists *only* for hypertext links in desktop Web browsers. When the user points to a hypertext link, the browser automatically changes the pointer to a pointing hand and changes the link's text formatting to the hover state. To indicate that the user is currently pointing to a hypertext link, employ a different color for hypertext links on hover. Use whatever other text formatting an information space applies to hypertext links in their available state as well, including underlining. In lieu of underlining associative hypertext links in their available state, many information spaces instead underline links on hover—using either a solid or a dotted underline.

- **Active**—When the user clicks a hypertext link, the Web browser automatically changes the link to its active state, changing its text formatting accordingly—that is, by default, rendering the text in red (#EE0000) and underlining it. Since, in the majority of cases, activating a hypertext link immediately traverses the link and displays a different page, information spaces do not generally implement any text formatting for this link state.

- **Selected**—Once the user clicks a link within an information space's structural navigation system, the browser changes the link's text formatting to the selected state. Within the navigation hierarchy, highlight *all* the hypertext links that the user has clicked to display the current page, thus indicating the user's navigation path. To highlight the selected links, use the same foreground and background colors as for the hover state and bold text. Do *not* underline the selected links.

- **Visited**—The browser automatically changes the color of visited links—by default, to purple (#551A8B), but alternatively, to another contrasting color that is unique to the information space's color palette. For example, you could use a lighter or more neutralized variant of whatever color the information space uses for available links. For associative hypertext links within a page's content, including links on article titles and author bylines, use a contrasting color to indicate *all* the links that the user has previously clicked during the current session to show that the user has already visited the corresponding pages or content on the same page. If associative links are in bold or are underlined when they're available, use the same formatting in their visited state.

- **Focus**—When a keyboard user moves the focus to a hypertext link using the *Tab* key—or, on a Mac, by pressing *Option + Tab*—the browser automatically changes the link to the focus state and renders an outline around it. This formatting is an essential aid to accessibility. [7]

Figure 12.2—Silicon UI Kit hypertext link states

The appearance of each of these link states should clearly distinguish it from all the other link states.

Action buttons

An action button is a basic user-interface element that the user can click or tap to initiate a specific action. On the Web, action buttons combine the following elements:

- **A visual element**—This is an HTML `<button>` element that is styled with Cascading Style Sheets (CSS).

- **A text element**—This is the button's label. However, this could optionally be a hypertext link.

- **Code**—In most cases, code initiates the button's action. However, clicking a hypertext link simply navigates.

A button's text label should be a verb that clearly indicates its resulting action. The label is typically in title case, bold text, and centered. Whenever possible, a label should comprise just one word and never more than three words. Always capitalize a button label's first and last words. To clarify the meaning of an important button and facilitate its recognition, some action buttons could also have an easy-to-interpret icon to the left of the text label.

Although action buttons should have a consistent appearance across an entire information space, the degree of visual emphasis on primary and secondary buttons can differ, typically through the use of filled and outlined buttons, respectively. The sizes of buttons can differ as well, depending on a button's context. However, an information space should use just a few standard button sizes—for example, a large call-to-action button within a marketing context, a standard size for most action buttons, and a small button size for use in contexts in which space is at a premium. Ensure that the sizes of all buttons provide easily tappable targets.

To provide affordance and, thus, make action buttons appear clickable or tappable, their appearance typically emulates that of a hardware button to some degree. Thus, action buttons are usually rounded rectangles. However, while the buttons from different user-interface frameworks could have a different corner radius and more or less dimensionality, designers no longer create the heavy, three-dimensional buttons that were once common.

Action buttons should consistently indicate their state, and the appearance of each of these states should clearly distinguish it from the other states. Figure 12.3 provides an example. A button's states might include the following:

- **Available**—For buttons that perform instantaneous actions—whose availability is neither conditional nor predicated on the user's previous actions—this is a button's initial state, and the button is clickable or tappable.

- **Unavailable**—For buttons whose availability is conditional and predicated on the user's previous actions, this could be a button's initial state.

- **Hover and click**—While hover and click could be separate states, combining these states when the user clicks a link is now common since touch devices don't support hover.

- **Processing**—If a button's action is not instantaneous because of the amount of processing that is necessary to complete an action, a processing or loading status indicator could appear to the left of the button's label. [8, 9, 10]

States

Figure 12.3— Silicon UI Kit button states

The same button states should also apply to icons.

Depending on their context, action buttons can serve several functions. Common types of action buttons include the following:

- **Process-completion buttons**—Such buttons are associated with the completion of Web forms or other processes. Some common examples include **Sign In**, **Send**, **Save**, **Search**, **Delete**, **Proceed to Checkout**, and **Place Your Order** buttons.

- **Call-to-action buttons**—These CTA buttons should provide clear calls to action, representing conversions that are valuable to an organization. Common examples of CTA buttons include **Sign Up**, **Add to Cart**, and **Buy Now** buttons.

- **Navigational buttons**—These buttons simply navigate to another page. Examples include **Read More** buttons on a directory page that encourage the user to read the rest of an article or other content.

- **Progressive-disclosure buttons**—These buttons could either expand or lazy load more content on the same page. For example, a **View More** button could expand more content, changing the button's label to **View Less**. Subsequently clicking **View Less** would then collapse the expanded content. Alternatively, a **View More** button could lazy load additional search results or more products belonging to a particular category.

Desktop navigation patterns

"The navigation designer must first be comfortable with the tools and elements [of] Web navigation and realize that, for any one navigation problem, there may be a number of mechanisms that solve it."—Jim Kalbach [11]

Let's review the navigation patterns that are in common use for information spaces on both desktop computers and larger tablets.

Horizontal navigation bars

On the desktop, the most common navigation pattern is the *horizontal navigation bar*, which resides near the top of an information space's pages, immediately below its header. You can use this pattern for the *global* horizontal navigation bar of an information space of almost any scope, *except* one that has a very broad navigation hierarchy.

To represent an information hierarchy that comprises multiple levels of categories, some information spaces stack two or occasionally three horizontal navigation bars near the top of all their pages. Thus, horizontal navigation bars can represent both global *and* local navigation. However, on many other information spaces that have a global horizontal navigation bar, local navigation links reside on either expandable menus or a vertical navigation menu.

Because an information space's *global* horizontal navigation bar must accommodate *all* the links that correspond to its top-level categories—and any *local* horizontal navigation bar must comprehend *all* the links that represent a section's next lower-level categories—the design of the information architecture and navigation bars should ideally occur concurrently. However, preserving the coherence and clarity of the information architecture should always be paramount when you're making decisions about whether to use a horizontal navigation bar and what links to place on it. For example, a horizontal navigation bar would *not* be the best choice for an information space whose scope is likely to expand significantly *or* where translating the link labels would result in long translations—for example, for translations into French, Spanish, German, Polish, Italian, or Dutch. [12] To enable users to easily discern which links to click to reach the information they're seeking, ensure that each category link on a navigation bar provides good information scent and is sufficiently clear and specific to make it easy to distinguish from all the other categories.

When designing a horizontal navigation bar, factors to consider include the scope of the information space, the links that are necessary at each level of the navigation hierarchy, the amount of horizontal screen real estate available for the navigation bar, the numbers of characters in the links' labels, the labels' font sizes and weights, and the amount of space separating the links. Typically, the number of links residing on a horizontal navigation bar ranges from between five and twelve links. But let's consider the Web sites and apps of big, well-established newspapers, whose print analogs have traditionally had large numbers of sections and, thus, have many more links on their primary navigation bar. For example, *The New York Times* Web site has 20 navigation links on its global navigation bar, as shown in Figure 12.4.

Figure 12.4—*The New York Times* Web site's global navigation

The Washington Post iPad app has so many links on its global navigation bar that it displays them in a carousel, and the user must scroll horizontally to bring all the links into view, as Figure 12.5 shows. Plus, tapping a plus-sign icon at the right of this bar displays a **Menu Preferences** overlay that lets the user configure which links appear on the global navigation bar.

The Washington Post

| For You | Latest | Politics | Opinions | Style | Investigations | Climate | Well+Being | Tech | Recipes | Lifestyle | World | D.C., Md. & Va. | Sports | Discover |

Figure 12.5—*The Washington Post* app's global navigation

Providing clear, albeit brief labels for navigation links is essential—as is providing adequate separation between them. Ideally, each link on a navigation bar should comprise just one word, but only *if* that single word would fully convey the link's meaning. If clearly labeling *any* links would require using multiple words, increase the amount of whitespace between *all* the links to adequately separate them from one another. Plus, to ensure good usability, the font size of the link labels must be sufficiently large for easy legibility. [13, 14, 15]

Expandable menus

Many information spaces have navigation bars that comprise multiple instances of a specific type of expandable menu such as drop-down menus or megamenus, which display navigation links at the next lower level of an information architecture. Some navigation bars consist only of navigation links, while others consist of a combination of links *and* menus. On a navigation bar, a tiny, downward-pointing triangle to the right of each menu title should convey its function and distinguish menu titles from simple navigation links. However, in this era of minimalistic design, many information spaces have eliminated these helpful indicators.

Using expandable menus for exploratory browsing requires more effort on the user's part because these menus initially hide an information space's local navigation links. However, their use does conserve screen real estate, and accessing an expandable menu requires only a single click.

On the desktop, pointing to a menu title on a navigation bar should just temporarily highlight it. On any platform, clicking a menu title should highlight it *and* display an expandable menu. Although many implementations of expandable menus on the desktop Web appear on hover rather than on click, this interaction model often has unexpected results, creating what usability professionals have referred to as *mouse traps*, with menus potentially opening unexpectedly and obscuring the information the user actually wants to view on a page. The interaction cost of the user's inadvertently opening an expandable menu, then recovering from this error would be especially high for weighty megamenus and navigation drawers. Menus that appear on hover can also present accessibility issues to users of screen readers and people who have motor impairments. Thus, displaying an expandable menu should ideally require the user to click a menu title.

Once the user opens an expandable menu on a navigation bar, the user can click a link on the menu to navigate to another page on the information space. The menu item that corresponds to the user's current page should appear highlighted and should *not* be a link. To close an expandable menu without clicking a link to navigate to another page, the user should be able to click outside the menu *or* click the same menu title again. [16, 17, 18, 19, 20]

Drop-down menus

Within the context of an information space's navigation system, a *drop-down menu* is a form of expandable menu that should reside either on the global navigation bar *or* any local navigation bar. The *Nielsen Norman Group (NN/g)* Web site employs some drop-down menus on its global navigation bar, as shown in Figure 12.6.

Figure 12.6—Drop-down menus on the *NN/g* Web site

Avoid placing *all* of an information space's global navigation links on a single drop-down menu. Doing so would obscure the information space's global navigation, impair discoverability, make it more difficult for users to ascertain the information space's structure and understand its scope, make switching between its top-level categories more difficult, and significantly increase the interaction cost of navigating the information space. [21]

Viewing *all* the navigation links on a drop-down menu should *not* require scrolling. Limit the number of navigation links on a drop-down menu to minimize the distance, as well as the time it takes for the user to move the pointer from the menu's title to its most distant navigation links. [22] On a drop-down menu, each menu item is clickable or tappable and is either a navigation link *or* a submenu title. Clicking a navigation link displays a different page on the information space and closes the menu.

Submenus

A *submenu*—also known as a *hierarchical menu* or *cascading menu*—descends from a submenu title on a drop-down menu and displays a list of related links at the next level of the navigation hierarchy. On a drop-down menu, a right-pointing triangle to the far right of a submenu title clearly indicates that the corresponding submenu will appear to its right when the user clicks the submenu title or, on the desktop, points to the title, as shown in Figure 12.7. When displaying a submenu on hover, add a 500-millisecond delay before displaying the submenu to prevent the user from inadvertently opening the submenu. [23]

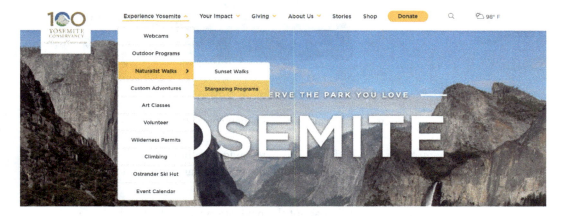

Figure 12.7—Drop-down menu with a submenu on Yosemite.org

Clicking a navigation link on a submenu displays a different page on the information space and closes both the drop-down menu and the submenu. To close a submenu without navigating to another page, the user should be able to either click outside the submenu *or* click the same submenu title again. On the desktop, the user can simply point to another item on the same menu.

Avoid placing frequently used navigation links on submenus. Using submenus buries an information space's local navigation links more deeply, adds significant interaction cost, and can present accessibility issues to people who have motor impairments. To avoid adding unnecessary complexity to an information space's navigation system, minimize the use of submenus, and always limit the number of submenu levels to just one. [24]

Megamenus

A *megamenu* is a large, rectangular, expandable menu that has a horizontal aspect ratio, can accommodate a large number of navigation links, and comprises columns of navigation links. A megamenu's size limits its usefulness to devices with larger screens such as desktop computers or large tablets. As Figure 12.8 shows, using a columnar layout gives a megamenu a coherent structure and minimizes the distance and, thus, the movement time from a menu's title to its most distant navigation links.

However, because a megamenu also comprises a large number of navigation links, the visual-search time the user needs to find a specific link on a megamenu is greater. [25, 26]

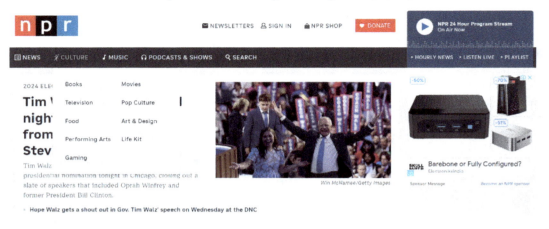

Figure 12.8—A columnar megamenu on *NPR*

You can use a megamenu to display *all* the subordinate navigation links at the next level or even multiple levels of the navigation hierarchy. To ensure that the user can view all of these navigation links at once, a megamenu should *never* scroll.

To prevent users from becoming confused about the navigation hierarchy to which a particular megamenu belongs, be sure to clearly represent the visual connection between the megamenu and its menu title.

If, in the course of moving the pointer diagonally to a navigation link on an open megamenu, the user incidentally points to an adjacent menu title, the megamenu should remain open for 500 milliseconds to let the user complete the interaction and prevent the user from inadvertently closing the megamenu.

Because a megamenu often covers much of a page, displaying a megamenu on hover is particularly egregious. Once the user opens a megamenu by clicking its menu title on a navigation bar, the user should click a link on the megamenu to navigate to another page on the information space. [27]

Group labels

One significant advantage of megamenus over other types of expandable menus is the ability to organize related navigation links in groups. While most megamenus organize navigation links under group labels that are not clickable, a megamenu could alternatively comprise two levels of navigation links. In both cases, a megamenu represents two levels of an information space's hierarchy.

To ensure that users can quickly scan a megamenu's group labels and easily choose a navigation link, each logical grouping of links must be neither too large nor too small, users must be able to easily distinguish the groups' labels from one another, and the most important or frequently used groups should reside in the upper-left side of the megamenu. Once users identify the relevant group, they can limit their perusal of navigation links to just the group that would most likely meet their need.

To clarify the structure of a megamenu, express its visual hierarchy using formatting such as typography—specifically, font weight, font size, and case—color, and perhaps indentation. As shown in Figure 12.9, the Nashville Zoo's Web site uses group labels. [28, 29]

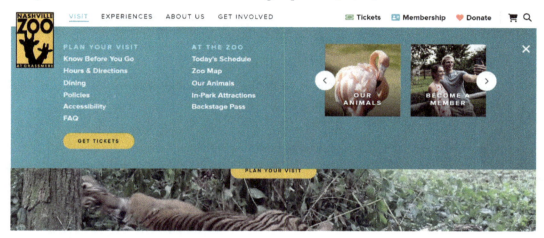

Figure 12.9—A megamenu with group labels on the Nashville Zoo's Web site

Images and scope notes

Another advantage of megamenus is that they offer sufficient space to present rich content, including images such as photos or icons and *scope notes* that describe the content in an information space's sections. Therefore, navigation links can either be simple text links or combine text and image links. For example, the use of photos depicting specific products is especially common on ecommerce stores, but image links could alternatively be icons that represent concepts. Such images can help users quickly and accurately identify navigation links.

Megamenus that comprise a fairly limited number of top-level links to key sections of an information space could include brief scope notes that describe the content in each corresponding section. Together, top-level links and scope notes could provide another grouping mechanism for each section's subordinate links. The megamenus on HubSpot's blog include both iconography and scope notes, as Figure 12.10 shows. [30, 31]

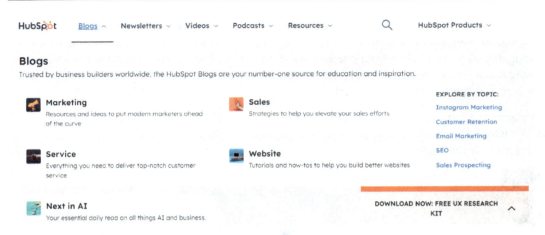

Figure 12.10—A megamenu on HubSpot with icons and scope notes

Vertical navigation menus

A *vertical navigation menu* is a fairly common navigation-design solution—whether for an information space's global navigation, local navigation, or a navigation menu that combines both—and can be an effective, flexible choice in the following cases:

- If an information space has a limited scope and a flat navigation hierarchy, a simple list of links can present *all* the links necessary for its *global* navigation. Such a vertical navigation menu would typically reside to the left of the main content area on all pages. However, on an information space's home page, the vertical navigation menu might alternatively be centered and occupy the entire page, with the links in a large font. In either case, the navigation links should be left aligned to ensure that they are easy to scan and read.

- If an information space's *global* navigation must include more links than a horizontal navigation bar could accommodate, display these global links in a vertical navigation menu that resides to the left of the main content area on all pages. Such a vertical navigation menu can accommodate as many global navigation links as are necessary to represent whatever information hierarchy would accurately convey the scope of the information space's content. Having more space for displaying links also ensures that their labels provide maximal information scent.

- If an information space has a global horizontal navigation bar and the *local* navigation comprises more links than would fit on a horizontal navigation bar, display the local links in a vertical navigation menu to the left of the main content area.

- If a vertical navigation menu comprises multiple levels of an information hierarchy, consider nesting the lower-level navigation links under the higher-level links in the form of a hierarchical list or tree navigation, as Figure 12.11 shows. Employ text formatting and indentation to help users distinguish the various levels of the hierarchy. Depending on the global navigation link and, thus, the section of an information space that the user is currently viewing, the local navigation links differ. Be sure to indicate which global and local links correspond to the current page.

- If an information space's scope could expand significantly in the future and, thus, require the addition of many navigation links without necessitating a complete redesign of its information architecture, a vertical navigation menu would let you simply integrate the new links and extend its vertical length. This need might exist in a large organization whose content is continually evolving or a high-growth startup. Place the most important links nearest to the top of a vertical navigation menu. This is especially important if the user must scroll to view the entire menu.

- If many of an information space's navigation links have long labels that require more horizontal space, increase the overall width of the vertical navigation menu to accommodate the links. If some links wrap onto two or more lines, be sure to vertically separate all the links with more whitespace.

- If translating the navigation links' labels into other languages would result in longer labels that need more horizontal space or would wrap onto multiple lines, display them in a vertical navigation menu.

Figure 12.11—A hierarchical vertical navigation menu on the *Apple Developer* site

Any vertical navigation menu should reside at the left of all pages on an information space whose language reads from left to right, where users are most likely to notice it and can easily distinguish its purpose. On such an information space, placing a vertical navigation menu on the right might cause users to confuse it with any links comprising a specific page's contextual navigation. In contrast, for an information space whose language reads from right to left, a vertical navigation menu should appear at the right of all pages.

An advantage of a vertical navigation menu is that it is easier and more efficient to scan than a horizontal navigation bar because it requires fewer fixations of the eye. The user can see more of the menu with less effort. However, a vertical navigation menu does take up more space on a page, which would otherwise be available for displaying more content. [32, 33, 34]

Accordions within vertical navigation menus

An *accordion* is a heading that, by default, has a downward-pointing caret or triangle icon to its right. Clicking the heading expands a panel beneath it, changing the icon to an upward-pointing caret or triangle. To collapse the panel, the user can click the heading again, changing the icon back to a downward-pointing caret or triangle. Within the context of a vertical navigation menu, the heading is a global navigation link that, rather than navigating to another page on an information space, expands a panel beneath the heading that contains a vertical menu of subordinate local navigation links, as shown in Figure 12.12. This accordion expands from a rail comprising labeled icons.

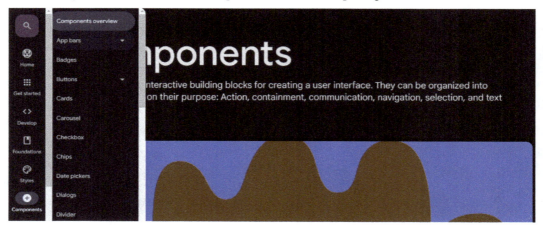

Figure 12.12—A vertical navigation menu containing an accordion

Thus, an accordion provides a form of progressive disclosure that supports the more compact display of the navigation options within a vertical navigation menu and reduces the amount of scrolling the user must do to view all the options. [35, 36]

Tree views in vertical navigation menus

In *tree-view navigation*, local navigation links are nested under the global navigation links, as shown in Figure 12.13. By default, a plus-sign icon or right-pointing triangle appears to the left of a global navigation link. Clicking a global navigation link expands the list of local links that are subordinate to it and changes the icon to a minus sign or the triangle to a downward-pointing triangle. To collapse a list of local links, the user can click the global navigation link again, changing its icon back to a plus sign or the triangle back to a right-pointing triangle.

Tree-view navigation can support any reasonable number of levels of local navigation and readily accommodate changes such as the addition of new links or translations that result in longer link labels. However, depending on the number of levels of local navigation and their indentation, tree-view navigation could potentially consume a lot of horizontal screen real estate and, thus, limit the space that is available for the main content area. [37]

Figure 12.13—Tree-view navigation in a vertical navigation menu

Navigation drawers and sidebars

Both Material Design's navigation drawers and Apple's sidebars can present all the global navigation links, as well as any local navigation links that are available from a given page of an information space, providing easy access to the links and indicating what navigation option corresponds to the current page. Once the user has navigated to a page using a navigation link, an active indicator shows that it is no longer a link. Do not use either a navigation drawer or a sidebar in combination with a horizontal navigation bar.

Key among Material Design's navigation components, which enable users to traverse an information space, is the *navigation drawer*, of which there are two types:

- **Standard navigation drawers**—Employ these in expanded, large, and extra-large window sizes—that is, on large tablets or the desktop. These drawers could have a fixed width of 360 dpi and be permanently visible on a page, as down in Figure 12.14, *or* be expandable and collapsible.

- **Modal navigation drawers**—Use these in compact and medium window sizes—that is, on phones, phablets, or small tablets. These drawers are typically expandable and collapsible to make the most of these devices' limited screen real estate. More about these later when we look at mobile navigation patterns.

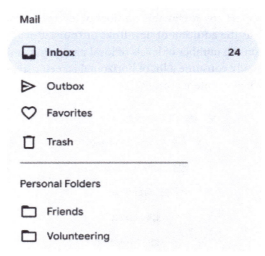

Figure 12.14—A standard navigation drawer in Material Design

For navigation drawers of either type which expand and collapse, the drawer can be expanded or collapsed by default, and all such drawers have a standard width of 360 dp. To expand or collapse a navigation drawer, the user can click a navigation menu icon, *or* hamburger menu icon. A *scrim*—a dark gray, transparent surface that covers the content of the page in the background behind the navigation drawer—blocks all user interactions with the elements on the page in the background, makes the page's content less prominent, and emphasizes the foreground content—that is, the links in the navigation drawer. [38, 39]

Similar to Material Design's navigation drawers, *sidebars* are navigation components in macOS, iPadOS, and iOS that display an information space's navigation in a small, medium, or large window and extend its full height. A sidebar typically appears within the primary pane of a *split view* and contains the information space's global navigation, as shown in Figure 12.15. Expanding a sidebar by default ensures that it is readily discoverable. If necessary to accommodate a window's primary content, changing the size of a sidebar's containing window could optionally collapse or expand the sidebar automatically.

Figure 12.15—A macOS sidebar in the primary pane of a split view

On the desktop or a large tablet, if an information space also has one or more levels of local navigation, they could appear in side-by-side panes within the sidebar that are laid out from left to right. For such a hierarchical navigation system, when the user clicks a link in the global navigation, the corresponding local navigation links would appear in a secondary pane in the split view; clicking a local navigation link would display the next level of local links in a tertiary pane; and so on throughout the navigation hierarchy. Such a split view could take up a large amount of horizontal screen real estate, so the user should be able to collapse and expand the sidebar using the appropriate interactions for the platform— for example, a **Show/Hide** button in macOS or an edge swipe gesture in iPadOS. [40]

Alternatively, either a navigation drawer or a sidebar could incorporate an accordion to display local navigation links that are subordinate to particular global navigation links, as shown in Figure 12.16. However, an accordion within a navigation drawer or sidebar should display only a single level of local navigation links.

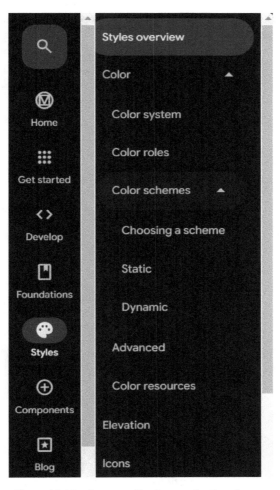

Figure 12.16—An accordion in a navigation drawer

In both navigation drawers and sidebars, place the most frequently used navigation links at the top of their vertical navigation menu and logically group related links.

Navigation rails

A *navigation rail* comprises a vertical array of icons that represent an information space's global navigation and, for languages that read from left to right, resides at the left side of an information space's pages when present. Each icon has a textual label centered beneath it. Clicking an icon either navigates to another page on the information space or displays a modal navigation drawer containing subordinate links.

The navigation rail component in Google's Material Design has a standard width of 80 dp, occupies a fixed position on the screen, and does not scroll. A rail typically includes between three and seven navigation icons. The icon corresponding to the current page appears selected and is *not* a link. The Material Design Web site itself employs a navigation rail, as well as a modal navigation drawer that displays local navigation links and can incorporate an accordion to display an additional level of local navigation, as shown in Figure 12.17.

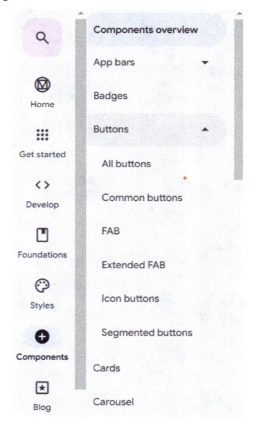

Figure 12.17—Material Design's navigation rail, modal navigation drawer, and accordion

The use of the navigation rail component is appropriate in medium, expanded, large, and extra-large window sizes—that is, on tablets of any size or the desktop. In large and extra-large windows, a navigation rail representing an information space's global navigation might be the *only* navigation component. Do not use a navigation rail and a navigation bar in the same user interface. An optional menu icon at the top of the navigation rail could display additional global navigation links in a modal navigation drawer. [41]

The navigation rail is *not* unique to Material Design.

Desktop navigation layouts

On the desktop, an information space's structural navigation system typically uses one of the following five common layouts, which are depicted in Figure 12.18:

- Horizontal navigation bar for global navigation

- Horizontal navigation bars for both global and local navigation

- Horizontal global navigation with a vertical local navigation menu

- Vertical global navigation

- Vertical global and local navigation [42]

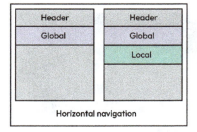 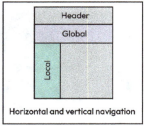 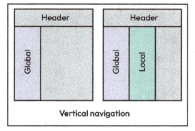

Figure 12.18—Common desktop navigation layouts

The breadth of an information space's global navigation, plus the breadth and depth of any local navigation determine which of these layouts would be optimal for that information space. Let's consider each of these desktop navigation layouts in turn.

Horizontal navigation bar for global navigation

For an information space that has a fairly limited scope and, thus, requires only a simple, flat navigation system, the use of a horizontal, global navigation bar is common. This navigation bar typically resides near the top of an information space's pages, immediately below the header. One advantage of this desktop navigation layout is its compactness. This navigation bar takes no horizontal space away from the main content area.

However, because a horizontal navigation bar can accommodate only a somewhat limited number of navigation links, the labels of these links must be concise—ideally, comprising only a single word, two words if necessary to convey a concept clearly, or a maximum of three short words and limiting the use of longer labels. To make it easy for users to visually distinguish these link labels, it is necessary to include adequate whitespace between them, especially if some labels comprise multiple words.

Many information spaces that have more complex information hierarchies use a horizontal, global navigation bar in combination with either drop-down menus or megamenus for its local navigation. The section "Expandable menus," earlier in this chapter, covers this in greater depth.

Another advantage of a horizontal, global navigation bar is that you can use fixed positioning to place the navigation bar at the very top of the information space's pages and scroll its content behind it. To make more of a page's content visible, you can automatically hide the navigation bar when the user scrolls down, then automatically show the navigation again when the user scrolls back up.

Horizontal navigation bars for both global and local navigation

For a simple, hierarchical navigation system, you can stack global and local navigation bars near the top of an information space's pages. This compact desktop navigation layout can support up to three levels overall and takes no horizontal space from the main content area. To make the user's location on an information space absolutely clear, highlight all the links that the user has clicked in navigating to the current page, as shown in Figure 12.19.

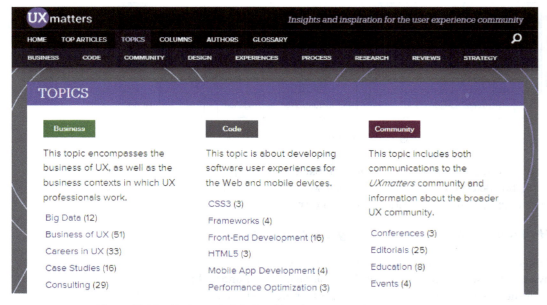

Figure 12.19—Horizontal global and local navigation bars on *UXmatters*

With the complex navigation hierarchies that are now typical of many information spaces, the use of drop-down menus and megamenus for local navigation has become increasingly prevalent. Other factors that have diminished the usage of multilevel navigation bars include the ascendancy of mobile devices and the prevalence of responsive Web design. As a consequence, the use of multilevel, horizontal navigation bars is now fairly rare. Nevertheless, an advantage of using horizontal, local navigation bars over using expandable menus remains their greater visibility and ease of use. Plus, by adequately addressing the design constraints that mobile design imposes with a mobile-specific design solution such as a hamburger menu, multilevel navigation bars can still be a viable design option on the desktop or tablet.

On the desktop, *tabbed navigation* that visually emulates the tabs of real-world file folders is a largely obsolete variant of two-level, global and local horizontal navigation bars. Because Web browsers use tabs to display all the different pages to which the user has navigated on various sites during the current session, displaying a Web site's tabbed navigation at the top of all its pages could potentially cause users to confuse the browser's tabs with a Web site's tabs.

Because the number of navigation links that a horizontal navigation bar can accommodate is somewhat limited, the labels of the links must be very concise—ideally, comprising just one word. If necessary to convey a concept clearly, two-word labels can work well, but link labels should comprise a maximum of three short words. Limit the use of longer link labels when possible. Ensure that users can visually distinguish the link labels on horizontal navigation bars by including adequate whitespace between them—especially if some labels comprise multiple words.

To conserve screen real estate when many local navigation links would require longer labels, you could instead place some or all of the local navigation links on either drop-down menus or megamenus. [43]

A key advantage of implementing horizontal navigation bars is the ability to place them at the top of all of an information space's pages using fixed positioning and scroll the pages' content behind them. With a larger number of navigation bars, this capability becomes more beneficial because it significantly reduces the interaction costs of using the navigation bars. To make more of a page's content visible, you can automatically hide the navigation bars when the user scrolls down, then automatically show the navigation bars when the user scrolls back up.

Another advantage of using horizontal navigation bars for both global and local navigation is that you can more clearly represent their hierarchical relationships visually by highlighting the user's selections on the navigation bars.

However, the use of horizontal navigation bars is suitable only when they can accommodate all the necessary navigation links, and it would unlikely be necessary either to add more new links than would fit on the navigation bars or to translate the link labels to other languages whose translations would result in excessively long labels. [44]

Horizontal global navigation with vertical local navigation

The global and local navigation together form an L-shape in this desktop navigation layout, as shown in Figure 12.20. This layout is in common use for large, hierarchical information spaces. Place a horizontal global navigation bar that can accommodate all the necessary links near the top of the page. [45]

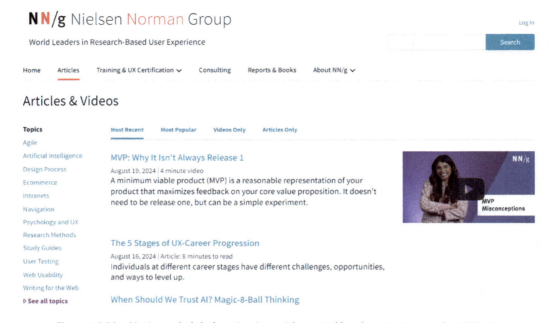

Figure 12.20—Horizontal global navigation, with vertical local navigation on the *NN/g* site

If an information space has a large number of local navigation links, lay them out vertically, on the left side of the page. This vertical layout can accommodate any reasonable number of levels of local navigation, as well as a large number of local navigation links at any level. However, because it is necessary to indent each level of local navigation links, with each level consuming additional horizontal screen real estate, this layout could limit the space that is available for the main content area. [46]

In the vertical navigation, the labels for all links should be left aligned. Placing each local navigation link on a new line makes it easy for users to distinguish the links from one another. To make the link labels easy for users to differentiate from one another, they should begin with unique words that convey their most important concept.

Although you should try to keep the labels of local navigation links as concise as possible, conveying their more detailed concepts clearly sometimes requires somewhat longer labels. Therefore, an advantage of vertical navigation is that, if necessary, you can wrap any longer link labels. However, doing so adds visual complexity. [47]

Vertical global navigation

For a broad, flat information space that has a large number of global navigation links, lay out the links vertically, at the left of the page—ideally, keeping all the links above the fold. Figure 12.21 provides an example of vertical global navigation from the Web site of the National Oceanic and Atmospheric Administration (NOAA). However, this layout could also work well for an information space that has a fairly limited scope.

Figure 12.21—Vertical global navigation on the NOAA site

Be sure to make it easy for users to identify the global navigation by setting it off visually from the main content area of the page. This global navigation layout does consume some horizontal screen real estate, limiting the space available for the main content area.

To make the labels for all vertical navigation links easy for users to scan, read, and visually distinguish from one another, these links should be left aligned and each link should start on a new line. Try to keep the labels of global navigation links as concise as possible. However, if any longer link labels are necessary, you can wrap them in a vertical navigation menu. Link labels must be easy for users to differentiate from one another, so should begin with unique words that convey a link's key concept.

Vertical global and local navigation

In cases where a horizontal navigation bar cannot accommodate all the necessary global navigation links, use *vertical navigation* in which the local navigation links are nested under the global navigation links, Hierarchical vertical navigation can support any reasonable number of levels of local navigation, as in the example from the U.S. Web Design System, shown in Figure 12.22. Plus, it can readily accommodate changes such as the addition of new links or translations that result in longer link labels. But, depending on the number of levels and their indentation, hierarchical vertical navigation could consume a lot of horizontal screen real estate and, thus, limit the space available for the main content area.

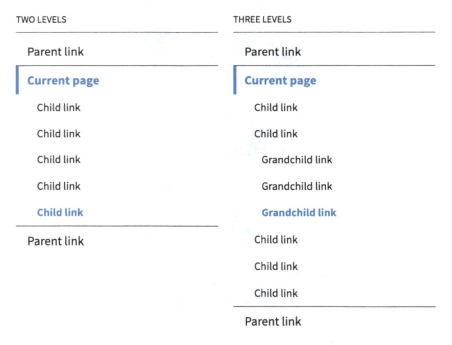

Figure 12.22—Hierarchical vertical navigation

Lay out a closely integrated hierarchy of global and local navigation links vertically, at the left of the page. This layout very clearly represents the hierarchical relationships between the various levels of links and can depict the information space's entire navigation hierarchy. Indicate the user's current location in the navigation hierarchy by visually highlighting the navigation links that the user has selected, making it easy for users to orient themselves to their current context. Each link label should start on a new line, making it easy for users to distinguish the links from one another. Indent each lower level of local links. Because this navigation layout can accommodate longer link labels whenever necessary, the labels for all navigation links can be sufficiently specific and, thus, consistently provide good information scent. [48, 49]

Mobile navigation patterns

"Mobile navigation must be discoverable, accessible, and take little screen space. Exposing the navigation and hiding it in a hamburger both have pros and cons, and different types of sites have different preferred solutions to the mobile-navigation quandary."—Raluca Budiu [50]

Essential characteristics of all mobile navigation patterns are that an information space's navigation components must be readily discoverable, consume minimal screen real estate, and be accessible to all people, while still prioritizing maximal visibility of the content on its pages. Balancing these countervailing goals is a key challenge of mobile navigation design. The navigation patterns that this section describes are in common use for information spaces on both mobile phones and smaller tablets. [51]

Mobile navigation patterns include various types of visible and hidden navigation components. Visible navigation components such as horizontal navigation bars and tab bars provide maximal discoverability, excellent usability, and strong information scent. In contrast, hidden navigation components are less discoverable, provide lower information scent, and users incur greater interaction costs in using them. Hidden navigation components in common use on mobile devices include navigation and hamburger menus, which could incorporate accordions, navigation drawers, or navigation sidebars. [52]

Horizontal navigation bars and tab bars

Although the *horizontal navigation bar* originated in the design of traditional desktop Web sites, it later became a fairly common navigation element on mobile devices and tablets. A horizontal navigation bar must accommodate an information space's global navigation links, arraying them across the top or bottom of a page. However, because of the navigation bar's constrained space and the limits this imposes on the number of links it can display, it is now difficult to find examples of the horizontal navigation bar on mobile Web sites. Nevertheless, this is still a useful navigation design pattern for an information space with just a limited number of global navigation links—typically only four links or, at most, five links with short labels. Although it might be possible to use a carousel to accommodate additional links, this solution would hide key navigation links, making them more difficult to discover and use, so would not be optimal. [53]

The Guardian Web site, shown in Figure 12.23, provides an example of the use of a horizontal navigation bar on a mobile device. This navigation bar scrolls with the page's content, which dedicates more screen real estate to the display of content, but makes it difficult to get back to the navigation bar. A better design solution would be a fixed-position navigation bar that collapses when the user scrolls down, then expands when the user scrolls up.

Figure 12.23—Navigation bar on *The Guardian* mobile Web site

In Material Design, a mobile app's *navigation bar* resides at the bottom of the screen and comprises three to five labeled icons that represent global navigation options. Tapping an icon navigates to the icon's destination view. The icon that corresponds to the current view should appear highlighted. Scrolling downward can collapse, or hide, the navigation bar, while scrolling upward can expand, or show, it. A predictive back gesture, or left or right swipe, on the navigation bar displays the previous view. [54]

The *tab bar* in iOS similarly resides at the bottom of a mobile app's screen, comprises a minimal number of labeled icons—up to five tabs in iOS or six in iPadOS—and is available globally. Tapping an icon navigates between different panes of content in the same view. The icon that corresponds to the current view appears highlighted. [55]

iOS also includes what Apple calls a *navigation bar* component that resides at the top of a mobile app's screen. However, other than a **Back** button in its upper-left corner that lets users navigate back through a hierarchy of views, this navigation bar doesn't provide navigation functionality. Its other key features are a large title area and an icon or command in the upper-right corner that operates on the content in the main view. [56]

Figure 12.24 shows the navigation bar in Material Design and the tab bar and navigation bar in iOS. On each platform, these components reside in the foreground on most pages of a mobile app, so the app's content scrolls behind them, and because they are persistent and, thus, always visible, they limit the amount of vertical screen real estate that is available for the app's content.

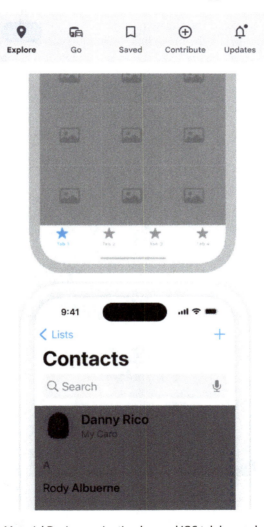

Figure 12.24—Material Design navigation bar and iOS tab bar and navigation bar

Therefore, when employing these persistent navigation components—especially when including both the tab bar *and* the navigation bar in an iOS app—try to avoid the persistent display of any other supplementary, or utility, navigation or search components on the screen.

Navigation or hamburger menus

Most mobile information spaces now employ a *navigation menu* or *hamburger menu* when displaying five or more global navigation links and any local navigation links. For an information space with multiple levels of navigation links, such a menu is the most effective way of presenting them. Typically, this vertical menu is hidden by default, and the user can tap a **Menu** link or hamburger icon to open it. For a user interface that reads from left to right, the **Menu** link or hamburger icon should be near the upper left of each page or screen and the menu should reside at the left of each page; for a user interface that reads from right to left, the **Menu** link or hamburger icon and the menu should reside at the right of each page or screen. Figure 12.25 shows an example of a hamburger menu.

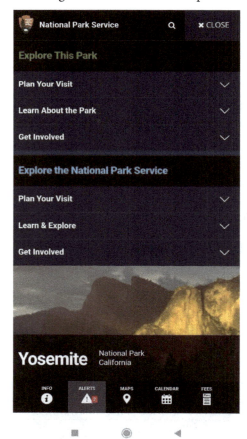

Figure 12.25—Hamburger menu on the NPS.gov mobile Web site

Although a navigation or hamburger menu often comprises just a flat list of global navigation links, they frequently employ progressive disclosure to present a single level of local navigation links using accordions or several levels of local links using navigation drawers. More about these shortly.

A navigation or hamburger menu consumes minimal screen real estate, but is a less discoverable form of mobile navigation than a navigation bar or tab bar. Although such menus are fairly easy to access, finding and displaying them requires greater user effort, or *interaction cost*. Therefore, these menus are most appropriate on a content-rich information space on which the user would be less likely to need to navigate directly to a specific section rather than browsing—especially on one that also provides a prominent search capability. [57, 58]

However, because hidden menus are less discoverable, providing additional, more visible means of supporting navigation is often helpful, especially when doing so is necessary to promote an information space's important business goals. Supporting means of navigation could include a home page that functions as a hub—and, in some cases, the main pages of an information space's major sections can be hubs as well. Hubs offer an overview of key content and give users direct access to it. Presenting lists of internal links to content about specific topics—typically, below a page's main content—is an effective way of surfacing fresh content within an information space's sections. Other common means of providing access to content that is more deeply buried within an information space include embedding internal links within a page's main content and offering lists of related links that accompany the main content, including links to other types of content. [59]

Repurposing the vertical navigation menu of a desktop information space for use on a mobile device requires little adaptation other than its placement on a navigation or hamburger menu. However, transforming a horizontal navigation bar into such a menu can require completely rethinking a navigation system's design and implementation. [60]

Accordions

"An accordion is a design element that expands in place to expose some hidden information. Unlike overlays, accordions push the page content down instead of being superposed on top of page content."—Raluca Budiu [61]

In a navigation or hamburger menu that comprises both global and a single level of local navigation links, accordions provide a means of progressive disclosure that lets the user display subnavigation links that are subordinate to specific global navigation links, as shown in Figure 12.26. The user can tap a global navigation link to expand, or show, then collapse, or hide, an accordion, which contains a submenu that typically comprises half a dozen or fewer local navigation links. If these submenus were to contain many more links, too few global navigation links would be able to appear on the screen simultaneously with a submenu's local navigation links, requiring the user to scroll too much and, thus, making information seeking overly laborious.

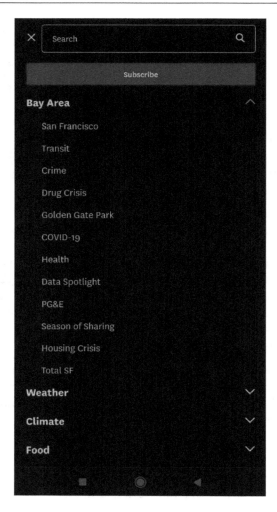

Figure 12.26—Navigation drawer with an accordion on the *San Francisco Chronicle* mobile Web site

The use of accordions in navigation or hamburger menus conserves space and fulfills the key design goals for subnavigation on mobile devices, as follows:

- **Minimal interaction cost**—With just two taps, the user can access any part of an information space whose navigation hierarchy comprises two levels, without excessive scrolling to reach the subnavigation links or needing to reload the page.

- **Support for any path through the information space**—When there is no particular path that users might typically take, they must be able to easily go to different pages in the same section or jump to different branches of the information hierarchy. Thus, regardless of whether the user would want to remain within a single section *or* go from a page in one section to a page in another section during the same session, an accordion would facilitate both interactions.

- **Maximal discoverability**—Accordions are integral parts of an information space's primary navigation system, which aids discovery and prevents user confusion. But their visual design must make it easy for users to distinguish between the global and local navigation links—that is, the main menu and the accordions—typically, by using color or value and indentation. [62]

Another variant of the accordion defines the navigable structure of an entire page and comprises links to the page's expandable subsections, which contain the page's content. Figure 12.27 provides an example. This type of accordion provides an overview of a page's content—enabling the user to build a mental model of the page—and gives the user direct access to specific sections of the page. An accordion is a very effective design solution for mobile pages, as long as each section of a page contains only a limited amount of content. A section that contains too much content requires excessive scrolling, would probably move the section's label beyond the viewport of the screen, and thus, could potentially cause the user to become disoriented.

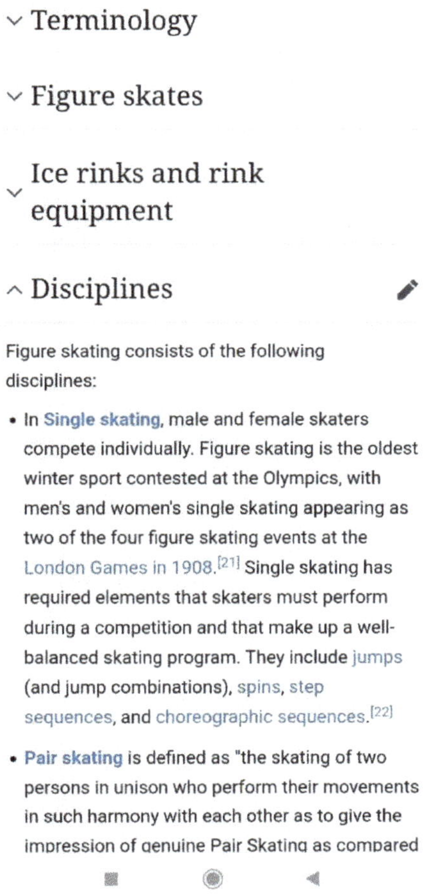

⌄ Terminology

⌄ Figure skates

⌄ Ice rinks and rink equipment

⌃ Disciplines

Figure skating consists of the following disciplines:

- In Single skating, male and female skaters compete individually. Figure skating is the oldest winter sport contested at the Olympics, with men's and women's single skating appearing as two of the four figure skating events at the London Games in 1908.[21] Single skating has required elements that skaters must perform during a competition and that make up a well-balanced skating program. They include jumps (and jump combinations), spins, step sequences, and choreographic sequences.[22]

- Pair skating is defined as "the skating of two persons in unison who perform their movements in such harmony with each other as to give the impression of genuine Pair Skating as compared

Figure 12.27—An accordion displaying content on the *Wikipedia* mobile Web site

The user should be able to tap the label for an expanded section of an accordion to collapse it. To prevent unnecessary scrolling, automatically collapse whatever section of the accordion the user had previously expanded when the user taps another label to expand a different section of the accordion. Alternatively, for a Web site, you could allow the user to return to a view of the accordion in which all sections are collapsed by implementing the browser's **Back** button to collapse the section of the accordion that is currently expanded.

If a Web page has extensive content, use fixed positioning to ensure that the label for the section the user has expanded remains visible once the user has scrolled its content. Otherwise, the interaction cost of scrolling to collapse the section of the accordion that is currently expanded would be excessive. [63]

Navigation drawers or sidebars and sequential menus

Through progressive disclosure, a navigation drawer or sidebar can present all the levels of an information space's navigation system in an overlay—whether there is just one level of global navigation or several levels of local navigation. The user can tap a hamburger icon or **Menu** label to display a navigation drawer that contains a menu of global links. In what Raluca Budiu of Nielsen Norman Group (NN/g) calls *sequential menus*, the user can successively display each level of the information space's navigation hierarchy—one level at a time—over the previous level of links, until the user drills down to the lowest level of the navigation hierarchy, as shown in Figure 12.28. [64]

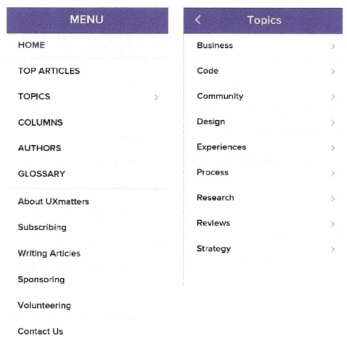

Figure 12.28—Navigation drawer with a sequential menu on the *UXmatters* mobile site

When the user initially expands a navigation drawer containing a sequential menu, it comprises only global navigation links. If there are any local navigation links below a particular global link, a right-pointing caret typically indicates their presence. Tapping the global link displays the local links that are immediately subordinate to it. If there were another level of local navigation links, another right-pointing caret would indicate that. Tapping a first-level local link displays the second-level local links beneath it—and so on, until the user has traversed all the levels of local links. If a local link at any level of the navigation hierarchy has no subordinate links, tapping the link displays another page on the information space. The user should always be able to navigate back up the navigation hierarchy by tapping a **Back** button in the upper-left corner of the navigation drawer.

Material Design 3 has brought greater standardization to the navigation drawer. Figure 12.29 shows the *modal* navigation drawer, which is used primarily for mobile devices and tablets, where it can be open or closed by default. However, it can also be used for windows of any size on the desktop, where it can appear permanently on the screen. To open a modal navigation drawer, the user can tap a hamburger menu icon that is on a navigation rail or on the page. Behind the modal navigation drawer, a gray transparent background, or *scrim*, often blocks interactions with the page's content. To close a modal navigation drawer, the user can tap a navigation link, the scrim, or outside the drawer, or swipe toward the anchoring edge of the drawer—that is, from right to left for a left-aligned drawer.

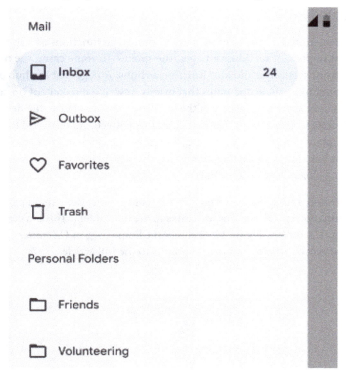

Figure 12.29—The modal navigation drawer in Material Design 3

A navigation drawer basically comprises a list of navigation links within a side sheet. To provide structure to a longer list of links, a drawer can optionally include a header, section labels above groupings of related links, and *dividers* that group related links. The links in a navigation drawer can consist either of an icon *and* a link label, with the icon preceding the label, *or* of only a link label. Within a given navigation drawer, icons should appear to the left of either *all* links or *no* links. Link labels should be concise. However, if a link label would exceed the width of the side sheet, truncate the label and append an ellipsis at its right rather than wrapping the label. If a navigation drawer contains more links than can fit in the vertical space within the drawer, the user can vertically scroll the list of links.

Once the user taps a navigation link and its destination page appears, an *active indicator*, a rounded rectangle with a contrasting background, appears behind the link. [65]

The navigation drawer with sequential menus has become the dominant pattern for mobile information spaces that must display multiple levels of navigation on small screens, is easy to use, and has only a moderate interaction cost. However, users who are spatially challenged are less efficient when using navigation drawers than users with strong spatial abilities and may become disoriented when using them. Some users might become frustrated by deep navigation drawers with sequential menus that require them to tap a long series of links before finally arriving at their destination page. [66]

Hub-and-spoke navigation

In hub-and-spoke navigation, an information space's home page functions as a hub and provides a list of links to its most important sections or pages, the spokes. In some cases, the main pages of an information space's major sections could also function as hubs. A page that is a hub often comprises a list of links in combination with scope notes that briefly describe the content on each destination page. On most information spaces, the clarity of the link labels should obviate the need to devote such valuable screen real estate to scope notes. Nevertheless, this pattern can be useful for an information space on which the content is very specialized and, therefore, the labels are rather esoteric. Using this design solution in combination with a navigation or hamburger menu provides an effective means of compensating for the deficiencies of such a hub.

Another variant of the hub-and-spoke pattern is a home page or a section's main page that comprises excerpts from an information space's content with links to the full content. This is a common pattern on Web magazines such as *UXmatters*. Figure 12.30 shows the home page of *UXmatters*, which comprises excerpts from its most recent articles and provides links to the full articles.

Figure 12.30—The *UXmatters* home page serves as a navigation hub

Although this can be an effective navigation pattern for a mobile information space, it requires the user to repeatedly traverse the navigation hierarchy in returning to the home page or a section's main page to navigate further. Therefore, the hub-and-spoke navigation pattern is most useful when the navigation hierarchy is very shallow, as is the case on *UXmatters, or* the user is focusing on a single task rather than browsing to find a variety of content on an information space during a single session. [67, 68]

Mobile navigation layouts

On the small screens of mobile devices and small tablets, information spaces that have a complex, hierarchical information architecture typically employ a hamburger menu icon and a navigation drawer, sidebar, or another type of overlay to display their hierarchical navigation, which might contain accordions.

In fact, this mobile navigation pattern has become so popular that some desktop Web sites now display their entire structural navigation system in the same way. However, even though using this pattern across all platforms would provide maximal cross-platform consistency, it would also greatly reduce the visibility and, thus, the discoverability of the global navigation options on the desktop and large tablets, making it difficult for users to appreciate the extent and nature of the information space's content and discouraging exploratory browsing. Therefore, employing this design approach on the desktop would be particularly inadvisable on an information space that has only a limited number of global navigation links. [69, 70]

Progressive disclosure

To economize on the amount of screen real estate an information space's local navigation occupies, most information spaces with local navigation that comprises multiple levels of links employ some form of progressive disclosure to display the various levels. The downside of using progressive disclosure is that the local links' lack of visibility discourages exploratory browsing. Each navigation layout that I've just described that comprises both global *and* one or more levels of local navigation can support some degree of progressive disclosure for both the desktop and mobile devices, as follows:

- **Horizontal navigation bars**—By default, display only a horizontal, global navigation bar, hiding all local navigation links. Once the user clicks a global navigation link, display the top-level local navigation links that are subordinate to that link on another horizontal navigation bar immediately below the global navigation bar. If an information space's local navigation comprises more than one level, the user can display each successive level of the navigation hierarchy by clicking lower-level links. To explore a different path within the same section of the information hierarchy, the user can click a different link on a local navigation bar. To explore a completely different part of the information hierarchy, the user can click a different link on the global navigation bar.

- **Combination of horizontal and vertical navigation**—Initially, display only a horizontal, global navigation bar, hiding the local, vertical navigation to reserve more space for content on the home page. Once the user clicks a global navigation link, display the local navigation in a sidebar at the left of the page. If the local navigation comprises multiple levels, initially hide all but the top level of local navigation links. By clicking a link at each level of the local navigation hierarchy, the user can expand its lower levels in a tree view or accordion. If the user clicks a different global navigation link or a different, higher-level, local navigation link, expand that branch of the local navigation and simultaneously collapse any other branch that was previously expanded.

- **Vertical navigation**—An accordion or tree view could initially display only the global navigation links in a navigation drawer, sidebar, or other type of overlay at the left of the page, hiding all local navigation links. If the user clicks a link in the global navigation, display the local navigation links that are immediately subordinate to that link. If an information space's local navigation comprises multiple levels, the user can click a link at each level to successively display the lower levels. If the user clicks either a different global navigation link or a higher-level, local navigation link, expand that branch of the local navigation and, at the same time, collapse any other branch that was previously expanded.

In vertical, tree navigation, indicate that a grouping of links is currently expanded or collapsed by placing one of the following elements in proximity to the immediately higher-level link:

- **Carets or triangles**—You can place carets or triangles either immediately to the left of a link, as Figure 12.31 shows, or at the far right of a navigation link's bounding box, as Figure 12.32 shows. On different information spaces, the initial placement and behaviors of these elements can differ somewhat. By default, a section of links that is collapsed might have a right-pointing *or* a downward-pointing caret or triangle. Once the user clicks a navigation link to expand its subordinate links, a downward-pointing caret or triangle might replace a right-pointing caret or triangle; or an upward-pointing caret or triangle, a downward-pointing triangle. The user can optionally click the top-level link again to collapse the section, returning the caret or triangle to its original state. Be consistent in your use of carets or triangles across an information space.

- **Right-pointing carets**—For structural navigation systems, carets are in common use across the desktop, tablets, and mobile devices. By default, a right-pointing caret is at the far right of each navigation link's bounding box, and its subordinate links are hidden. On the desktop, once the user clicks a global navigation link, its subordinate local links expand in another pane, as Figure 12.33 shows. *If*, as on some sites, the global link remains visible, the user can click the link again to collapse the pane. On a mobile device, a panel of lower-level links typically replaces the level of navigation in which the user tapped a link. Typically, in the upper-left corner of the navigation panel's chrome, a **Back** button or left-pointing caret—ideally, with a label that indicates the user's prior destination—lets the user navigate back up one level at a time in the navigation hierarchy.

- **Plus (+) and minus (-) signs**—The use of tree navigation with tree-view controls to expand and collapse the nested levels of navigation, as shown in Figure 12.34, is much more common in applications than on Web sites or other information spaces. By default, a plus sign typically appears to the left of a navigation link, and the corresponding section of subordinate links is collapsed. Once the user clicks a navigation link to expand a section of subordinate links, a minus sign replaces the plus sign. The user can optionally click that link to collapse the section, replacing the minus sign with a plus sign. On some Web sites, the plus and minus signs appear to the right of a navigation link, as shown in Figure 12.35.

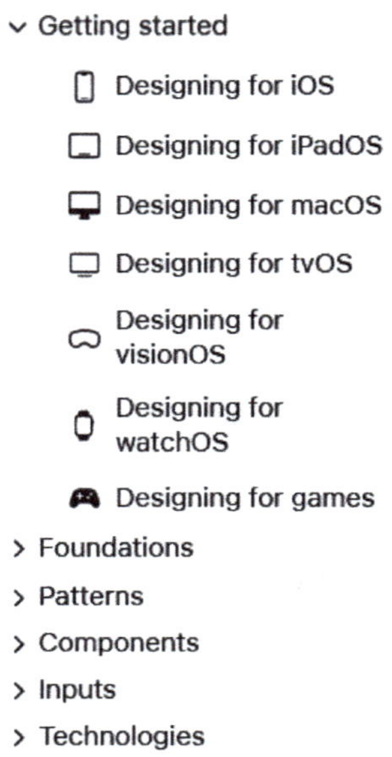

⌄ Getting started

    📱 Designing for iOS

    ⬜ Designing for iPadOS

    🖥 Designing for macOS

    📺 Designing for tvOS

    🥽 Designing for visionOS

    ⌚ Designing for watchOS

    🎮 Designing for games

> Foundations

> Patterns

> Components

> Inputs

> Technologies

Figure 12.31—Carets to the left of links display accordions on the desktop

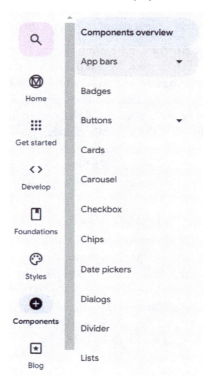

Figure 12.32—Triangles to the right of links display accordions on the desktop

Figure 12.33—Right-pointing carets to navigate the SiriusXM mobile app

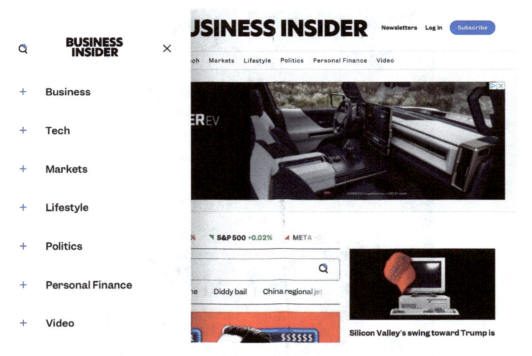

Figure 12.34—Tree-view controls on the *Business Insider* Web site

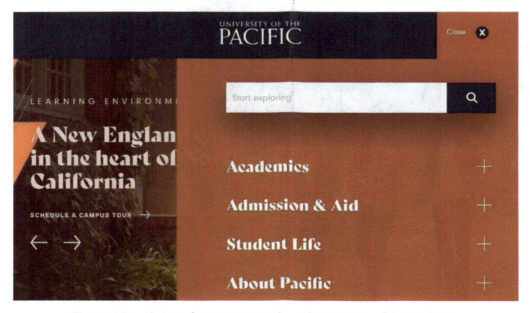

Figure 12.35—Variant of tree-view controls on the University of the Pacific site

Some other desktop navigation design patterns that support progressive disclosure include drop-down menus and megamenus. For more information about these patterns, refer to the section "Navigation design patterns," earlier in this chapter.

Visualizing navigation design

"Page layout and graphic design give the navigation its final form. … Aspects such as the order of options, their arrangement on the page, the font type and size used, and color can be critical elements. They can make or break the navigation system." —Jim Kalbach [71]

The navigation design patterns and layouts that you learned about earlier in this chapter—in the sections "Desktop navigation patterns," "Desktop navigation layouts," "Mobile navigation patterns," and "Mobile navigation layouts"—play a foundational role in determining the overall structure of an information space's pages. Incorporating these navigation patterns and layouts into an information space's page layouts results in a consistent structure across pages.

Designing an information space's navigation system and page layouts and creating the design deliverables that represent them is typically a highly iterative process. You need to revise and optimize your designs until they satisfy all essential business and user requirements.

Throughout your navigation-design process, you might visually represent design solutions at progressively greater levels of detail and design fidelity. When laying out an information space's pages, including its navigation system, you'll initially sketch several possible design solutions to determine the optimal layouts for specific types of pages. Then, when documenting the navigation system's design, you'll create low- to high-fidelity wireframes that depict the navigation system within the context of specific page layouts or page templates. If necessary, your wireframes could evolve into high-fidelity mockups that depict your final, detailed designs of page layouts or a click-through or fully functional prototype.

Depending on your design deliverables' purpose and audience, the refinement of their presentation could differ greatly. You could create simple sketches, low-fidelity wireframes, or visually refined, high-fidelity wireframes. For example, to win new clients or secure funding for your in-house design efforts, you might need to invest in creating beautiful, highly refined deliverables that would impress your stakeholders. In contrast, if your goal is merely to convey your design ideas to your product team efficiently and effectively, throughout an iterative design process during which your designs might frequently change, create simple, low-fidelity sketches or wireframes. You must understand the needs of your design deliverables' audience to ensure that you adequately address them.

While producing design deliverables is an essential part of the design process, on today's highly collaborative agile and Lean UX projects, the creation of elaborate design deliverables has become much less common. Most teams now rely more on direct communication.

However, in cases where you do need to progress from your lower-fidelity design deliverables to higher-fidelity design deliverables—for example, from wireframes to mockups or a prototype—freeze the final version of your prior deliverable to prevent your having to maintain multiple deliverables over time. You'll learn more about the appropriate fidelity of design deliverables later in this chapter, in the section "Choosing the right level of fidelity." [72, 73]

Laying out the navigation system on pages

"The visual representation of your design should be no more polished than the thinking behind it."—Kim Goodwin [74]

The layout of the various components of an information space's navigation system conveys much about their function. When designing page layouts, you must visually represent the navigation system's structure—both to codify and refine your design ideas and to convey them to your teammates.

Initially, working collaboratively with your core product team, you'll probably create sketches of possible navigation designs on a whiteboard or on paper, as shown in Figure 12.36. Incorporate your teammates' feedback into these sketches as you work.

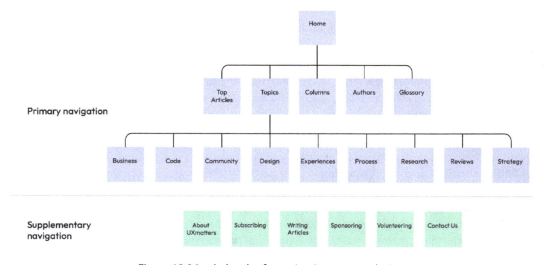

Figure 12.36—A sketch of a navigation-system design

Alternatively, your team could write the label for each navigation-system link on a Post-it note, then experiment with various layouts of these notes on a whiteboard to determine the navigation system's structure. Figure 12.37 provides an example.

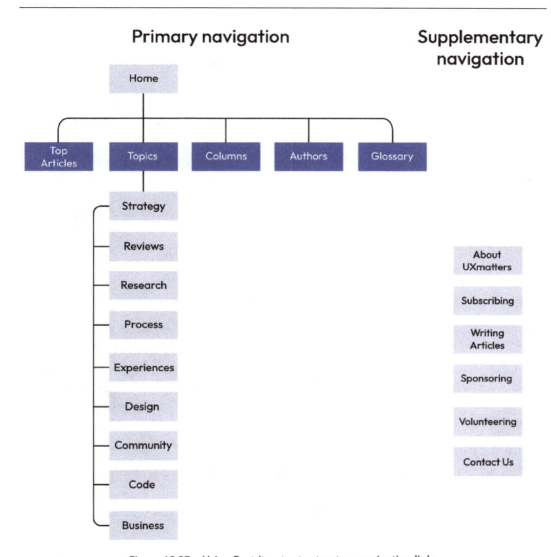

Figure 12.37—Using Post-it notes to structure navigation links

Be sure to take photos of your sketches or layouts of Post-it notes as they evolve to preserve the full history of your decision-making process.

Share your proposed design solutions with your broader product team, gather and synthesize their feedback, then incorporate their feedback into your designs. By taking a collaborative approach to ideation and design, you can ensure that your teammates and stakeholders will buy into your design solutions. [75, 76, 77]

Creating wireframes

"A wireframe ... is a basic outline of an individual page...."—Christina Wodtke [78]

Wireframes are simple digital representations of an information space's proposed page layouts. They define the structure of an information space's *key pages*—such as the home page and its sections' main pages—as well as reusable *page templates* that enable your team to achieve consistency across many pages of a given type—such as content pages, product pages, or search-results pages.

One key reason for creating wireframes is that they can help you understand all the components and content that a page must accommodate. Thus, wireframes depict the overall layouts of the pages' main content elements and the various components of the information space's navigation system. However, they do *not* initially represent any visual-design details or include much content other than labels. As the fidelity of your wireframes increases over time and pages' actual content becomes available, you should add that content. [79, 80, 81, 82]

Next, we'll look at the process of creating wireframes in greater depth.

Overview of the process of creating wireframes

"Developing wireframes ... helps the information architect decide how to group content components, how to order them, and which groups of components have priority."—Peter Morville and Lou Rosenfeld [83]

If your team has not yet created a wireframe template for use across all projects, create a wireframe template that you can use when creating all the wireframes for the information space's pages. Then identify all the pages for which you need to create wireframes. Follow these steps when creating the wireframe for each individual page:

1. Choose the page for which you want to create a wireframe.

2. Choose or create an appropriate template for the page that includes any existing reusable navigation or other global components.

3. Create any new reusable navigation or other global components that the information space requires and add them to the wireframe for the page.

4. Add all page-specific interactive components and content elements, including textual labels and any other available text.

5. Add numbered callouts for components and elements of the wireframe that require annotations.

6. Add numbered annotations to a sidebar.

You could follow a similar process when creating any page mockups. [84, 85]

Creating wireframe-page templates

To facilitate consistency across the wireframes for a particular project and make your sets of wireframes more usable—especially when multiple designers are working on them—develop wireframe-page templates. Add a header to each wireframe-page template that comprises the project name, the designer's name, a unique wireframe number, and the latest revision date, as shown in Figure 12.38. You might also include a page title and a version number.

Project: UXmatters Redesign	Designer: Pabini Gabriel-Petit
UXmatters Article Page: W-2.2	Revised: November 15, 2014

Figure 12.38—The header for a wireframe-page template

You can use numbered callouts on a wireframe to identify a page's specific elements, choosing a high-contrast background color for the callouts to clearly set them apart from the wireframe. Such numbered callouts often take the form of a circle or a square. Provide annotations regarding each numbered element in a sidebar to the right of the wireframe. If a wireframe depicts the final design for a page or screen, these numbered callouts could correspond to the numbered sections of a detailed functional specification. See the section "Creating a template for your specifications," later in this chapter, for descriptions of some of the administrative information you might want to include with your sets of wireframes. [86, 87, 88]

Creating page templates and reusable components

To ensure consistency in an information space's page layouts, create *page templates* that you can reuse across many pages of a given type—for example, content pages, product pages, or search-results pages.

When creating wireframes from page templates that represent multiple pages of an information space, maintaining consistency across all of the pages is essential. While this can sometimes be challenging over multiple design iterations, failing to keep all your wireframes up to date could result in implementation errors. Plus, on the rare project that requires you to archive all your design documentation, you'll need to ensure that your final wireframes are completely up to date and reflect the pages that actually get built.

To ensure consistency in an information space's navigation components and other global components— such as the header and footer that appear on most pages—design and implement reusable components. Then employ these components in creating all your wireframes for both pages and page templates.

A Design Lead should be responsible for establishing procedures for the design and implementation of an information space's page templates and reusable components, as well as guidelines for their use, maintaining these resources, and ensuring their consistent use across all projects.

To facilitate the use of an information space's page templates, global components, and navigation components, devise a standard numbering format for them—for example, for page templates (PT-1 and so on), for global components (GC-1 and so on), and for navigation components (NC-1 and so on). Then, number each of the page templates, global components, and navigation components that you create.

Provide design guidelines for the page templates to ensure their consistent usage, appearance, and placement of all standard and optional elements across all the pages that derive from each template. Provide design guidelines for each reusable global and navigation component as well to ensure its consistent usage, appearance, placement, and behavior across all pages and page templates.

Be sure to allow sufficient flexibility in your template and component designs to accommodate variations in layout, appearance, and behavior. Note any possible variations in page templates and components—for example, mandatory versus optional elements. [89, 90]

You could optionally create storyboards comprising multiple wireframes that depict common workflows through an information space, illustrate the progressive states of a navigation system's interactive components, or show various views of a page in different states.

Choosing the right level of fidelity

"Adopt the lowest fidelity necessary to get the job done. … However, the most appropriate level of fidelity for your deliverables will depend on several factors [such as audience, time, and scope]." —Cennydd Bowles and James Box [91]

A wireframe's fidelity describes its level of detail. Depending on the needs of your audience and your wireframes' stage of development, their fidelity could differ. For example, during the earlier stages of a navigation-design project, you might create several lower-fidelity wireframes to explore different design solutions, while during its later stages, you might create high-fidelity wireframes depicting only what you hope would be your final design solution. You'll revise your wireframes as you add and iteratively refine your navigation system's design details.

Low-fidelity wireframes typically are in grayscale, include no graphic elements, and other than key labels, no actual content. Rectangular boxes typically represent the various areas of a page, which hold different types of content, while boxes that have an X through them represent images. *High-fidelity* wireframes might employ some color, include some graphic elements and actual content, depict the position and size of the navigation components, and simulate the actual widths of pages. Because of high-fidelity wireframes' greater level of design detail, creating them is more time consuming. The fidelity of most wireframes probably lies somewhere in the middle of this range. Wireframes should convey only whatever details are necessary at a given stage of a specific project.

Wireframes should include any important images on a page. Incorporate any actual images that already exist—such as an organization's logo in a page header or social-media icons. For images that are not yet available, add image placeholders to the wireframe, showing their placement and approximate dimensions, a brief description of the image within the placeholder, and a detailed description of the image's purpose or content in the annotations.

Some designers use dummy, or *greeked*, text in wireframes, especially during the earlier stages of design. This practice can help other team members to focus on structure and page layout rather than on content when reviewing the wireframes. However, many designers prefer to use representative or real content, actual labels, and real data in wireframes. When you're creating navigation-design deliverables, the best approach is to use the actual labels for the navigation links. This lets you identify any layout problems that long labels or dynamic text might cause during design rather than after implementation. Plus, your stakeholders and usability-test participants can see the draft copy and actual labels in context.

When you're wireframing an information space's navigation components, render the actual labels of their navigation links using the actual fonts, weights, and sizes—thus, ensuring that the available space can accommodate them. Figure 12.39 shows a fairly high-fidelity wireframe of an article page on *UXmatters* and the global navigation bar. The wireframe includes some graphic elements and uses the site's actual fonts. Its purpose was to convey the page's design to our Web developer.

Project: UXmatters Redesign   Designer: Pabini Gabriel-Petit
UXmatters Article Page: W-2.2   Revised: November 15, 2014

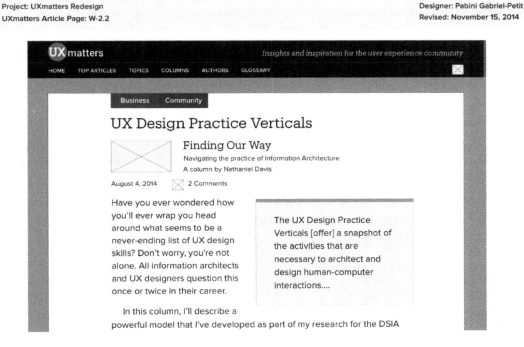

Figure 12.39—A high-fidelity wireframe of an article page on *UXmatters*

[92, 93, 94, 95, 96]

Annotating wireframes

A thoroughly annotated, unambiguous set of wireframes could provide adequate documentation for the design of an information space's navigation system in the following cases:

- The navigation system is fairly simple.
- The navigation system is based on well-understood design patterns.

- The project employs an existing design system that includes detailed visual design and behavior specifications.

- Most importantly, designers and developers work in close collaboration with one another. Achieving the necessary level of collaboration generally requires relatively small, collocated teams. While remote teams can maintain the level of direct communication that is necessary for collaboration, this requires that all team members be highly committed to collaborating and share common work hours. [97]

Creating mockups

"There are few communication techniques available to interface designers that speak with the immediacy and relevance of the mockup."—Luke Wroblewski [98]

As a navigation-design project progresses, your team could optionally choose to create increasingly realistic, digital renderings of page or screen designs, making your designs for the navigation system increasingly tangible. Similar to wireframes, mockups generally represent only an information space's key page layouts and page templates for specific types of pages.

Mockups of page layouts represent all the visual-design details of your final page designs, incorporate detailed designs for all components of the navigation system, and depict these details as the Development team would actually implement them. Mockups of a navigation system focus primarily on the appearance of that navigation system—that is, how it looks.

The process of creating mockups of pages or screens is similar to that of creating wireframes, which is outlined in the section "Overview of the process of creating wireframes," earlier in this chapter. If you're documenting your navigation-system design by creating detailed navigation design specifications, you could incorporate your mockups into the specifications as figures.

If you're creating a new design system, including navigation-design components, the importance of the project makes creating high-fidelity mockups absolutely essential. However, if you're using an existing design system, creating mockups is unnecessary because the designers and developers who created that design system have already done the detailed visual-design and interaction-design work for you—thus, medium-fidelity wireframes are all you need. [98, 99]

Annotating wireframes and mockups

Add annotations that describe the components and elements of a wireframe or mockup to a sidebar at the right. When annotating your wireframes or mockups, provide brief descriptions that indicate such salient details as the following:

- Identifiers for the elements of a page

- Descriptions of the content in particular areas of a page

- Link destinations

- Functionality of components or elements

- Behaviors of interactive components or elements

- States of components or elements

- Whether components or elements are dynamic

- Earlier design deliverables that inform your designs

- Sources of content

- Sources of data elements

- Numbers of images in a carousel or array of product images

- The rationale behind and justifications for particular design decisions

[99, 100, 101, 102]

Figure 12.40 shows an annotated wireframe of an article page on *UXmatters*. You could alternatively use numbered callouts to identify page layouts' various elements or provide numbered annotations, as shown in Figure 12.41, then link them to the detailed specifications.

Figure 12.40—An annotated wireframe of a page design

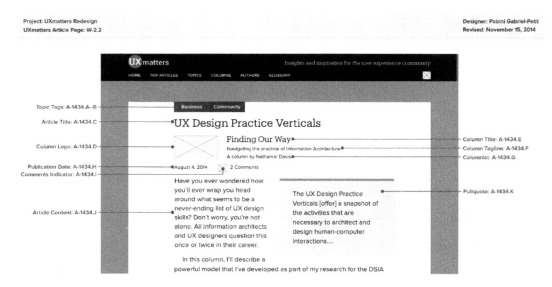

Figure 12.41—An annotated wireframe with numbered annotations

Provide a high-contrast background for all numbered callouts, using a color that clearly sets them apart from the page designs, which could be quite colorful. You could optionally add redlines to copies of your mockups as well, showing elements' measurements and spacings in pixels.

Creating prototypes

It's worthwhile to develop a behavioral prototype if you're designing custom widgets, interactive data visualizations, [or] animated behaviors.... Such a prototype doesn't need to be complete—all it has to do is demonstrate the specific behavior....."—Kim Goodwin [105]

A *prototype* accurately represents the navigation system's final visual design, emulates the user experience of navigating through the information space's pages and the users' resulting workflows, and demonstrates the behavior of the system's navigation components. A prototype of a navigation system focuses primarily on expressing the behavior of that navigation system—and thus, shows how it *feels*.

Although creating a prototype is an optional step of the design process, the level of effort that is necessary to produce a prototype of an information space's navigation system is relatively low in comparison to the huge value of being able to test the navigation system's actual behaviors with users.

You can easily implement a clickable or tappable Web prototype representing an information space's navigation system by using HTML to create navigation links that connect its pages and content elements. Another significant advantage of creating a Web prototype using HTML/CSS or an existing Web framework is that you can readily implement responsive page designs that adapt to different devices' screen sizes and Web browsers. Plus, if you build a prototype using HTML/CSS and JavaScript, you can be certain that your team can build your design for a Web site or Web application as designed. [106, 107, 108]

Specifying navigation-system designs

"Creating detailed interaction design specifications that clearly communicate your design intent makes it possible for developers to make your design a reality."—Pabini Gabriel-Petit [109]

Because the primary purpose of detailed functional specifications, user-interface design specifications, or form and behavior specifications is to guide development, the specifications for a navigation system document every aspect of its structure, appearance, and behavior. These detailed specifications describe how the navigation system *works*, including the functionality of its components and the user's workflows or usage scenarios.

Creating detailed, well-illustrated specifications also helps ensure the completeness of an information space's navigation-system design and prevents the ambiguities that could result in developers' misinterpreting the specifications and, thus, their implementing them inconsistently. Detailed specifications are essential in situations where designers might not be available to answer developers' questions.

A navigation specification's overall structure typically reflects that of an information space's navigation hierarchy and the pages it comprises. [110, 111]

Creating a template for your specifications

"Including the names of contributors, who are often peers in other disciplines, is very important to establishing shared ownership of a specification, which ultimately describes what a product team has agreed to build."—Pabini Gabriel-Petit [112]

Regardless of the format of your specifications, creating a template for them can reduce your effort across projects. One option is creating an HTML/CSS template. A template for a detailed specification might typically include sections for the following information:

- **A title page**—Include the authors and contributors to the specification and their contact information.
- **Executive summary**—A brief overview of the information in a specification that would be most important to key stakeholders, who might not have time to read the entire specification, but must approve the specification. Provide a brief summary of information about the project's context, including its scope and the overall timeline; an overview of the information space's features and capabilities and their value to users; and a review of key and unique elements of the design.
- **Conventions**—A brief list of formatting conventions the specification uses
- **Table of contents**—A numbered list of the headings of the sections and subsections that the specification comprises
- **Version history**—A table containing a chronological listing of all numbered versions of the specification, their authors, and their dates—from the most recent version to the oldest version
- **Change log**—A table containing a list of all tracked changes to the specification that is ordered by the numbers of the sections in which they occur, indicates the date and author of the change, and provides brief notes regarding the changes. Providing a change log helps ensure your teammates don't miss any important changes.

- **Approvals**—A table containing a list of all approvers of the specification and the dates on which they approved the specification

- **Glossary of terms**—An alphabetical list of any uncommon terms that are specific to an information space or its technology and are necessary to the readers' understanding of the specification. Providing a glossary prevents unnecessary repetition of information in the specification.

- **Personas**—A set of personas that represent an information space's key user groups and describe their relevant characteristics and behaviors; define key user requirements to provide the basis for your design solutions, ensuring that they'll deliver value and satisfy users' needs; and provide the rationale for your design decisions. Introduce the personas by defining their role in the design process.

- **Usage scenarios**—Sets of key-path scenarios that describe each persona's main tasks and step-by-step workflows. Express them as storyboards that comprise images of pages or screens or numbered lists of steps. Introduce the scenarios by describing their utility.

- **Page templates**—Reusable page layouts for an information space's common page types and their variants, showing the navigation system in context and including detailed interaction-design and visual-design specifications for all variations in layout, appearance, and behavior. Other detailed specifications should refer to these page-template specifications rather than repeating the same information.

- **Reusable components**—Detailed interaction-design and visual-design specifications for an information space's navigation components and other global components, including specifications for all their variations in layout, appearance, and behavior. Other detailed specifications should refer to these component specifications rather than repeating similar information.

- **Detailed specifications**—These specifications constitute the main body of the specification. Each subsection provides detailed interaction-design and visual-design specifications for a key page or a navigation component, along with your rationale for your design solutions and images that depict them.

- **Color palette**—This section summarizes all the color specifications for an information space, shows all the colors in the color palette—along with their hexadecimal, or hex, codes—and specifies the usage of each color.

- **Type specifications**—This section summarizes all type specifications for an information space and specifies the usage of each font for particular content types such as levels of headings (H1, H2, H3, and so on), body text, and the navigation system's links.

- **Icons**—This section comprises a collection of digital-image assets for any icons the information space uses. Provide images for each icon, representing all of their states. Include the file name for each image.

All detailed specifications comprise numbered sections and subsections that make it easier to refer to and use them.

One advantage of creating detailed specifications as HTML/CSS documents and publishing them on a server, as I usually do, is your ability to present full-screen mockups and annotations side by side, unconstrained by the limitations of a print document. To ensure visual accuracy, you should ideally view mockups on a screen anyway. Plus, you can build a table of contents comprising links, letting readers easily navigate to any part of the specification; link entries in the change log to specific changes within the specification; create cross-references to link any content within the specifications, as necessary; and provide links to digital assets such as icons.

Other types of design deliverables—such as sets of wireframes or mockups—might require similar administrative information. [113, 114, 115, 116, 117]

Specifying a navigation component

"A good spec is prescriptive, clear, explanatory, efficient to read, and appropriately formatted. … What goes in the spec needs to be exactly what gets built…."—Kim Goodwin [118]

The detailed specifications for a navigation component typically include the following information:

- The name of the navigation component
- Elements it comprises and their behaviors and states
- The links it comprises and their states and destinations
- A map of the navigation hierarchy for structural navigation components
- Visual design specifications such as fonts and colors
- Sources of any data elements
- References to earlier design deliverables that inform your designs
- Any business rules that apply—for example, differences that exist depending on who the user is, the role of the user, whether the user is or is not signed in, *or* whether the user is in the midst of a particular workflow such as a checkout process
- Branding requirements
- Accessibility requirements

Because these navigation components are present on most of an information space's pages, provide detailed specifications for them in only one place, then refer to these specifications when describing a particular instance of a navigation component on a specific page. [119, 120, 121]

Specifying a page or screen

The detailed specifications for each page or screen typically comprise the following information:

- The page's title
- A brief description of the page's function or content

- Any page template on which the page's structure is based

- The navigation components on the page

- The states of the links on a page

- Detailed specifications for all the functional elements on the page, including their states and other behaviors

- Visual design specifications such as fonts, colors, dimensions, and spacings

- Sources of images on the page

- Categories to which the page's content belong

- Guidelines for creating page titles and Web addresses, or URLs (Uniform Resource Locators)

- References to earlier design deliverables that inform your designs

- Any business rules that apply to the page—for example, whether the page is available when the user is or is *not* signed in

[122, 123, 124]

Numbering sections, figures, and tables in specifications

To facilitate teammates' finding and referring to specific information, specifications should comprise numbered sections, figures, and tables. Figures typically represent site maps, page flows, and screen images. Such images often express specifications more concisely than words could, so enable you to minimize the amount of written documentation. Tables might outline all the elements that a navigation component or page comprises and could let you write their specifications more concisely. [125]

Tailoring specifications to their audience

Members of your specifications' target audience include people in the following roles:

- **Developers** are the primary audience for your specifications, which must provide all the information that is necessary to build the user interfaces you've specified for an information space.

- **Software-quality assurance (SQA) engineers** base their test plans and ad hoc and automated test scenarios on your specifications.

- **Business stakeholders**—for example, product managers, product owners, or business analysts— ensure that your specifications satisfy all business and user requirements and evaluate your specifications to determine whether the product team can build the system they describe within the necessary time and resource constraints.

- **Content owners**—such as business leaders, content strategists, information providers, content creators, and other members of the Content Management and Content Development teams—rely on your specifications in assessing an information space's content requirements and providing content that meets them.

- **Authors of user documentation** rely on your specifications to understand users' needs and the user interface for which they'll provide documentation.

- **Future team members**—whether designers, developers, product managers, or managers of teams—must rely on your specifications to become familiar with your historical design decisions.

Structure and format your specifications to make it easier for the various members of your target audience who have different needs to find the information they need.

If you're working on a project with a globally dispersed team, you might need to provide very detailed specifications. However, a key tenet of agile development calls for minimizing project documentation and instead relying on constant, face-to-face communication. So you must tailor your specifications to the needs of your project team. [126, 126, 127]

Presenting your design deliverables

"Once you're confident that your designs meet both business and user goals, it's time to gather feedback from your team. … Taking control of the feedback process is one of your hardest challenges, but get it right and you can turn negative, arbitrary criticism into valuable critique."—Cennydd Bowles and James Box. [128]

You'll create various design deliverables to capture your initial, evolving, and final design concepts, clearly communicate them to your teammates and stakeholders, and facilitate getting their feedback and, ultimately, their buy-in and approval.

Presenting your design deliverables at each stage of a design project also gives you the opportunity to manage people's expectations of them and clarify their purpose. Plus, receiving feedback on your design deliverables from both your team and stakeholders enables you to understand how well they're meeting their needs.

Never throw deliverables over the wall, distributing them to teammates who work in other functions or to stakeholders without adequate context or explanation. To make sure this doesn't happen, hold a design-review meeting whenever you complete a new deliverable or a significant new version of a deliverable. [129, 130, 131]

Preparing for a design-review meeting

Whether you're planning to share wireframes, mockups, or a prototype, present your design deliverables to their intended audience either in person or online to prevent any misunderstandings about what you intend them to convey. Although a design review should ideally be a face-to-face meeting with your key teammates and stakeholders, with modern online-meeting software, you can make a remote design review work well. However, it can be harder to hear everyone in an online meeting, so you should invite fewer participants to a remote design review.

Book a small conference room that is large enough to comfortably accommodate all invited participants. The room should have plenty of whiteboards and wall space on which you can tape up printouts of the agenda, personas, usage scenarios, existing designs, and sketches of new design concepts that participants can annotate. You'll also need a large-screen television or a high-quality screen projector on which you can display your presentation, designs, or detailed design documentation. Ensure that there is adequate connectivity to enable everyone in the room or online to view whatever you're displaying on their notebook computer or tablet. Make sure there are plenty of electrical outlets, too.

Schedule two to three hours for the meeting, but book the conference room for two additional hours—an hour before the meeting to ensure that you have enough time to set up the conference room and an hour afterward for collecting your data and design artifacts. Make sure that the layout of the room provides good visibility and is conducive to both collaboration and discussion. If a large project's scope requires additional meetings, schedule a series of design-review meetings over a period of several days.

While you must invite all the people whose participation in the meeting would be absolutely necessary to explore and resolve design issues, try to limit the meeting to only four to six people—certainly no more than eight. Only a small group of people can sustain the sort of focused, collaborative discussion that is necessary to generate the feedback you need and come up with creative solutions to design problems.

If including all your teammates *and* stakeholders in one meeting would result in an overly large, unwieldy meeting, consider first doing a design-review meeting with only your core team members, including your Product Manager, Lead Developer, UX Designers, SQA Lead, and people from the Content Development team. Incorporate their feedback into your design deliverables, then present your final design to key stakeholders and your boss in another meeting.

Well in advance of your design-review meeting, send an email invitation to the participants. To ensure that everyone is aware of the meeting's purpose, briefly describe what you'll be showing during the meeting and include the agenda. Also include links to your presentation, design deliverables showing page or screen designs, or detailed design documentation, ensuring that any remote participants can still follow along if they experience any connectivity issues. Request everyone's full attention and participation during the meeting, without distractions from phone calls, texts, email messages, or social-media posts. Ask participants to ask clarifying questions, share their candid feedback, and stay until the very end of the meeting when you'll be making some big decisions as a team—for example, to resolve trade-offs between requirements or design details and implementation costs.

Once you've completed your design deliverables, share them with your team a day or two before your design-review meeting. Sharing your design solution in advance of the meeting reduces everyone's stress going into the meeting. Therefore, when you begin the meeting, participants will be more willing to give you the time you need to clearly convey the appropriate context for the design review.

If you're concerned about any particular team members' reactions to your work, ask whether they would like to discuss your design solution before the meeting or have any questions about it. Your discussion might even result in your getting their buy-in in advance of the meeting—and going into the design review with the buy-in of one or more key team members can help your meeting go more smoothly.

Consider doing a practice run of your presentation with your UX teammates before the design-review meeting—especially when you'll be presenting to executives or important clients. Your teammates might be able to suggest additional information that you should include or identify points at which where your pacing is too fast or too slow. If there are multiple presenters, you can also decide who should present what and practice transitioning between them. [132, 133, 134, 135, 136, 137]

Planning the agenda for your design review

Set a clear agenda for your design-review meeting, according to its purpose, as follows:

- **Discussing alternative design concepts**—Early during a design project, when you're presenting several alternative design concepts to your team, be sure to provide a clear, descriptive designation for each design concept that emphasizes its most salient characteristic. Put together a presentation that shows all of your alternative design concepts, presenting each concept in its entirety before moving on to the others. Then lead a discussion, comparing the pros and cons of each design concept, and get your teammates' input on them. Never try to cobble together a design solution from several different concepts. Instead, choose a single coherent design concept, then refine it.

- **Obtaining feedback on your designs**—When setting the agenda, indicate any specific questions to which you need answers and design elements on which you want feedback. *Good feedback* is constructive and actionable because it honors the context of the existing design vision and the needs of users. Constructive feedback provides clarity on the design problem your team is addressing. In contrast, *bad feedback* is negative criticism that is based only on stakeholders' personal tastes, offers no design rationale, and fails to consider the needs of actual users.

- **Getting stakeholders' approval of your final design**—The goal of this meeting is to get stakeholders' buy-in and approval of your final design solution, so the Development team can move forward with implementation. Therefore, this meeting really is all about presentation. You'll explain your learnings from user research, present your design rationale, and walk your audience through your design solution. Describe how you've incorporated their previous feedback *or*, if you decided against a suggestion, describe your efforts to explore the suggestion, explain why you rejected it, and show what you've done instead. [138]

The Design team should plan the presentation's narrative collaboratively. Decide who should present each specific part of the presentation.

Provide your agenda on a whiteboard or a large sheet of paper at the front of the conference room. Include the agenda at the beginning of your presentation as well, then again at the beginning of each of its major sections, showing your current point within your agenda. [139]

Creating an effective presentation

Even though you've created a design deliverable to meet your Development team's needs, you still need to prepare a presentation for your design-review meeting to make consuming your design solutions easier for executives and stakeholders, who probably won't have time to read your design deliverables, as well as for others who might not be able to attend your meeting.

Start by presenting your agenda to provide an overview of what your meeting will cover. Then, at the beginning of each major section of your presentation, display the agenda again and show your current point within it. Before beginning your design walkthrough, communicate your ground rules for offering critique.

Plan the order in which to present your designs for navigation components and page layouts. Prepare slides that show each navigation component and page layout, along with your design rationales. Base your design rationales on a solid foundation—for example, findings from UX research, business requirements, design principles and guidelines, the task scenario for which you're designing a solution, or feedback on your designs. Be sure to share any user-research findings that inform the creation of your personas, usage scenarios, and requirements in detail. Convey your personas' viewpoints by speaking in their own voice. By clearly communicating your design rationales, you can help anyone viewing your presentation on their own to better understand your designs.

For each design solution, your slides should consist of the following:

- A clear definition of the design problem

- Relevant, research-based personas and their usage scenarios

- Business and user requirements that the design must satisfy

- An overview of the design solution's key concepts

- A visual representation of the design

- Your rationale for the design

- Storyboards showing personas' key-path scenarios and workflows

- A review of relevant findings from UX research

Plan for the delivery of your presentation to take at least an hour and try to limit it to no more than an hour and a half. Present only the highlights from your deliverables. If you limit your presentation's length and focus primarily on the topics that are of particular interest to your audience, you should be able to hold everyone's attention throughout your presentation. Minimize your coverage of research activities and other project information.

The content on your slides should be concise and take the form of bulleted lists, not include everything you'll say when making each point. Illustrate your presentation with screen drawings and diagrams. If you're communicating complex processes, consider building some slides one element at a time.

The Design team should review and finalize all slides of the presentation the day before the design review. Be sure to review all scenario slides especially carefully to ensure that all page or screen states are correct. [140]

Facilitating a design-review meeting

The essential value of presenting your designs to your teammates and stakeholders is having the chance to gather actionable feedback from them. They'll help you identify design issues and missed opportunities and, ultimately, improve the user experience.

Other benefits of facilitating a successful design-review meeting are building your team's shared understanding of the design problems you need to address and the solutions you've devised, building trust among your teammates and stakeholders, and gaining their alignment behind and commitment to your designs. Ensuring these positive outcomes requires creating an environment for your design review that is conducive to everyone's equal participation, eliciting high-quality feedback from *all* essential team members and stakeholders, and achieving alignment on optimal design solutions.

As the designer, you should be responsible for facilitating the design-review meeting. Only you know exactly what you need to cover. Plus, nobody else understands, can explain, answer questions about, and sell your design decisions as well as you can. If more than one designer is presenting design solutions, designers should present their own designs and respond to any questions or feedback about them. [141]

Setting expectations for the design-review meeting

"Have a shared set of design principles by which your team evaluates all work. Allow only feedback that is couched in terms of your design principles. There should be no more 'I like' or 'I don't like'."—Traci Lepore [142]

At the beginning of your design-review meeting, clearly communicate its objectives and establish both the scope of the critique and its ground rules. Tell participants that you want to complete your design walkthrough before taking any questions or discussing their feedback, so you can make your full case for your design solution before they offer their critique. (Your well-prepared teammates and stakeholders probably already have some feedback.) Ask participants to note any questions or suggestions they think of during your design walkthrough—along with the titles of the pertinent pages or screens—but to defer discussing them until you've concluded your presentation. Reassure participants that you'll probably answer most of their questions during your design walkthrough and that they'll be able to ask any remaining questions and share their feedback with you during your second pass through the walkthrough.

Encourage participants to share constructive, actionable feedback—whether positive or negative. Be sure to convey what level of feedback would currently be helpful. The fidelity of the design deliverables you're presenting should correspond to the level of feedback that you're seeking.

Establishing these ground rules before you begin your design review can help keep everyone on track with your presentation. [143, 144]

Conducting a structured design walkthrough

Delivering a presentation is the best means of structuring a design-review meeting and keeping it on track. Doing a structured design walkthrough and explaining your design rationale for each navigation component and page or screen that you present ensures that you'll be able to gather well-ordered, well-informed feedback.

Present your wireframes or mockups for the pages or screens of key workflows first. For example, when leading a design walkthrough for a navigation system, show the user's path through the navigation experience. Relate the story behind your design's evolution, providing a quick overview of any design directions that you rejected along the way and the reasons for your design decisions. After presenting the key workflows, use any remaining time to give a quick overview of other pages or screens.

Your presentation should do the following for each design problem and solution that it covers:

- Clearly define the design problem that you're solving, including the business and user requirements that the design must satisfy.

- Provide an overview of your design solution to establish the context for the design, as well as to put everyone on an equal footing, regardless of their recent level of engagement with the design process.

- Present your detailed design solution.

- Communicate your design rationale for your solution.

- Provide validation for your design rationale by sharing relevant findings from user research, presenting relevant research-based user personas and their usage scenarios, or leveraging relevant design principles and guidelines.

During your design-review meeting, you'll likely have the opportunity to explain your design rationales in greater depth than in your design deliverables or presentation. [145]

Eliciting feedback through a team discussion

"Feedback is so vital to the process, failing to provide the environment in which everyone's feedback can be of optimum quality is a dereliction of our duty as UX designers."—Martina Hodges-Schell and James O'Brien [146]

Once you've completed your initial design walkthrough, ask participants about their questions and concerns and lead them in a team discussion. Structure the discussion by doing a second walkthrough, giving all participants in the meeting a chance to ask their remaining questions, provide any feedback they want to share, discuss design issues, and offer possible solutions for them. Focus your detailed discussion primarily on the design issues that have the greatest impact or are unique in some way.

As the facilitator, it's your responsibility to encourage candid discussion and make sure that all participants feel you're really listening to their feedback, respect their role in the process, and are open to their ideas. Therefore, when responding to participants' questions or concerns, be sure that you fully understand them before answering. If you don't know the answer to a question, admit it, promise to find out, then get back to them with the answer. Or, if you can, refer the question to another team member who is present in the meeting. Make sure that you've adequately addressed the question before moving on.

At the beginning of your team's discussion, set aside three areas on a whiteboard, as follows:

- **Key issues**—These are issues that the team *must* resolve during the discussion—such as important questions, concerns, requirements, design ideas, or trade-offs. Being confronted with such a list of issues helps stakeholders focus their discussions on the big issues.

- **Parking lot**—This lists important questions, concerns, assumptions, feedback, and design ideas that fall outside the scope of the current discussion and, thus, would interrupt its flow. Loop back to them as they become relevant to your discussion.

- **Open issues**—Once the team has discussed its key issues, compile a list of all remaining open issues. At the end of the meeting, capture any design issues that remain unresolved in a list of open issues. Also, note any open questions that your UX research has not resolved.

Whenever participants start having discussions on the side rather than engaging in the current topic of discussion with the entire team, ask them whether there is an important issue that the team as a whole needs to address. If the issue is relevant to your current discussion, go ahead and discuss it. If not, add the issue to your parking lot. [147]

Working collaboratively to resolve design issues

"Focus more on hearing people and gaining agreement than on trying to be right. … Approach the discussion not as a battle you need to win, but as an exchange of views to build shared understanding."—Kim Goodwin [148]

When discussing your design decisions with your teammates and stakeholders, be prepared to explain them—especially if they provide feedback or make suggestions with which you strongly disagree because acting on them would degrade the user experience. However, you should always be open to new ideas that might improve a design solution. Don't become so attached to your ideas that you defend them too aggressively and alienate your teammates. While user-centered design (UCD) should be your goal, there are always business requirements and technical constraints that you must consider as well. [149]

Be sure to make any necessary changes to the personas to get everyone's agreement on them before you progress to seeking alignment on key-path scenarios and user requirements. Work as a team to prioritize the personas and scenarios according to the business value they represent. You'll base business and user requirements on the personas and scenarios, then make decisions about the features and functionality to design and implement based on the priorities you've set.

Try to resolve important issues during your discussion whenever possible. However, some issues might require additional research or thought. Therefore, at the end of the meeting, add any remaining open issues to your list of open issues, assign each open issue to a specific person, and set a deadline for its resolution. Schedule a follow-up meeting to discuss your learnings and make your final decisions.

Hopefully, you're working with a multidisciplinary team that is highly collaborative, and those participating in your design-review meeting can quickly work through the design issues, come up with creative solutions for them, and achieve alignment. The essential outcome of any design-review meeting is that your team knows what to do next. [150, 151]

Wrapping up your design-review meeting

At the end of the meeting, take photos to create a permanent record of the parking lot, lists of action items, and any other notes or sketches that you captured on the whiteboard during the meeting. When concluding the design-review meeting, thank everyone for their participation and offer to follow up individually with anyone who wants to discuss particular design issues with you in greater depth. [152]

Guidelines for providing constructive design critiques

"To make design-critique sessions productive and worthwhile, [you must] clearly communicate and socialize design-critique ground rules across teams."—Atul Handa [153]

All participants in a design critique must observe basic ground rules for effective communication and collaboration, such as the following:

- **Focus on goals.** Establish and work toward shared goals. In what ways does the design meet business requirements and user needs? Where does it fall short? Critique should be based on how well a design satisfies these goals, *not* people's personal opinions. Accommodate the goals and concerns of colleagues working in other disciplines.

- **Be respectful.** Assume that your colleagues are competent, have positive motivations, and want to achieve the best outcomes possible—just as you do. Always be mindful of your colleagues' feelings. Never denigrate or be dismissive of others' views. Your criticism should never be personal. Criticize ideas, *not* the people sharing them. Critique the design, *not* the designer.

- **Engage in active listening.** Listening to understand rather than to respond requires that you be fully attentive and demonstrate that you've heard what the speaker has communicated to you, either by paraphrasing, clarifying, or summarizing it. Active listening increases engagement, promotes mutual understanding, helps the speaker feel heard, and improves the listener's comprehension and retention of information.

- **Be open-minded.** When trying to refine a design and align teammates and stakeholders behind a design solution, everyone must keep an open mind and be ready to consider others' ideas—especially when the participants represent different disciplines. Give everyone an equal opportunity to provide feedback.

- **Be user-centered.** To do full justice to a design, all the participants in a critique must evaluate it from the users' point of view rather than their own. The designers must help those who work in other disciplines to adopt a user-centered mindset and share personas representing the users and their usage scenarios with the team.

- **Be collaborative.** Adopt a collaborative rather than a competitive mindset and work with your teammates and stakeholders to achieve a common purpose: designing a better product or service for users. Assume that everyone on your team is making their best effort to do good work. Whenever possible, build on one another's ideas.

- **Be responsible.** The participants in a design review have shared goals and obligations. To meet these goals, the product or service being designed must satisfy users' needs and deliver business value. Achieving an optimal design solution requires that all participants be present and fully attentive during the design walkthrough and the discussion that follows, be open to new ideas, and offer constructive, actionable feedback.

When *eliciting feedback* on your navigation-system designs from teammates and stakeholders during your design-review meeting, it's important that you first align everyone's expectations regarding the types of feedback you need and how best to provide meaningful feedback rather than having participants jump right into sharing their opinions.

To ensure that the design feedback you gather is constructive and actionable, do the following:

- Communicate your ground rules for critiques.

- Present your user personas and their usage scenarios to provide the proper context.

- Explain the business context within which you've created the design.

- Convey the intent of the design solution that you're presenting.

- Request feedback on how well you've achieved that intent.

- Reflect on the feedback you receive. What value does it provide? How can you incorporate the feedback into the design?

When *offering design feedback*, all teammates and stakeholders should do the following:

- Familiarize themselves with and reflect on the context of the design-review meeting: understanding the users and satisfying their needs, comprehending the business context, and solving the design problem.

- Ask clarifying questions about the designer's intent in making a particular design decision. Rather than relying on assumptions, try to ascertain the reasons for a decision.

- Consider whether the designer's intent adequately addresses business requirements and user requirements.

- Evaluate whether the design actually meets the designer's intent.

- Provide meaningful, constructive, actionable criticism that takes the full context of the design into account.

- Provide support for the good aspects of a design—what would work well for users—as well as criticizing its flaws.

- Focus on clarifying, critiquing, elaborating on, and improving design solutions rather than proposing different solutions.

- If necessary, explore alternative design ideas.

- Consider the consequences of making design changes, including the scope of the changes and their business and technical impacts.

- Make a concerted effort to resolve open issues and help make actual design decisions during the design-review meeting.

- If the designer disagrees with your feedback and rejects it altogether, make sure that the notetaker records your feedback and finds out what other stakeholders think about it. [154, 155, 156, 157]

Responding to stakeholders' feedback

To moderate your stakeholders' expectations, you must respond clearly to their suggestions and requests, as follows:

- **Reject a suggestion outright.** One of your goals is to avoid scope creep so you'll need to say *no* when an idea falls outside the scope of your project's goals. Note valuable ideas for possible inclusion in a later release. With your deeper understanding of your overall design, some ideas simply won't make sense to you. However, whenever you reject a stakeholder's suggestion, you should provide a good reason for doing so. If a stakeholder continues to make unconvincing arguments, you might try a few additional counterarguments. But, if you cannot fairly quickly resolve an issue, suggest moving on and agree to look into it further after the meeting.

- **Consider a suggestion.** If you're unsure about a suggestion and need time to consider it further or want to do some research to better understand it, say you'll look into the issue and take an action item to do so. Once you've agreed to consider a suggestion, you must follow through and get back to your stakeholders with your response in a timely manner.

- **Accept a suggestion.** If a stakeholder makes a good point or offers a useful suggestion on how to improve your design solution, make sure that you're aligned on how to move forward, then agree to make the change. Take an action item to revise your design accordingly. Incorporating your teammates' feedback makes it more likely that you'll get their buy-in when you need their approval. [158]

Capturing feedback

A designated notetaker should capture a thorough record of the entire design-review meeting—from the objectives you set at the beginning of the meeting to a summary of the action items you've assigned to participants. Throughout the meeting, the notetaker should capture *all* the feedback your teammates and stakeholders share, along with your responses to their feedback; details about all the design decisions you make and any design directions that you've rejected; and the action items to which you've all agreed, including the people assigned to them and their deadlines.

Post any of this information that participants need to review during the meeting on a whiteboard, then add it to the meeting notes later. To make preparing the meeting notes for distribution later as easy as possible, the notetaker should try to structure the notes for easy reference from the beginning, then refine the notes' structure after the meeting to make them more coherent. The notes' topical coherence is more important than their chronology.

If necessary for clarity, draw clean sketches of your final design solutions on a whiteboard. Take photos of design sketches and any other important information your team has captured on a whiteboard. You should also create an audio or video recording of the meeting that you can review later to ensure that you don't miss anything important. Capture as much of the information digitally as possible to prevent wasted effort when compiling your notes.

Each participant in a design-review meeting who has action items should document their own action items in detail, so everyone is clear on their next steps following the meeting.

Provide online access to all notes, photos, and audio or video recordings of the meeting and share the links with everyone who participated in the meeting, as well as teammates whose work would benefit from reviewing them. Ask your teammates to review the notes as soon after the meeting as possible, resolve any open issues, and add any missing information. [159, 160]

Testing navigation design

"Create paper or HTML prototypes for users to navigate. … Think carefully about what you want to test and how you can construct the test to yield trustworthy results. … When presenting hierarchies, … the presentation of sample second-level categories can substantially increase users' abilities to understand the contents of a major category, [by increasing] the scent of information."—Peter Morville and Lou Rosenfeld [161]

To evaluate and refine your navigation-system designs, you'll engage in iterative cycles of design and testing with users at the various stages of the design process. Early on, when sketching page layouts or creating low-fidelity wireframes that depict an information space's navigation system on specific types of pages, you might do paper prototyping using your sketches or wireframes. Later on, you might conduct usability testing using high-fidelity wireframes or mockups. Your team could even create a click-through or more fully functional prototype of the navigation system to facilitate more rigorous usability testing prior to the actual implementation of the information space's user interface.

You can use a variety of methods to validate the design of a navigation system, including A/B and multivariate testing, tree testing, and data analytics. Chapter 5, *UX Research Methods for Information Architecture*, describes tree testing and data analytics.

Because the designs for an information space's global and local navigation are so critical to its usability, it is essential that you thoroughly test their designs. [161] Ideally, you should create test scenarios for different personas' key navigation and task scenarios. [162]

Testing your wireframes with users

"Participants tend to feel more comfortable providing critical feedback on designs that feel unfinished, while they feel more hesitant to criticize something that looks like it's already finished or has taken a lot of time and effort to create."—Jim Ross [163]

Regardless of the fidelity of your wireframes, you can print them out and use them in conducting paper prototyping, which is a very easy and effective means of testing early iterations of your designs for a navigation system. Using paper-prototyping techniques, you can even represent interactivity. Observe how test participants interact with the information architecture and navigation system within the context of an information space's pages.

When you're testing the usability of your navigation design, use the actual labels for links and page titles, so usability-test participants can make sense of them in context. [164, 165]

Testing a prototype with users

Your team could create a Web or other prototype of a navigation system's design, then conduct usability testing to evaluate the design. Even if you'll ultimately be implementing a platform-specific mobile app, creating a Web prototype can be helpful for use during usability testing.

Because of the realism of the prototype's user experience, testing the prototype can reveal previously unidentified usability issues. However, there is less opportunity to remedy any issues that you discover late in the development cycle. Although your team should work to address any critical issues, it might be necessary to defer the resolution of less important issues until the next release. [166, 167]

The navigation stress test

Keith Instone devised a quick, easy method that you can use to test the design of an existing information space's navigation system: the *navigation stress test*. Using this method involves posing some challenging questions to participants that can help you discern the effectiveness of the navigation system, then identify opportunities to improve it. To conduct a navigation stress test, follow these steps:

1. Recruit a small number of actual users to participate in the stress test, as well as other members of your team and some people who are completely unfamiliar with your information space.

2. Choose any low-level page on your information space—*not* the home page or a section's main page.

3. Print out the page in black and white and make a copy for each participant.

4. Have participants do the following:

    - Pretend they've landed on the page by clicking a link on another information space or in Web-search results.

    - Answer the following questions:

        • What information space am I on?

        • Where am I on the information space? In what high-level section? What is the parent page of this page? What are the information space's other sections?

        • What is on this page? What is its topic?

        • Where can I go from here? Are the labels of associative links sufficiently clear and well differentiated that I know where they'll take me? How can I get to the main page for this section? How can I get to the home page from here?

        • What is the navigation path to this page from the home page?

    - Mark up the printout with their answers to these questions.

5. Compare and discuss everyone's answers, looking for similarities and differences.

6. Identify opportunities for improving the navigation system.

7. Choose another low-level content page and repeat this process. [168]

Summary

In this chapter, you learned about navigation design patterns for organizing and representing groups of hyperlinks that enable users to traverse an information space's navigation hierarchy, progress through linear sequences of interactions, or navigate directly to specific content. The detailed descriptions of these patterns comprehend everything from fundamental navigation elements such as hyperlinks and action buttons to desktop navigation patterns and layouts to mobile navigation patterns and layouts, and a discussion of progressive disclosure.

This chapter also provided in-depth information about visualizing navigation-system designs by creating a variety of design deliverables, including sketches, wireframes, mockups, prototypes, and navigation design specifications. You also learned about presenting your design deliverables in design-review meetings and participating in design critiques. Finally, I provided a brief overview of some methods of testing navigation-system designs.

Next, in Chapter 13, *Designing Search*, you'll learn about various approaches to designing and implementing internal search systems, as well as detailed design patterns for search user interfaces, search-results pages, filtering systems, and faceted-browsing user interfaces.

References

To make it easy for readers to follow links to the references for this chapter, we've made them available on the Web: `https://github.com/PacktPublishing/Designing-Information-Architecture/tree/main/Chapter12`

Get This Book's PDF Version and Exclusive Extras

UNLOCK NOW

Scan the QR code (or go to `packtpub.com/unlock`). Search for this book by name, confirm the edition, and then follow the steps on the page.

Note: Keep your invoice handy. Purchases made directly from Packt don't require one.

13
Designing Search

"People just want to find the information they need. Integrating navigation and search ... better supports how people really look for information. ... Whether a person searches or browses is situational. ... Information needs dictate the method of seeking."—Jim Kalbach [1]

When on an information space, most people engage in both of these information-seeking behaviors: *browsing* by using its navigation system and *searching* using its search system. Most information spaces should support both of these information-seeking behaviors. However, an exception might be providing *only* a well-designed, easy-to-use navigation system for an information space with a very limited scope, whose content is fairly stable and unlikely to expand much. In contrast, providing only an internal search system is generally inadequate because the lack of a navigation system would prevent users from easily building a mental model of the information space and learning the language of the domain, in which users should posit their search queries to get optimal results.

An information space's internal search system lets users perform searches and view lists of search results that comprise links to specific information that most closely matches their need. Thus, depending on the user's information need, searching can provide an efficient and valuable alternative to browsing to navigate an information space. But searching is a truly effective information-retrieval strategy *only* in cases where the user is looking for known items of information that are actually available on the information space *and* the user is capable of crafting a search query that accurately describes a specific information need. [2, 3]

Searching can be an iterative process that might require users to refine their search queries or, more often, follow several links to discover all the information they need. However, according to Jakob Nielsen, only about 10% of users revise their original search query when it fails to find what they need. [4]

This chapter covers the following topics regarding the design of search systems:

- Challenges of searching an information space
- Assessing the need for an internal search system
- The value of implementing an internal search system
- Optimizing the quality of search results

- Optimizing the search experience

- Search user-interface design patterns

- Some specialized types of search systems

Challenges of searching an information space

"Users are incredibly bad at finding and researching things on the Web. ... Most users reach for search, but they don't know how to use it."—Jakob Nielsen [5]

Many users lack effective search strategies to help them overcome the challenges of searching. [6] Users typically encounter the following challenges when searching an information space:

- **Users must have considerable knowledge of an information domain to search successfully.** To formulate a productive search query, the user must have an understanding of an information space's domain and knowledge of the terminology that describes it. The domain defines the nature and scope of the content that users might find on the information space. For example, on an ecommerce store, an experienced user would be familiar with the categories and subcategories of products on offer, while a novice user would not.

- **Searching is a more cognitively demanding activity than browsing.** While browsing relies on users' *recognition* of the navigation links they should follow, searching increases users' cognitive load because it requires them to *recall* information from their memory to formulate a search query. To devise a successful search query, users must employ relevant terms that accurately describe their information-seeking goal. All this mental effort increases the users' cognitive load.

- **Searching has a higher interaction cost than browsing.** Users must type a search query—an activity that is both time-consuming and subject to errors—submit the search query and wait for results to load, then scan a list of search results to identify the links that might lead to useful information. Providing autosuggestions can reduce this interaction cost somewhat because the user might need to type only part of the search string, then rely on recognizing a query in the list of suggestions rather than recalling exactly what terms to use in devising a search query.

- **Users must have some expertise in the use of search systems and experience formulating search queries.** Using an information space's search system and expressing search queries can be challenging. Many users have difficulty devising effective search strategies that meet well-defined information-seeking goals and formulating search queries that would deliver useful search results. The user's information-seeking goals could range from looking up very specific facts to seeking general information about a broad topic, from searching for particular items to purchase to looking for a gift for someone without any definite ideas in mind, or to embarking on a complex journey of discovery through which users continually learn, analyze the search results, and refine or broaden their search goals.

- **The varied contexts within which users search could present particular challenges.** The physical challenges of searching an information space within different contexts such as on a desktop computer in the workplace, a notebook computer or tablet at home, or a mobile phone or wearable device out in the world differ significantly. Conducting searches using devices that have screen sizes ranging from very large to very small requires the use of search user interfaces that display optimally on specific screen sizes, so the user's context differs across devices. When users search at their workplace, they tend to search information spaces that relate to their industry domain and their searches are typically more task oriented. In contrast, searching on a mobile phone lets users find information that is relevant to their current real-world experiences, but they must often contend with connectivity issues or the glare of the sun on their screen. [7, 8]

Assessing the need for an internal search system

"[Do not] assume … that a search engine alone will satisfy all users' information needs."—Peter Morville and Lou Rosenfeld [9]

Users generally expect to be able to search an information space when they don't immediately see the information they need. Plus, users just might not be willing to spend time browsing. So most information spaces should provide an internal search capability. Still, when considering whether to implement search for an information space, you should assess the following:

- **Users' domain knowledge and search expertise**—For users to successfully search an information space, they must have a clear goal for searching and possess sufficient knowledge of the information domain to formulate useful search queries. Users must also have some expertise in using search systems to be able to employ a search user interface to search the information space, to revise their search queries to obtain more relevant search results, and to identify the most relevant links among a list of search results.

- **Users' dominant mode of information-seeking**—Depends primarily on their information-seeking goal and the type of content that an information space provides. When users know exactly what information they're seeking and are confident an information space offers that information—as when they're looking up a fact, searching for a specific document, or refinding some content that they've previously seen—they can simply search for it. In contrast, if users *don't* know exactly what information they need, exploratory browsing might be their prevalent mode of information seeking. Thus, a search could become a journey involving the serendipitous discovery of information along the way that could actually change what users are seeking. In such a situation, users might struggle to express their search queries.

- **Scope of the content**—If an information space lacks sufficient content to require a search system, users must be able to find the information they need by relying on a well-designed navigation system and browsing. Because the design and development of a navigation system is essential in any case, you should consider what additional value a search system would deliver versus the effort that implementing and maintaining it would require. But whenever the scope of an information space becomes too great to easily find specific content by browsing, it's definitely time to add an internal search system.

- **Content's focus**—More transactional applications comprising a user's own information are less likely to support search than content-rich information spaces. Plus, the types of content on an information space determine the ability of a search system to wholly satisfy users' information-seeking needs. On an information space with more technical, domain-specific content, users need to be able to search, as well as browse for information. But which of these complementary methods of information seeking users would be more likely to choose depends on their domain expertise, as well as the specificity of their information needs.

- **Support for a domain's terminology**—Defining metadata that describes the content objects on an information space—including its pages; images, videos, and other media; and documents—improves both the navigation and search systems. Plus, the implementation of a controlled vocabulary of the preferred and variant terms that describe a domain *or* a thesaurus that also defines links to broader, narrower, and related terms often aids the effectiveness of an internal search system by improving the relevance of its search results. Determine whether an organization should prioritize investing in the development of these capabilities for use in categorizing an information space's content. If not, you should question whether implementing a search system would be the optimal decision and consider building a site index instead, comprising a comprehensive collection of links to the information space's content, which would also help users look for known items of information.

- **Structural coherence**—In cases where an organization has failed to align behind a coherent structure and navigation system for an information space, the implementation of an internal search system becomes especially important. If an information space has an incoherent navigation hierarchy, users would have difficulty orienting themselves and finding the information they need without an internal search system.

- **Content that changes frequently**—The expansion of many information spaces' content tends to be more organic than planned, which can make maintaining a navigation system challenging. For an organization that frequently publishes or syndicates new content on its information space or generates dynamic content, it can become impractical to maintain an index, table of contents, or catalog of the content manually. In contrast, an internal search system's automated, full-text, or other more sophisticated indexing capability can reliably enable users to access all of an information space's content.

- **Learnings from search-log analytics**—By studying the search queries that users employ, you can better understand what information users are seeking on an information space and, thus, assess what information might be missing from that space. You can also learn what terms people are using to express specific concepts and adapt your style guide accordingly. Leveraging these learnings lets you improve an information space's navigation and search systems *and* discover what content your team needs to create. [10, 11, 12, 13]

The value of implementing an internal search system

"When your site is easy to search, customers are more likely to stay on the site, consume your content, make a purchase, and even become repeat buyers. Plus, the more people searching on your site, the more data you can collect on their behavior and needs to inform [the] business."—Louise Vollaire [14]

Typically, between 30% and 60% of an information space's visitors conduct a search using its internal search system, and visitors who search are two to four times more likely to convert and become customers than those who don't. [15] In fact, most information spaces can derive business value from implementing an internal search system. Whatever an organization's business model, an effective search system that delivers useful search results helps visitors find the information they need, convert, and become customers. Of course, what constitutes a conversion depends on the type of organization. [16, 17]

For publishers of content, the whole point of an information space's existence is to engage visitors in consuming their content, so possible conversions include reading articles, sharing links to articles on social media, commenting on articles, following the information space or an author on social media, or signing up for newsletters. For businesses that need to persuade visitors to read their marketing content and learn about what they offer, possible conversions might include reading pages that describe the business and its products and services, reading blog posts, engaging in conversations with a chatbot, making an appointment to meet with a salesperson, sharing links on social media, following the business on social media, or signing up for newsletters or notifications.

Ecommerce businesses garner business value primarily from their Web and mobile applications, whose internal search systems play a significant role in their success. Among shoppers on ecommerce stores, 43% prefer to rely solely on search rather than navigation, while 49% use search more often than navigation. [18] According to an *Econsultancy* study, people who visited an information space to search for something specific and successfully found what they sought were 80% more likely to convert. Plus, ecommerce businesses were more likely to retain customers who successfully found the products they needed, thereby increasing the business's revenues. But about 41% of searchers failed to find what they were seeking on their first try and only half of them were willing to try again. [19]

When the Nielsen Norman Group (NN/g) conducted extensive testing of the search systems on ecommerce sites, they learned that users' first search attempt found what they were seeking 64% of the time. When users reformulated their queries, an additional 10% of searches were successful—resulting in an overall success rate of 74%. Thus, a well-designed search system can contribute to any business's success. [20]

To derive the greatest business value from their internal search system, ecommerce businesses should endeavor to personalize search results—especially for users on mobile devices—so they can better satisfy the needs of individual users and increase the business's revenues. However, personalizing search results can be challenging, especially for infrequent users and those who aren't signed in, for whom an information space has only very limited data. [21]

Optimizing the quality of search results

"The Google experience— … an experience that is relevant, personalized, and anticipatory—is what your customers demand and expect and … pays off."—Andrea Polonioli [22]

Users of information spaces' internal search systems are increasingly demanding that their performance and user experience attain the quality to which they've become accustomed from using Web-search engines such as Google. Several factors largely determine the quality of the search results that an internal search system delivers, as follows:

- **What text the search system searches**—The text that an information space's internal search system actually searches depends on the capabilities and configuration of the system. For example, it might simply search the full text of each page. However, a search system should preferably search specific elements of indexed content on each page such as its title and Web address, or URL (Uniform Resource Locator). Alternatively, a search system could search for indexed terms from the information space's controlled vocabulary—that is, meaningful terms within the page's main content. An index's scope could be either comprehensive or limited to content within a specific zone of the information space. For more information, see the subsection "Determining what to search."

- **What search algorithm the system uses**—Different internal search systems use a great variety of different search algorithms. Most of these algorithms rely on some type of pattern matching, comparing users' queries to text strings in an information space's index, then adding matching pages to a list of search results. The precision and recall of the search system's results depend on what search algorithm the system uses. For more information about search algorithms, see the subsection "Balancing a search algorithm's precision and recall."

- **How the search algorithm interprets queries**—Depending on a search system's capabilities and configuration, the approaches it employs in refining search queries to improve the precision of the results it generates can vary. For example, a search system could use a spelling checker; normalize queries using stemming or lemmatization; employ a controlled vocabulary or stop words; or search for similar pages. For more information about how a search system might refine a user's search queries and, thus, its search results, see the subsection "Automatically refining search queries."

- **How the system orders or ranks search results**—Different search systems order or rank search results in different ways—for example, by automatically sorting the results alphabetically, chronologically, or numerically *or* ordering the results by their relevance, recency, popularity, ratings, or document type. The ways in which users would use the search results should determine the optimal way in which a search system should order or rank search results by default. For more information about how a search system orders or ranks search results, see the subsection "Ordering and ranking search results." [23, 24]

Determining what to search

"Indexing is one of the first steps—after crawling—in a complex process of understanding what [pages] are about for them to be ranked and served as search results by search engines."—Dave Davies [25]

A disadvantage of full-text searches is that, in addition to searching a page's main content, the search system would also search irrelevant content elements on the page such as navigation links and other lists of links or categories that appear on many pages. In contrast, assigning indexing terms to an information space's pages; then determining the relevance of specific pages to these indexing terms; and scoring, sorting, and ranking search results based on a variety of factors enables searches to be both more precise *and* more efficient than simply relying on full-text search.

Key approaches to indexing include the following:

- Indexing semantic elements in content

- Indexing search zones [26]

- Latent semantic indexing [27]

By indexing semantic elements in content

An information space's search system crawls the content and links on its pages and indexes the content to ensure that the search system can retrieve relevant results, deliver precisely the information users are seeking, and thus, provide greater value to users. [28] The search system should index specific semantic content elements on the information space's pages. To determine which content elements to index, you can conduct user research or, for an existing information space, analyze the search-query logs.

Some semantic content elements that organizations commonly index include pages' titles, authors' names, and pages' body text; Web addresses, or URLs; and metadata such as page descriptions and categories. By indexing only semantic content elements whose inclusion in searches would provide value to users, you can exclude many irrelevant search results. Indexing semantic content elements also lets you design search results that provide more meaningful information to users. [29]

Since the World Wide Web Consortium (W3C) recommended the adoption of Hypertext Markup Language 5 (HTML5) in 2014, the use of structural semantic markup such as <main>, <section>, <article>, <summary>, <header>, <footer>, <aside>, and <nav> has made identifying specific elements of pages to either include them in *or* exclude them from a search easier. These tags describe the purpose and meaning of specific elements on a page and help a search system better understand a page's structure and content. The <main>, <section>, <article>, and <summary> tags identify unique content elements that a search system might index, while the <header>, <footer>, <aside>, and <nav> tags identify page elements that a search system should likely exclude when indexing a page's content. [30, 31]

By indexing search zones

A *search zone* defines a subset of an information space's content as indexable. While decomposing an information space's content into search zones can offer different views of the content to users, indexing such collections of content can be expensive. However, creating search zones makes it easier for users to find the content they need on an information space.

For example, you could index specific, large collections of content by category, type, format, topic, audience, author, business unit, or department; or by content that is related chronologically—such as by recency—or geographically—such as by nearby places. Although you should try to include *only* destination pages that provide relevant content in search zones—*not* navigation pages—distinguishing between them can sometimes be difficult. To improve performance, minimize overlaps between these indexes.

Creating search zones can enable users to focus their searches on content that is of interest to them by setting filters before conducting a search and, thus, improve the relevance of search results. However, using filters increases the user's interaction cost, so their use should always be optional. Therefore, you must provide useful defaults that search and display relevant results from all of the information space's content. [32] For more information about filtering, see the section "Filtering search results," later in this chapter.

By employing latent semantic indexing

A newer approach to indexing that leverages the power of artificial intelligence (AI), the natural-language processing (NLP) technique *latent semantic analysis* (LSA), or *latent semantic indexing* (LSI), analyzes the relationships between an information space's pages and the terms in their main content, then generates a set of related concepts by using this *distributional hypothesis*: words with similar meanings occur in similar content and, conversely, words occurring within the same contexts, or with similar distributions, tend to have similar meanings.

This technique creates a document-term matrix of the word counts on each page, in which the rows represent unique words and the columns represent pages. An element's weight in the matrix corresponds to the number of occurrences of terms on a page. Rare terms are upweighted to reflect their relative importance. This technique can mathematically determine how similar two pages are.

By using latent semantic indexing to automatically index an information space's pages, a search system can understand the meaning and context of each page's content and, thus, deliver more relevant search results with greater precision. Semantic indexing also provides greater efficiency by reducing processing time, as well as the consumption of computational resources, while quickly and accurately delivering the information users need. [33, 34]

Balancing a search algorithm's precision and recall

"Precision is … the proportion of items retrieved that are relevant to the query; recall is the proportion of relevant items that are retrieved."—Tony Russell-Rose and Tyler Tate [35]

In information retrieval, precision and recall are inversely related performance metrics that assess the results an information space's search algorithm retrieves in response to a user's search query. Definitions of these two ways of measuring the effectiveness of a user's search strategy and the quality of the results the search system finds are as follows:

- **Precision**—This refers to the percentage of relevant search results among all the results that the search algorithm retrieves in response to a user's search query and, thus, measures the overall accuracy and relevance of an entire search-results set. To determine *precision*, divide the number of relevant pages a search algorithm retrieves by the total number of pages it retrieves. According to this definition, each result the search algorithm returns has equal weight. Thus, it disregards relevance ranking. [36]

- **Recall**—This refers to the percentage of *all* relevant pages on the information space that the search algorithm retrieves in response to a user's search query and, thus, measures the comprehensiveness or completeness of the search results, which should include all significant pages. To determine *recall*, divide the number of relevant pages a search algorithm retrieves by the total number of relevant pages on the information space. [37, 38, 39, 40, 41]

Depending on the context of the search tasks that users might perform on a specific information space, its search system might prioritize either precision or recall, as follows:

- **Precision**—Prioritizing precision produces fewer, highly relevant search results and could cause a search algorithm to fail to retrieve some relevant pages. If the goal of a user's task is to find a useful answer to a specific question *or* to read an excellent article on a specific topic, a high level of precision would be necessary, so a search should produce only a few highly relevant results. For example, a search algorithm can attain greater precision if it is capable of searching pages' metadata or specific fields in a content-management system's database *or* indexing an information space's content using its controlled vocabulary. Refer to Chapter 7, *Classifying Information*, for more information about controlled vocabularies.

- **Recall**—Prioritizing recall produces a large number of search results that have various levels of relevance—likely even some results that are wholly irrelevant. Research tasks typically demand a high level of recall, so search results should include many pages with differing levels of relevance, including some pages that have little relevance. [42]

In devising a search algorithm, it is necessary to balance these countervailing forces to best meet the needs of users. While the goal of any search system is to attain an equilibrium between optimal precision and recall, a search system cannot deliver high levels of both precision and recall. Prioritizing precision sacrifices recall, while prioritizing recall sacrifices precision. [43] Two other metrics that are useful in measuring the precision of search results are as follows:

- **Average precision**—This is a measure of relevance from the perspective of the user sequentially scanning the results in a set of search results.

- **Cumulative gain**—This is a measure of the amount of information the user could potentially gain from an entire set of search results. [44]

Traditional approaches to achieving greater precision

Search algorithms have traditionally achieved precision in search results by using the following approaches:

- **Employing a graded-relevance model**—This ranking model produces relevance rankings for search results by predicting graded-relevance scores for query-page pairs, then weights and sorts results according to their score. Thus, top-ranked results have greater weight and appear higher in a results set. [45, 46]

- **Using a weighted keyword-matching score**—The search algorithm can retrieve and rank search results by using a weighted keyword-matching score, leveraging the *bag-of-words (BoW) model* to represent text as an unordered collection of words, or another *term-vector model* to represent pages or items of text as vectors, in which the distance between vectors indicates their relevance. This approach computes token weights in matches, assigning higher weights to more important tokens. Machine-learning (ML) algorithms can operate only on numeric data, so the *tokenization* of natural language breaks ordered sequences of symbols into *tokens*—that is, *n-grams* representing letters, punctuation marks, spaces, syllables, or, more rarely, words. [47, 48, 49, 50]

- **Employing a learning-to-rank model**—Training machine-learned ranking models enables the search algorithm to list more desirable search results before those that are less desirable. However, if the rankings derive from clicks or other measures of searchers' engagement, these techniques might conflate relevance with desirability. Plus, although a learning-to-rank (LTR) model can improve precision, its purpose is to improve ranking, *not* relevance. Most LTR models use a combination of query-dependent, relevance-ranking factors and query-independent, desirability-ranking factors, which can sometimes result in poor tradeoffs between relevance and desirability. [51, 52]

- **Mapping query tokens to more meaningful entities**—Using *named-entity recognition (NER)* improves precision by classifying named entities of specific types that occur within an information space's content—that is, proper nouns such as the names of organizations, people, places, and products—then mapping search keywords or query tokens to the corresponding predefined categories in the information space's index. Thus, NER can align the search algorithm's understanding of users' queries with the index's representation of the information space's content. NER can be based on rules or machine learning (ML). [53, 54]

Traditional approaches to achieving higher recall

Search algorithms have traditionally achieved higher recall in search results by using the following approaches *before* scoring and ranking search results:

- **Query expansion**—The search algorithm can expand the scope of matching results by including certain tokens—such as synonyms or abbreviations—or phrases, *or* by using spelling correction, stemming, lemmatization, and other approaches to automatically refine users' search queries. Alternatively, the algorithm could support the use of the Boolean operators AND and OR in search queries by users to to broaden or narrow their searches, respectively. Because query expansion can change the meaning of a query, deprioritize the resulting matches and place them below the results that match the original query. Query expansion could reduce the context of the overall query, leading to a loss of precision.

- **Query relaxation**—While the search algorithm's goal in query relaxation is still to increase recall, the algorithm can exclude certain tokens or phrases or make them optional. For example, a sophisticated query-relaxation strategy could exclude tokens that are redundant or provide no useful information. Alternatively, a simpler form of query relaxation could retrieve search results that match all but one query token, but that might risk excluding a token that is critical to the query's meaning. [55]

However, both of these methods carry risks to precision because of their inability to consider query context holistically. In fact, query relaxation is so risky that many search systems use this approach only as a last resort—for example, if a search would otherwise return no results.

Using AI-powered search to achieve greater precision

Artificial intelligence (AI) can improve the precision of search results through *query understanding*—that is, determining the user's intent in expressing a query—which occurs *before* the search algorithm scores and ranks search results. Methods of understanding the meanings of queries using AI-powered search include the following:

- **Query classification**—This method maps a query to one or more descriptive categories. Category types might include topics, types of products, locations, or colors. Query classifications can optionally be hierarchical. They apply to the entire query rather than specific words or phrases within the query. The training of a query-classification model relies on the same engagement data as that for training a ranking model. Query classifiers can use large language models (LLMs) to understand queries. For more information about LLMs, see the section "Improving precision *and* recall by leveraging both LLMs and NLP," later in this chapter.

- **Assessing the relevance of individual results to a query**—*Synonyms*, different words with the same meaning, and *polysemes*, words that have various meanings, can prevent search algorithms that rely on bags-of-words representations of search queries and content from successfully assessing relevance. But search algorithms that use natural-language processing (NLP) and *dense-embedding representations*—an array of vectors that encode the meanings of words so words that are closer in the vector space are similar in meaning—can better understand and align the meanings of both queries and content. Therefore, they can do a better job of assessing the relevance of individual results to a particular query and return the most similar results based on distance in the vector space, even without exact keyword matches. [56, 57, 58, 59]

Using AI-powered search to achieve higher recall

Artificial intelligence can improve recall by taking the following approaches to AI-powered search:

- **Holistically considering query context**—This approach leverages dense-embedding representations of the meanings of queries and content through similarity-based retrieval from a vector database and improves recall without risking the imprecision of query expansion or relaxation. However, it is still necessary to ensure alignment between the representations of queries and content.

- **Rewriting queries based on query similarity**—Similar queries often simply represent different ways of expressing the same search intent. While word order, stemming, lemmatization, tokenization, stop words, and spelling errors can cause variations in query strings, it is easy to identify them as similar or equivalent queries. Understanding and recognizing the semantic equivalence of synonyms or paraphrasings in queries requires that an AI-powered search algorithm be capable of improving recall through whole-query expansion or relaxation. For detailed explanations of stemming and lemmatization, see the section "By normalizing text using stemming or lemmatization," later in this chapter. For additional information about tokenization, see the section "Traditional approaches to achieving greater precision," earlier in this chapter. For detailed information about the use of stop words, see the section "By specifying stop words," later in this chapter. For detailed information about handling spelling errors in search strings, see the section "By using a spelling checker," later in this chapter.

Although AI-powered search can help improve precision, its ability to leverage a more semantic approach to retrieval makes AI most useful in improving recall. [60]

Improving precision and recall by leveraging both LLMs and NLP

"Broadly speaking, search applications focus on three concerns: precision, recall, and desirability. AI mostly helps with the first two...."—Daniel Tunkelang [61]

When users' search queries are less than optimal, both precision and recall can suffer. However, with the integration of *large language models* (LLMs) into some modern search systems, AI can now play a role in processing search queries. *LLMs* are machine-learning (ML) models that use deep-learning algorithms to understand and process natural language and have learned natural-language patterns and entity relationships through training on massive amounts of text data. The use of LLMs and natural-language processing (NLP) in AI-powered search systems can improve both precision and recall in several ways, as follows:

- **Natural-language query expansion**—Most search algorithms rely on the use of a combination of *term frequency–inverse document frequency*—that is, *tf–idf*—which weighs the importance of each word on each page of an information space's entire *corpus—and* the Boolean operators AND, OR, and NOT—which modify search terms to broaden or narrow searches. *Term frequency*, the frequency of query terms on a page, can be misleading because the algorithm considers all terms equally when assessing relevance. To solve this problem, reduce the weight of terms that occur too frequently in the collection to be meaningful in determining relevance by a factor that increases with a term's collection frequency. Calculate the *inverse document frequency* by dividing the total number of pages by the number of pages containing the term to determine whether a term's occurrence is common or rare across all the pages, then take the logarithm of the quotient. An LLM can understand a natural-language query by decomposing its natural-language phrases into sets of terms and synonyms, then expand the query contextually and semantically by creating a structured Boolean query from those terms and synonyms. Because the LLM understands the context and semantics of the query, the result is improved precision and recall.

- **Content summarization and differentiation**—When a search retrieves a very large number of results, automatically generated summaries of each page's content can reduce the time it takes the user to choose the most relevant results. Plus, identifying similar and unique pages within the search results and explaining what makes them similar or different can help the user differentiate between the results.

- **Entity recognition**—LLMs can facilitate more specific and accurate searches through the recognition and extraction of data entities that represent sets of information about concepts from both pages and queries. This improves both precision and recall.

- **Query reformulation**—When the user's initial search query fails to provide useful results, LLMs can either suggest alternative queries or reformulate the initial query, enabling the user to explore additional results and improving recall.

Because of their ability to understand natural language and support semantic search, LLMs can more accurately deliver relevant search results *and* adapt to users' needs and evolving data sources, improving both the precision and recall of AI-powered search. [62, 63, 64, 65, 66, 67]

Automatically refining search queries

"Spelling correction is a must-have for any modern search engine. Estimates of the fraction of misspelled search queries vary, but a variety of studies place it between 10% and 15%. For today's searchers, a search engine without robust spelling correction simply doesn't work."—Daniel Tunkelang [68]

There are many approaches that a search algorithm could employ in automatically refining search queries to improve the precision of the results it generates, including the following:

- Checking and correcting the spelling of queries

- Normalizing text using stemming or lemmatization

- Using a controlled vocabulary

- Specifying stop words to exclude from queries

- Searching for similar pages

The utility of a particular approach in a given situation depends on the types of content an information space comprises and the information needs of its users. Understanding users' information needs is paramount when choosing what query-refinement tools to implement as part of a search system. [69]

By using a spelling checker

A spelling checker that can automatically correct the spelling of search queries is an essential component of any modern search system. Users commonly misspell search terms or inadvertently introduce typos into them. [70, 71]

According to Andrea Polonioli, "Another way of helping customers is to embrace typos and understand the term a user most likely meant to enter. [Typos] have increased since screens got smaller and now around 15% to 25% of searches will feature at least one misspelt character." However, only 29% of information spaces' internal search systems can handle queries with just a single-character misspelling, and only 39% support synonyms. [72] Thus, the lack of a spelling-correction system is a common and preventable reason for search systems' failing to find *any* results for a search query.

Implementing an optimal search experience requires investment in a high-quality spelling-correction system. While today's off-the-shelf spelling-correction systems are sufficiently customizable to meet most needs, choosing the best spelling-correction system for any given search system requires having an understanding of how these systems work, including both their configuration and the process by which they check the spelling of users' queries.

The ultimate goal of spelling correction is to identify a correction that has the highest probability of being correct, from among all possible candidate corrections, including the user's original query. [73]

Configuring the search system's spelling checker

Some advanced spelling-correction systems now employ AI by leveraging machine-learning (ML) algorithms and natural-language processing (NLP) techniques. A spelling-correction system that uses an ML algorithm requires several configuration steps prior to its use in checking the spelling of queries, as follows:

1. **Building an index of tokens for use in generating candidate spelling corrections**—Because ML algorithms can operate only on numeric data, preprocessing natural language using tokenization is necessary. *Tokenization* breaks sequences of adjacent symbols that occur in a particular order into *tokens*—that is, *n-grams* representing letters, punctuation marks, spaces, syllables, or, more rarely, words. A spelling-correction system requires an index that supports approximate string matching, identifying misspelled tokens by retrieving tokens using their substrings. Misspelled tokens typically differ from intended tokens by at most a few characters. Thus, a small *edit distance* quantifies the dissimilarity between two strings as the minimum number of operations necessary to transform a misspelled token into an intended token. The system generates the index by identifying the unique tokens within the information space's content, or *corpus*, then inserting the appropriate substring-to-string mappings into an n-gram index.

2. **Building a language model to estimate the probability of an intended query**—In other words, the language model calculates the probability that the searcher would type a given search query. By normalizing the historical frequencies of unique queries, the language model determines a probability distribution. To determine prior probabilities, the model combines token frequencies in the query logs with token frequencies in the information space's content, or *corpus*. However, the corpus frequencies should *not* overshadow those of the query logs, which demonstrate the searcher's actual intent.

3. **Building an error model to estimate the probability of a misspelling in an intended query**— Misspellings can comprise one or more errors such as an *insertion* of an extra letter—for example, a repeated letter; a *deletion* of a letter—for example, leaving out a repeated letter; a *substitution* of one letter for another; or a *transposition*—that is, swapping the order of consecutive letters. Using ML, the system trains the model on a collection of example spelling mistakes. [74, 75, 76, 77]

The process of checking a search query's spelling

When checking the spelling of an intended search query, the spelling-correction system employs a three-step process, as follows:

1. **Generating spelling-correction candidates for the query**—The spelling-correction system leverages the spelling-correction index in generating a set of candidates. For example, it might retrieve all tokens within an edit distance of one of the misspelled queries. Although increasing the edit distance would grow the set of candidates exponentially, the probability of producing the correct candidate would decrease. Using a substring index, retrieve all the candidates, concatenating them with ORs or ANDs.

2. **Calculating a probability score for and ranking each spelling-correction candidate**— Determine whether spelling correction is necessary by using the error model to compute the probability that the query is correct. Choose the optimal spelling-correction candidate by ranking all candidates, including the query itself, based on their probability, then use the top-scoring candidate.

3. **Determining whether to present a spelling correction**—If the original query is the top candidate and has a high probability, consider the query to be spelled correctly. If the top candidate is *not* the original query, but has a high probability, automatically correct the spelling of the query, notifying the searcher of the spelling correction by displaying a prominent message above the search results, including a link to display the results for the original query. If the top candidate is the original query, but the next candidate has almost as high a probability, *or* the top candidate is *not* the original query and has a lower probability, display the results for the original query and ask the user ***Did you mean…?***, proposing an alternative spelling. Try to correct a query that would retrieve no results, but never rewrite a query in a way that would retrieve no results. [78]

By normalizing text using stemming or lemmatization

"Stemming and lemmatization are text preprocessing techniques that reduce word variants to one base form. ... They reduce the inflected forms of words across a text data set to one common root word, or ... lemma."—Jacob Murel and Eda Kavlakoglu [79]

Normalization, stemming, and lemmatization are natural-language processing (NLP) techniques and, thus, are foundational to the larger discipline of AI.

The *normalization* of text enables search systems to recognize different words as variants of a single root word. Some ways of normalizing text include correcting spelling errors, transforming uppercase to lowercase letters, translating acronyms and abbreviations to their spelled-out forms, changing numbers to words, and removing hyphens from compound words.

Understanding search algorithms requires a foundational understanding of language, or *linguistics*. Two particularly useful means of normalizing text for search systems are stemming and lemmatization. The use of these NLP approaches can significantly improve the relevance of search results by evaluating the meaning and syntax of a search query—for example, whether the intent of a query is to find definitional information or how-to information.

Users frequently employ search queries that include some form of a word with the expectation that the search results will include pages that include various inflected forms of that word. These grammatical variants of words result from *declension*—the inflection of nouns, pronouns, and adjectives' according to their case, number, and gender—and *conjugation*—the inflection of verbs by their tense, number, person, voice, and mood. [80, 81, 82, 83]

By using a stemming algorithm

An information space's search system should provide a *stemming* capability, which normalizes text by automatically removing any suffixes to reduce search terms to their root form, or *stem*, then expands the search query by conflating the search terms' variants, or *inflections*. However, stemming cannot normalize irregular verb forms or handle prefixes.

The search system uses the same stemming algorithm in indexing the information space's pages, ensuring that the index includes all the various inflections of roots. Because the search system can search for inflections of search terms, search results can include the pages that match them. A search system with strong stemming can significantly expand a user's query to include pages that match all the inflections of the search terms and, thus, greatly improve recall, while a search system with weak stemming might simply expand search terms to include both their singular and plural forms. A search system that has no stemming capability or only weak stemming has greater precision, but lower recall.

Various stemming algorithms exist, and the rules, or *heuristics*, on which they are based differ from one stemmer to another, as well as across languages. Although the performance and accuracy of these stemming algorithms can differ, stemming is generally a fast, efficient approach to normalizing search terms.

A stemmer's rules might stem words based on their case, gender, number, tense, person, mood, or voice. For example, stemming might reduce the word *introduction* to *introduc*, so the search would comprehend the plural form *introductions*; the verb *introduce* in all its tenses; and the gerund *introducing*.

Two problems are characteristic of stemming, as follows:

1. **Under-stemming**—This issue occurs when a stemming algorithm fails to map words that have similar meanings to a single root, with the result that a search query does *not* retrieve some highly relevant pages.

2. **Over-stemming**—This problem occurs when a stemming algorithm maps semantically distinct words to the same root, with the consequence that a search query retrieves some irrelevant pages. [84, 85, 86, 87, 88]

By using a lemmatization algorithm

Lemmatization reduces the morphological variants of a search term to a single base form that must be a valid word in the dictionary. *Morphology* refers to the structure and formation of words in a language through the following:

- *Inflection* of nouns, pronouns, and adjectives by their case, number, and gender *or* verbs by their tense, number, person, voice, and mood. For example, *sang* is an inflection of *sing*.

- *Derivation* of words from existing words or roots by adding a noninflectional affix or changing the form of the word or root. For example, *personalize* derives from *personal*.

- *Compounding* of words by combining two or more existing words or roots. For example, *firefighter* is a compound comprising *fire* and *fighter*.

In determining each search term's base form, lemmatization first uses a vocabulary to identify its part of speech (POS), then assigns a POS tag to the term to indicate its syntactic function within a sentence. Then, according to the search term's specific POS, lemmatization applies different normalization rules. A lemmatization algorithm can apply more precise normalization rules to related terms and even modify a stem by reducing it to a *lemma*, or valid word. Thus, lemmatization can conflate both irregular verb forms and prefixes.

Although lemmatization can help solve the problems of under-stemming and over-stemming, it is a slower process than stemming. Plus, erroneous identification of a search term's POS would obviate any benefit of lemmatization over stemming's simpler suffix-stripping algorithms. [89, 90, 91, 92]

By using a controlled vocabulary

"Controlled vocabularies are expensive to build, use, and maintain, and they may contribute to clutter in a search interface. [Although] domain experts benefit from controlled vocabularies, ... results have been mixed for ordinary users."—Ying-Hsang Liu, Paul Thomas, Jan-Felix Schmakeit, and Tom Gedeon [93]

As Chapter 7, *Classifying Information*, describes, a *controlled vocabulary* comprises the standard, natural-language terms that an information space uses for particular concepts. Using terms from the controlled vocabulary when identifying, indexing, categorizing, and labeling an information space's pages enables the information space's search system to better understand the meanings of the terms that users employ in their search queries and, thus, improves the relevance of search results. The controlled vocabulary should also identify synonyms and related terms for the terms it comprises, enabling the search system to recognize their relevance, thereby broadening searches, and retrieve additional results.

Using a content-management system (CMS) in concert with a controlled vocabulary supports a metadata-driven approach to automatically organizing an information space's pages. *Descriptive metadata* identifies attributes of an information space's pages, provides the basis for their categorization, and defines relationships between them. Manually tagging pages using semantic metadata describes the categories to which the pages belong rather than their placement within a predefined information architecture. Descriptive metadata clearly distinguishes an information space's pages from one another; therefore, this metadata has the greatest impact on information retrieval. *Retrieval metadata* includes a page's title, subject, author, date, keywords, and security level and also facilitates pages' retrieval. The consistent application of metadata to pages is essential in clearly conveying their purpose and meaning and, thus, improving information retrieval.

The design of an information space's search user interface could support users in employing terms from an information space's controlled vocabulary in their search queries. For example:

- Autocompleting users' search queries could help users refine their queries using terms from an information space's controlled vocabulary. See the "Autocomplete" section later in this chapter.

- Autosuggestions that appear in a drop-down list beneath the **Search** box as the user types could help users recast a search query using terms from an information space's controlled vocabulary. See the "Autosuggest" section later in this chapter.

- A list labeled **Suggested terms**, comprising terms from an information space's controlled vocabulary, could appear immediately beneath the **Search** box—either as the user types *or* once the user clicks **Search** and the search-results page appears. These suggested terms might include related terms such as broader or narrower terms *or* synonyms. Coupling a checkbox with each suggested term could enable users to easily add terms to a search query by clicking an **Expand Query** button, thus expanding the query and, in most cases, enlarging the result set, which could help users avoid getting no search results at all.

Thus, the use of a controlled vocabulary can guide users to choose the same terms in describing or referring to a particular topic: in the case of the information architect—by proposing indexing terms for use when tagging pages with metadata—*and* for the searcher—by proposing query terms from the controlled vocabulary. So descriptive metadata that leverages an information space's controlled vocabulary can support effective information retrieval both by search engines and by users. [94, 95, 96, 97, 98, 99, 100]

Two purposes for which the development and use of controlled vocabularies can be especially useful are as follows:

1. Supporting information retrieval in scientific domains by automatically suggesting search terms that the search system has used in indexing the information space, including related terms and synonyms, thus bridging any gap between the language of the information space and that of the users

2. Supporting information retrieval across domains by automatically suggesting search terms that have sufficiently similar meanings in complementary domains [101]

By specifying stop words

Traditionally, the default or configurable settings of search systems have specified lists of *stop words*, or *stop lists*, comprising common, low-information words that searches should ignore. By omitting stop words that convey little semantic value from their index, search systems can filter out and prevent the analysis of these stop words before processing natural-language queries. This improves the relevance of search results, reduces the amount of data search systems must process by 35–45%, and thus, improves their data-processing performance.

However, while a stop list for a particular purpose could comprise any words and even some phrases, there are no universal rules for selecting them. Generic lists of English stop words have typically included words such as the following: *a, about, all, am, an, and, another, any, are, as, at, be, been, being, both, but, by, can, certain, did, do, does, doing, each, every, few, for, from, had, has, have, having, he, he'd, he'll, her, here, here's, hers, herself, he's, him, himself, his, how, how's, I, if, I'd, I'm, I'll, I've, in, into, is, it, its, it's, itself, let's, may, me, might, more, most, must, my, myself, no, nor, not, of, on, or, other, our, ours, ourselves, shall, she, she'd, she'll, she's, so, some, such, than, that, that's, the, their, theirs, then, them, themselves, then, there, there's, these, they, they'd, they'll, they're, they've, this, those, to, us, was, we, we'd, we'll, were, we're, we've, what, what's, when, when's, where, where's, which, while, who, whom, who's, why, why's, will, with, yet, you, you'd, you'll, your, you're, yours, yourself, yourselves,* and *you've.* As you can see, stop lists typically include many contractions.

Depending on an information space's industry domain, a search system's stop list could differ. Domain-specific stop lists can be beneficial in excluding words that are in common use in specialized domains such as medicine, science, education, sports, and politics. However, because certain words might appear on most of a given information space's pages, their inclusion could greatly reduce the relevance of search results.

Over time, search systems have generally progressed from typically using long stop lists, comprising 200–300 terms; to very brief stop lists of 7–12 terms; to having no stop list at all. The use of stop words can be problematic in cases where the user searches for phrases that include them, particularly in names.

Therefore, modern search systems that employ natural-language processing (NLP), which is an integral component of AI, consider more than just individual words. Rather than simply excluding a list of stop words when processing queries, such a search system can analyze *all* of the words in an information space's content, or its **corpus**, including those that are *not* specific to any topic. Therefore, a search system that leverages NLP might be better able to perform certain tasks such as the following:

- Clearly summarizing the text on destination pages for display in search results and, thus, making it easier for the user to distinguish between results

- Analyzing the user's sentiment, which requires comprehending the subtleties of a query's emotional tone and, therefore, must consider adjectives and negations such as *no* and *not*

- Discerning the user's intent in posing a query, which requires understanding the context of the query. For example, a search system could determine the context of a search query by analyzing such phrases as *What is…?* and *How to…?* to ascertain whether the user is seeking the definition of a term *or* information that describes how to perform a task and, thus, rank relevant results accordingly. [102, 103, 104, 105, 106, 107, 108]

By searching for similar pages

Within individual search results, some search systems provide a link to **More content like this** or a link with a similar label. When the user finds a particular search result that provides useful information, the user can click the link to obtain additional similar information. This approach assesses document similarity.

There are several ways of deriving a secondary set of similarly relevant search results from an individual search result for the user's original query. In one approach, the search algorithm converts the content on the page in question to the equivalent of a search query through the following process:

1. The algorithm excludes all stop words from the search.

2. The algorithm builds a query comprising only meaningful terms that might be representative of the page's content.

3. The algorithm searches for these terms.

As a result, the algorithm will retrieve relevant search results that are similar to the result that the user originally found useful.

Alternatively, such a link could present search results that include links to pages that either were indexed using similar metadata *or* pages that either were indexed using similar metadata or pages that the user found of interest cite or were cited by the original page. [109]

Ordering and ranking search results

"There are two common methods for listing retrieval results: sorting and ranking. … Sorting is especially helpful to users who are [making] a decision or [taking] an action. … Ranking is more useful when there is a need to understand information or learn something."— Peter Morville and Lou Rosenfeld [110]

Depending on the types of content that an information space comprises, its users' information needs, and the ways in which users prefer to use search results, the sorting *or* ranking and ordering of results on a search-results page can differ. [111] Once the search system indexes the information space's content, it assigns relevance to each page, then scores and sorts or ranks search results using various factors. [112]

Automatically sorting search results

A search algorithm could automatically sort search results by executing a default or configuration setting that would cause it to sort a results set by any specific content element or type of metadata.

The ability for users to sort search results manually can be especially helpful when the user is in the process of making a decision or preparing to take an action such as making a purchase. Conducting such a search on whatever content element or type of metadata would be appropriate in a given context could enable the user to either initiate or complete a particular task.

A search algorithm or user could sort search results in the following ways:

- **Alphabetically**—The order in which search results appear in an alphabetical list of results depends on the content element or type of metadata that provides the basis by which the search system sorts them. For example, a search system might sort a set of search results alphabetically by a page's title or author, *or* by metadata such as meta keywords, categories, or colors. Sorting search results alphabetically is especially useful for sorting people, places, or products by name. In most cases, you should omit initial articles such as *the, a,* and *an* from names to ensure that users can find things where they're most likely to look for them.

- **Chronologically**—Sorting a set of search results by date—for example, an article's or blog post's publication date or the date of a historical, recent, or upcoming event—displays the results in chronological order. On most information spaces, metadata automatically specifies the date and time when each page was created and last updated. It's usually best to display a list of search results regarding news stories or press releases in *reverse chronological order*, with the most recent results listed first, while lists of historical or upcoming events should be in chronological order.

- **Numerically**—Displaying a set of search results in numerical order can sometimes be useful. A search system might sort a set of search results by amounts of money or some other numerical data. For example, if a user is comparing products on an ecommerce app or site, the user might want to sort a list of search results by price or size. [113]

Ranking search results

"Ranking should consider more [than] just the relevance of the results. Query-independent considerations such as popularity, quality, or recency often determine which relevant results to present to searchers on the first page and in what order. The combination of query-independent signals reflects the desirability of the results."—Daniel Tunkelang [114]

Using a ranking algorithm is best when the user is conducting research, with the goal of discovering new information for the purpose of learning. A search algorithm could alternatively rank search results by leveraging a variety factors, including any of the following:

- **By relevance**—This is the most common approach to ranking and orders search results from the most to the least relevant. Whenever the user needs to find the most pertinent information on an information space to answer a given query, relevance ranking is essential. However, determining the right approach to ranking relevance is necessary to ensure that a ranking algorithm prioritizes the search results that optimally meet the user's particular needs. For more detailed information about ranking the relevance of search results, see the next section, "Relevance ranking." [115]

- **By recency**—Recency ranking considers the freshness of the information on pages in addition to ranking their relevance, giving more recent pages higher priority rather than ordering pages strictly by relevance or by date. To order search results for *recency-sensitive queries* such as those regarding breaking news, for which the user's intent can vary temporally, a search algorithm must support recency ranking. Recency ranking can consider publication and most recent modification dates and times for pages, as well as any other temporal data. For example, any page published or updated in the past 30 days might be considered fresh. The automatic identification of recency-sensitive queries requires the use of a highly precise classifier. Responding to such queries leverages a machine-learned recency-sensitive ranking model that has been trained on such queries. [116, 117, 118, 119]

- **By popularity**—Popularity ranking promotes the relevance of pages with which all users have collectively interacted most frequently. An algorithm might leverage an information space's search analytics to determine the number of times users have previously clicked the same search result for a particular query, *or* Web analytics that show the numbers of visits pages have received and the time each visitor has spent on a page. Other factors that indicate the popularity of the pages that appear in search results include the number of internal links to those pages and the popularity of pages linking to them. The assumption is that, because many users have interacted with a page, a product listing, or another item in the past, other users might also find it useful. Amazon is an example of an ecommerce app that considers popularity in ranking search results. [120, 121, 122, 123]

- **By document type**—A search system might elevate HTML pages over other file types such as PDF (Portable Document Format) documents. [124]

In most cases, assessing the desirability of search results does *not* require artificial intelligence (AI). It is possible to determine a search result's recency or popularity directly from the content itself or from search analytics. However, some search systems leverage AI in assessing the quality of search results—for example, the text's writing quality. [125]

Relevance ranking

"The more heterogeneous your documents are, the more careful you'll need to be with relevance ranking."— Peter Morville and Lou Rosenfeld [126]

Relevance is the degree to which a search system's results accurately match a user's query, intent, and expectations. The relevance of search results has a huge impact on the search user experience. An effective, efficient search system that delivers highly relevant results permits users to spend less time and effort manually refining their search queries—typing new search queries and revising those that haven't produced satisfying results. Therefore, optimizing the relevance of an internal search system's results is essential to meeting both business goals and the needs of users.

Different internal search systems could employ a great variety of relevance-ranking approaches and algorithms, depending on whether they are traditional, keyword-based search systems or modern search systems that leverage artificial intelligence (AI) in ranking search relevance. Additional factors in choosing an optimal search algorithm for an internal search system include the types of content an information space comprises and users' needs. [127, 128, 129]

Traditional factors in ranking search relevance

On information spaces with traditional, keyword-based, internal search systems, the search system parses the words in the user's search query, then the search algorithm analyzes the search index to determine which pages are relevant to the query. Key factors in ranking the relevance of search results might include the following:

- **Keyword occurrence and density**—The search algorithm determines which of an information space's pages include the search query's keywords, as well as the number of instances of each keyword on those pages.

- **Proximity of query terms**—The search algorithm might consider whether specific keywords are adjacent to one another, forming text strings that are similar to those in the user's search query; keywords appear in the same sentences or paragraphs; or keywords do *not* appear in close proximity to one another on a page.

- **Prominence of query terms**—The search algorithm might consider where the keywords appear on a page. Keywords that appear prominently within a page's title or Web address, or URL; *or* within metadata fields such as a page description, keywords, and subject categories might be more relevant than those that appear only in a page's main content.

- **Manually indexed query terms**—Manually indexing an information space's pages using its controlled vocabulary provides additional human-mediated information about each page's content by defining metadata such as a page's description, keywords, and subject categories. The search algorithm can then search these fields of metadata in retrieving matching pages. [130, 131, 132]

Relevance-ranking factors for AI-powered search

"The process of matching query terms to the content of articles, pages, documents, and product listings is critical for establishing relevance."—Elastic [133]

On an information space with a modern, AI-powered, internal search system, the search system uses AI and machine-learning (ML) models to score search results' relevance, then ranks them accordingly.

The search system might employ a variety of query-preprocessing techniques to derive maximal information from the user's search query to inform the search system's retrieval of relevant results, including the following:

- **Natural-language processing (NLP)**—Using a large language model (LLM), an AI-powered search algorithm can understand the user's natural-language query by decomposing the query into a set of terms and synonyms, then generating a structured Boolean query from them, thus expanding the search query both contextually and semantically.

- **Contextual techniques**—An AI-powered search algorithm can analyze the query terms within the context of the user's location, past search queries, and other user-specific information.

- **Semantic techniques**—An AI-powered search algorithm can deduce the user's intent for the search query by analyzing the semantic metadata that describes the categories to which the information space's pages belong. Manually tagging the information space's pages with descriptive, semantic metadata clearly distinguishes the pages from one another.

This preprocessing of search queries is essential to ensure the optimal performance of a search system and its ability to produce highly relevant search results. [134]

Key determinative factors in a ranking algorithm's ability to rank the relevance of search results could include the following:

- **Query-term occurrence and density**—The search algorithm determines which of an information space's pages comprise content that includes the search query's terms, as well as the number of instances of each query term on those pages. Content that matches the user's query terms could include articles, marketing information, profiles, product descriptions, or any other textual information that an information space indexes to support a search capability.

- **Linguistic analysis**—The need to accurately interpret natural-language phrasings, synonyms, and variants of words such as singular and plural forms, typos, and misspellings, as well as regional spelling variants and phonetic spellings, complicates the search algorithm's analysis of search-query terms. After performing a linguistic analysis, the search system can better identify relevant content based on its holistic understanding of the query and, thus, the user's intent.

- **Authoritativeness of content**—The search algorithm might assess certain content as having greater credibility and accuracy and, thus, being more authoritative and relevant.

- **Freshness of content**—For certain types of queries, the search algorithm might consider recent content to be more relevant.

- **Weighting of query terms**—The search algorithm might determine which query terms to prioritize by assigning a numerical value to each term.

- **The user's intent**—The search algorithm might analyze the semantic context of the user's search query to discern the user's goal in conducting a search. This is one of the most practical and cost-effective ways of employing AI in search—in part because it also comprehends a linguistic analysis.

- **Contextual factors**—The search algorithm might consider the context of the user's search—for example, factors such as the user's geographical location, language, device, Web-browsing history, and search history—in personalizing or localizing search results. These contextual factors are user specific and might derive from user behavior analytics (UBA) and other data analytics.

- **Engagement metrics**—The search algorithm could employ certain metrics in assessing users' engagement and satisfaction with particular search results, which can be highly indicative of their relevance. A high *clickthrough rate (CTR)*, the percentage of users who click a search result once they've viewed it, indicates high relevance. A search system can also leverage users' clicks and conversions to improve the relevance of search results. *Dwell time*, the amount of time the user spends on the destination page that appears after clicking a link in the search results, is another metric that can show relevance. Plus, search analytics provide insights into how frequently users search, the types of information for which they search, and whether they are successful in finding the information they need.

- **Business goals**—A search system must address an organization's business goals for both the information space and its search system. Businesses can refine their ranking algorithms to address specific business needs.

Consistently delivering highly relevant search results in response to users' queries is essential to keeping them engaged with an information space over longer periods of time, achieving higher conversation rates, and thus, increasing revenues. [135, 136, 137]

Improving the relevance of search results

Over time, administrators implementing AI-powered, ML-based search algorithms can refine and improve the relevance of their search results by taking the following approaches:

- **Broadening the scope of search results**—Defining related terms and synonyms in a controlled vocabulary expands query comprehension and, thus, broadens the scope of results, as does supporting partial matches of search queries. Plus, automatic recognition of the language in users' queries enables the system to autosuggest queries that are similar to previous search queries.

- **Improving tolerance of spelling errors**—Implementing spelling checking and autocorrection ensures that users get useful results despite their queries including spelling errors or typos and reduces users' frustration by making it unnecessary for them to continually refine their queries. Other approaches to handling spelling errors include automatically suggesting similar queries or providing a *Did you mean?* feature when correcting search queries.

- **Implementing facets, or smart filters**—On information spaces that provide large volumes of content such as ecommerce sites and apps, filters let users narrow the scope of their search queries. Depending on the types of content that an information space comprises, useful filters might include category, topic, date, location, price, color, or size.

- **Personalizing search results**—Personalization of search results according to the user's preferences, location, browsing history, search history, purchase history, or other behaviors, then ranking the personalized results as most relevant enhances the overall relevance of search results. Personalization is also a key factor in predictive search, which anticipates users' future needs based on past searches. Plus, personalization can be a key factor in displaying relevant autosuggestions.

- **Prioritizing high-value content**—By weighting search results from higher-quality parts of an information space as more relevant, you can ensure that the content that would be most valuable to users appears near the top of search-results pages. The analytics for high-quality pages indicate that they have strong engagement, high authority, many backlinks, and high conversion rates.

- **Leveraging user research and feedback**—Conducting user research on the search experience and gathering users' feedback on the relevance of their search results enables teams to learn about users' needs and how well the search system is satisfying them. User feedback is the best indicator of how relevant the search results actually are. The product team can then provide inputs to the system to help it continuously learn and improve.

- **Optimizing the user experience**—In addition to refining the overall user experience of the information space, optimize the search and navigation experiences. Ensure that the search bar is easy to find and search-results pages are easy to use. Refine the information architecture to enable the search system to crawl and index the information space's content more efficiently and effectively. Clean up the information space by removing or updating any outdated pages and eliminating redundant content, unused keywords or categories, and broken links. [138]

Optimizing the search experience

"Search begins with an information need that is articulated in some form of query. … [In some cases,] search is simply looking up a fact or finding a particular document. But search is also a journey. It's an ongoing exploration where what we find along the way changes what we seek. It's a journey that can extend beyond a single episode … and be conducted on all manner of devices."—Tony Russell-Rose and Tyler Tate [139]

To design an optimal search experience for the people who will use an information space's internal search system, consider the following factors:

- **What are the needs of the information space's users?** What information-seeking goals would users have when searching? Would they be searching to satisfy their personal needs or to complete tasks at work? Would they need to find a particular document or obtain a single correct answer to a specific question or want to seek in-depth information about a particular field of research? Should search results be brief or very detailed? What information would the search results need to provide to help users decide which results are relevant to their information needs and, thus, would be worthwhile to explore?

- **What level of expertise is characteristic of the users?** Are they novice or experienced searchers? Would they be likely to type brief queries comprising just a few words or are they motivated searchers who would specify detailed queries to ensure that they yield useful results? Would users appreciate the simplicity of specifying their queries using natural language, as most do, or would they be willing to create complex queries using Boolean operators or by filling out advanced-search forms? Would users be likely to refine their queries if they didn't initially get the results they need?

- **What is the context within which users would search?** What factors might influence the user's experience when using the information space's search system? What impacts would the user's location and immediate environment have on the user, if any? Would users likely be searching at work, within the context of their organization? Would they be searching within the comfort of their own home? Or would they be using their mobile device to search for what they need out in the world? Would other people participate in or direct their searches?

- **What types of content does the information space comprise?** Does the information space's content include both navigation pages and destination pages? Does the content primarily consist of pages of unstructured text in HTML, documents in particular formats such as PDF documents or spreadsheets, images, or video or audio recordings? Does well-defined metadata describe all types of content? Is the content dynamic or static? How frequently does the search system's index get updated?

- **What is the scope of the information space's content?** Would the users' searches be more likely to retrieve only a few search results or many pages or screenfuls of useful results? How successful would the search system be in delivering the most relevant results first? How well would the search user interface support users in quickly scanning the results to identify those that would best meet their needs?

- **What assumptions would users make when conducting a search?** Would users assume that they should type keywords into the search box, formulate a natural-language query that describes the information they're seeking, or pose a specific question? Would users assume that they need not create complex queries using Boolean operators, but could optionally include them? Once users have clicked the **Search** button, would they assume that the search system would search the entire information space, then display a list of relevant search results, or simply provide a useful answer to their question in natural language? [140, 141]

Search user-interface design patterns

"The patterns of search … have emerged as repeatable solutions to common problems. … Search tasks of any complexity require an iterative approach, involving the creation and reformulation of queries. … What we find along the way can change what we seek."—Peter Morville and Jeffery Callender [142]

Let's look at some user-interface design patterns for the common components of a traditional search system. While there are many variants of the functionality and user-interface designs for internal search systems, a traditional search experience typically comprises the following steps and the user-interface design patterns that support them:

1. Expressing and submitting a search query

2. Displaying search results

3. Optionally refining the search query

The search bar

A search bar typically consists of a search box and a **Search** button, which enable the user to express and submit a search query. Consistently place the search bar in a prominent location near the top of all pages on an information space, in proximity to the main navigation system—especially if users would likely be seeking known items. The visual design for a search box and **Search** button varies across information spaces, according to their design standards. Once the user conducts a search, the search bar should remain at the top of the search-results page. Figure 13.1 shows the search bar on the Etsy Web site.

Figure 13.1—Etsy search bar

The visual design of the search bar and its placement on an information space's pages can strongly influence the user's decision whether to search or browse. The more prominent the search bar, the more likely the user is to search. Depending on the search bar's prominence, users might interpret the design of the search bar as guiding them to either search or browse.

Designing user interfaces for mobile devices has resulted in several design patterns for displaying the search bar on mobile devices, in order from the most to the least prominent placement, as follows:

- **Displaying the search bar by default**—The most prominent placement of the search bar consistently displays the bar near the top of the page, as on Amazon.com, which is shown in Figure 13.2.

- **Displaying a Search icon**—This relatively prominent placement requires the user to click the **Search** icon to display the search bar, which should appear immediately beneath the icon, as on the *UXmatters* mobile Web site, as Figure 13.3 shows. Apple.com employs a variant of this pattern that displays a Search page comprising *only* a **Quick Links** list rather than a search bar, as Figure 13.4 shows, clearly pushing users to browse rather than search.

- **Displaying a Menu icon or button**—This least-prominent initial access point to an information space's search functionality requires the user to click the **Menu** icon or button to display one of several progressively less prominent variants of search: immediately displaying the search bar at the top of the navigation menu, as on *The New York Times*, as shown in Figure 13.5; or displaying a **Search** link or icon on the navigation menu that the user must click to display the search bar. While the **Search** link or icon is typically at the top of the navigation menu, on the *Baymard Institute* Web site, it is at the very bottom, as Figure 13.6 shows, making it much less likely that users would search the site. Regardless of the **Search** link or icon's placement on the navigation menu, once the search bar appears, it should be at the top of the page, as on the *Baymard* site.

The prominence of an information space's search functionality should align with the user's needs and reflect its strategic importance to a business. [143]

Figure 13.2—Amazon.com search bar on mobile

Figure 13.3—*UXmatters* search bar on mobile

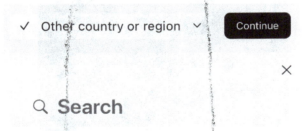

Figure 13.4—Search on Apple.com, on mobile

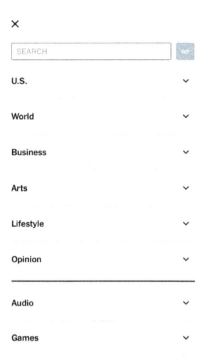

Figure 13.5—Search on *The New York Times* mobile navigation menu

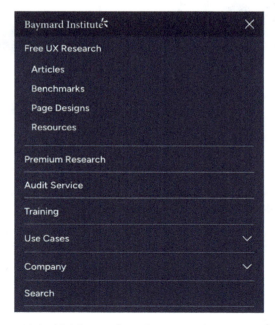

Figure 13.6—Mobile search on the *Baymard Institute* Web site

The search box

The search box lets the user type a search query and should be sufficiently wide to display most queries in their entirety. Longer queries are becoming increasingly common as searchers become more sophisticated. [144] According to *Statista*, in January 2020, about 4% of search queries comprised five words; 10%, four words; and 23%, three words. But, as has always been common, 40% of search queries are still two words in length and 20%, only one word. [145] By providing more space in which to display search queries, you can encourage users to type longer queries and, thus, potentially obtain more relevant search results. [146]

By default, an insertion point should be selected in the search box, enabling the user to immediately begin typing a search query, and the box should contain placeholder text that either indicates its purpose—for example, *Search for…*—or provides guidance to help the user formulate a meaningful query. The placeholder text should be in italics to ensure that the user does not mistake it for a value that the user has previously typed in the box or a prefilled value.

Once the user begins typing a search query, remove the placeholder text from the search box and display a **Clear** button at the far right of the box. Clicking the **Clear** button should remove any text the user has typed from the search box.

According to Forrester Research, 43% of users immediately type a query into a Web site's search box rather than using its navigation system to find what they're seeking, and searchers are up to three times more likely to convert than navigators. [147]

Note—Searchers sometimes overconstrain their searches by typing too many keywords in the search box, preventing them from finding what they need and potentially causing the search system to find no results at all. [148]

The Search button

A **Search** button should always reside immediately to the right of the search box, where users expect it to be even when space is tight on mobile devices. The button's proximity to the search box is especially important in cases where the user needs to search iteratively by revising a search query to obtain more useful results. To ensure maximal clarity, the **Search** button should be labeled **Search** or contain a **Search** icon, whose standard image represents a magnifying glass. If the latter is the case, make sure that the button is a sufficiently large tap target. The lack of a **Search** button in the user interface could hinder the successful completion of a search and potentially introduce errors such as the user's inadvertently clearing the search box.

Clicking the **Search** button or pressing **Return** or **Enter** submits the search query that the user has typed and displays a search-results page. To facilitate searching on a mobile device, the label for the corresponding button on its alphanumeric, touch keyboard should also be **Search**, *not* **Return**. However, users would most likely look for a **Search** button in the information space's user interface rather than the mobile operating system's keyboard. [149, 150]

Autocomplete

"Embrace type-ahead suggestions, where a list of potential query options can anticipate what a person would likely be looking for. This is not only helpful, particularly on a mobile device, it also reduces how hard a shopper has to think. Offering options taps into the human brain's ability to recognize a term, if offered to them, rather than have to think up the correct combination of words, which may be industry terms they are unfamiliar with."—Andrea Polonioli [151]

Autocomplete can provide up to about ten suggestions for completing the search query that the user is typing into the search box or six on a mobile device, displaying them in a drop-down list that automatically appears immediately beneath the search box. The user might find a list comprising too many suggestions overwhelming and either ignore them or spend an unwarranted amount of time considering the suggestions. Plus, a mobile device's touch keyboard could obscure an overly long list. As the user types additional characters into the search box, the list of suggestions continually changes. In each query suggestion, use color or text formatting such as bold or italics to highlight the suggested text, differentiating it from the text that the user has already typed, making the items in the list easier to scan. Selecting a suggested search query in the list highlights the selection, closes the list, and initiates a search. Alternatively, the user could simply finish typing the query, then click the **Search** button or press **Return** or **Enter** to initiate the search. Figure 13.7 shows autocomplete on Amazon.

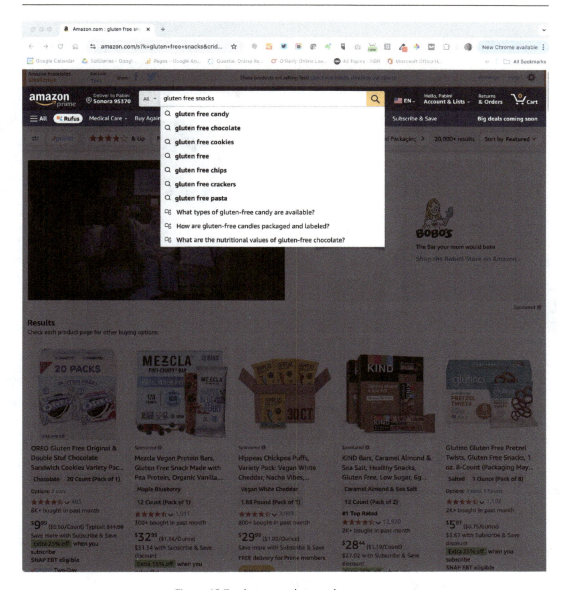

Figure 13.7—Autocomplete on Amazon.com

An autocomplete list can optionally include *scoped suggestions*—that is, groups of query suggestions that have category labels, as shown in Figure 13.8. Always use distinctive formatting to emphasize the category labels and make them easier to distinguish from query suggestions that are *not* scoped—for example, by indenting the query suggestions under the category labels and rendering the category labels in bold and perhaps in a different color.

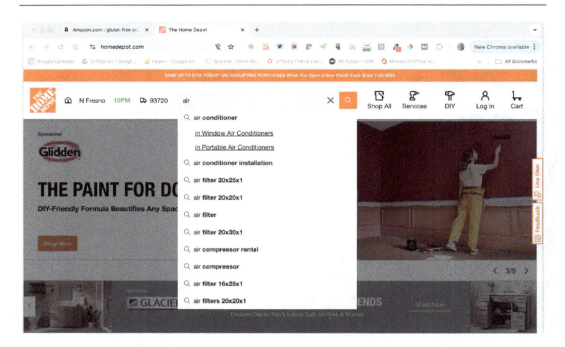

Figure 13.8—Autocomplete with scoped suggestions on The Home Depot Web site

Autocomplete, a form of predictive search, makes expressing queries more efficient by reducing the amount of typing the user must do—which is especially important for users on mobile devices on which typing is more difficult—*and* more accurate by preventing spelling errors and presenting specialized terms that are unfamiliar to the user. Recognizing terms is much easier for users than recalling them from memory. Using the correct terms ensures that the user can find useful results. Autocomplete searches a controlled vocabulary for character strings that match a partial query, making it particularly helpful for simple information-seeking tasks such as known-item searches. [152, 153, 154]

Autosuggest

"Search suggestions are … a popular way to help people overcome their limited generative abilities by showing a drop-down of fully formed potential queries as soon as users type in a few characters that hint at their needs. Although helpful, search suggestions can also be limiting; … if something isn't included in the search suggestions, users might never bother to search for it."—Jakob Nielsen [155]

Autosuggest provides up to about ten query recommendations or six on a mobile device, displaying them in a drop-down list that automatically appears immediately beneath the search box. The list of query recommendations continually changes as the user types characters into the search box. The query recommendations suggest popular queries that might include the user's text anywhere within the query and often include queries from the user's recent searches. Highlight the user's query text in each query recommendation, using color or text formatting such as bold or italics to differentiate the user's query text from the suggested query text and making the items in the list easier to scan. Selecting a recommended search query in the list highlights the user's selection, closes the list, and initiates a search. The user could alternatively finish typing the query, then click the **Search** button or press **Return** or **Enter** to initiate the search. Figure 13.9 shows a form of autocomplete on the Eileen Fisher Web site, which interactively displays the suggestions directly on the search-results page.

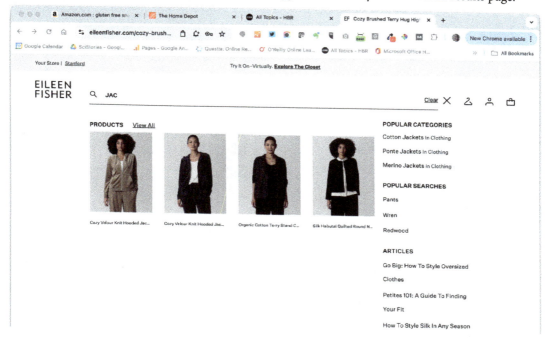

Figure 13.9—Autosuggest on Eileen Fisher

Autosuggest provides many of the same benefits of efficiency, usability, and accuracy as autocomplete, by minimizing the amount of typing the user must do, preventing spelling errors, and presenting specialized terms with which the searcher might be unfamiliar. Autosuggest could search the index for related keywords and phrases that do not precisely match the user's query or leverage users' query-reformulation data. In the context of complex information-seeking or exploratory-search tasks, offering recommendations that guide users to popular alternatives might inspire them to pivot and search for related information they might not otherwise have discovered. [156, 157, 158]

Combining autosuggest with expanded content recommendations

Autosuggest features sometimes include more than just query recommendations. Best Buy, which is shown in Figure 13.10, displays a typical autosuggest list in the column on the left, along with images depicting individual products in the selected catgory of products on the the right.

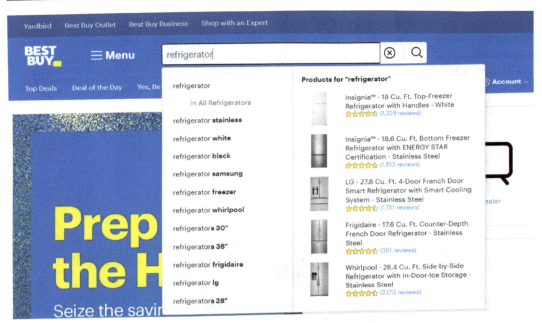

Figure 13.10—Autosuggest menu on Best Buy

Rather than just displaying query recommendations, as in a simple autosuggest list, an autosuggest megamenu might provide links to a variety of related information. While some related links could derive from the query the user is typing—such as related products or product-category pages—others might represent marketing promotions. As for any other megamenus, make sure that they are well-structured. Provide headings that convey their structure and make them easy to scan. Megamenus can consume considerable screen real estate and obscure the content on a page. So, if you display an autosuggest megamenu on the desktop, provide a simple autosuggest drop-down list on mobile devices.

According to NN/g, the expanded content recommendations in an autosuggest megamenu often combine text and image links and might include the following:

- Popular or trending searches or items for which users have frequently searched

- The user's recent searches

- Featured search results

- Links to related products or product-category pages

- Links to articles and other related content

However, users rarely take advantage of such expanded content recommendations. During an NN/g study of ecommerce sites, users encountered such links 60 times, but clicked them only seven times, regardless of the specificity of the user's search. Expanded content recommendations are problematic for the following reasons:

- **Their use doesn't align with the search mindset.** When users decide to search, they usually have a specific query in mind, want to find what they need as quickly as possible, and don't want to explore irrelevant information. They might even not notice links that do not relate directly to their query or, if they do, might find them distracting and become annoyed.

- **They load too slowly, especially when they include images.** Often, users submit their search query before the expanded content recommendations even appear. Also, since the user is continually modifying the query, if the content recommendations do finally appear, they might not match the user's ultimate query.

- **The user's attention is focused elsewhere.** When conducting a search, the user focuses on the search user interface, *not* on extraneous content that is competing for the user's attention. Users often ignore product images—even those that depict precisely the product they want—assuming they might be part of promotional content.

- **Even frequent users won't learn to use them over time.** Even after conducting multiple searches, users still don't notice and don't use expanded content recommendations.

Because of the many inherent problems of expanded content recommendations, you might decide that the investment necessary to implement them properly would be unwarranted. However, if you do implement expanded content recommendations, observe the following design guidelines to ensure that they are helpful to users who are conducting searches rather than unnecessarily imposing an additional cognitive load on them:

- **Always feature simple text autosuggestions prominently.** Users would be most likely to click them even when a megamenu also includes expanded content recommendations.

- **Limit the number of images, as well as instances of dynamic content.** Displaying too many images results in slower loading times. Plus, the images could compete for the user's attention. Displaying dynamic content that can change depending on user interactions such as inputs or selections requires additional processing that can create lags in loading times.

- **Provide clear labels for groups of expanded content recommendations.** Such groups might include **Recent Searches**, **Popular Searches**, **Trending Searches**, **Recently Viewed Products**, **Recommended Products**, **Top Products**, or **Categories**. The labels should make it clear to users why you're displaying each group.

- **Use consistent locations for expanded content recommendations.** Displaying specific types of expanded content recommendations in consistent locations within the megamenu rather than varying their location depending on the user's specific query makes it easier for users to find what they need. [159, 160]

Constraining the scope of searches

"Scoped search allows users to limit their search to a section or type of content … instead of searching everything…. [While] restricting search to a specific area of a Web site can provide better results, faster, … users [often] overlook, misunderstand, and forget about the search scope."—Katie Sherwin [161]

Some information spaces that organize their content into categories allow users to narrow the scope of a search to a specific category before conducting the search. Typically, a drop-down list of categories resides to the immediate left of the search box from which the user can select a category. An **All Categories** option should be selected by default because users expect that a search should comprehend an entire information space unless they've specified otherwise. Never select a specific category by default. When the user selects a category to constrain the scope of a search, clearly display the selected category in the list box. The user can then type a query in the search box and click the **Search** button or press **Return** or **Enter** to initiate the search.

eBay's product taxonomy comprises categories that appear in a drop-down list to the right of its search box. The **All Categories** option is selected by default. By selecting a category in this list, users can constrain the scope of a search to a specific product category. Figure 13.11 shows eBay's scoped-search feature, in which the drop-down list of categories appears to the right of the search box.

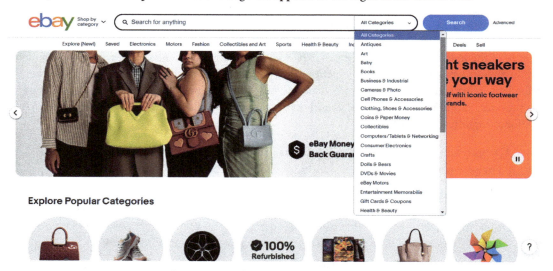

Figure 13.11—Scoped search on eBay.com

Narrowing the scope of a search to a specific category or section at the beginning of the search process accelerates the query-refinement process and, thus, might help users find what they're seeking more quickly. The drop-down list that provides this functionality might be labeled **Categories** or **Sections** or have no label. A scoped search also lets the search system tailor its query-refinement options to the selected category or section. But most users wouldn't want to bother constraining a search until they've already conducted a search and found the results to be unsatisfactory. Therefore, they are more likely to want to narrow a search as part of their query-refinement process.

However, if the user fails to realize that a search is constrained to a specific category or section, chaos and frustration ensue. The user won't be able to find the desired information and might assume that the information space simply doesn't provide it. So prominently display the category or section that the user has selected to limit the scope of a search. At the top of the search-results page, clearly state any constraint on the scope of the search—for example, *Searching only books. Do you want to search all categories? Select All Categories.*

When users initiate a new search, be sure to clear any category they've selected when conducting a previous search. If a category inadvertently remained selected, users might overconstrain their searches, which might then produce no results.

Providing scoped search can be beneficial in applications for which users must have domain expertise—especially if users would be conducting known-item searches. Scoped search offers users the benefit of being able to specify contextual queries. However, users who lack domain knowledge might not know what category they should select. Selecting the wrong category could overconstrain a search, making it more likely that it would produce no results.

Other permutations of scoped search include the following:

- **Scoped search in autocomplete and autosuggest**—If an information space comprises categories or sections *and* the search box includes an autocomplete or autosuggest feature, you can include scope suggestions within the autocomplete or autosuggest drop-down list beneath the search box. Apply distinctive text formatting to emphasize the user's search term, rendering it in bold or a color, and use plain text for any suggested text. To make it easier to distinguish scoped query suggestions from suggestions that are *not* scoped, use the following text formatting—light gray text in a smaller font size—to indicate the scope:

[Search term] in [Category or Section]

- **Providing scoped search for specific sections**—Avoid displaying multiple search boxes representing different search scopes across all the pages of an information space, which can be extremely confusing to users. Instead, limit scoped searches to specific sections of an information space that provide a specific type of data—for example, on an intranet, an employee directory, or an organization's address book—and allow users to conduct searches within that specific section using a search bar just for that section. When providing a search box for a scoped search, be sure to indicate the scope of the search by clearly labeling the search box—for example, **Search the directory**. Amazon provides scoped search for its *Your Orders* pages, allowing the user to search only her orders, as shown in Figure 13.12. [162, 163, 164, 165]

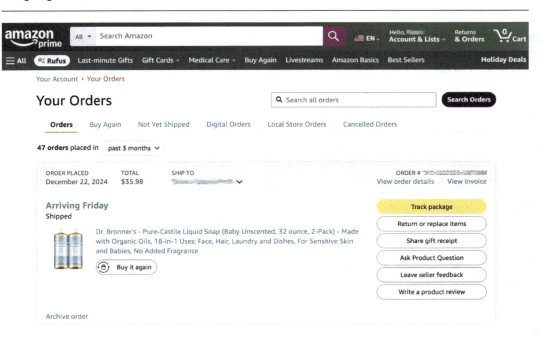

Figure 13.12—Amazon's scoped search for the user's previous orders.

Saving search queries

Users might occasionally want to save fruitful search queries, especially in domains for which the available data is likely to update frequently over time. An information space could optionally support manually saving search queries by providing a **Save Query** button in proximity to the search bar. [166]

Presenting search results

"There are two main issues to consider: which content components to display for each retrieved document and how to list … those results. … Display less information to users who know what they're looking for and more information to users who aren't sure what they want."—Peter Morville and Lou Rosenfeld [167]

In response to the user's search query, the search system generates and displays a set of search results, typically displaying them on multiple search-results pages. Determining what content elements to include in each result and how best to display the list of search results depends on the types of content an information space comprises and the needs of the user. [168]

The content elements search results comprise

"Show users who are clear on what they're looking for only representational content components, such as a title or author, to help them quickly distinguish the result they're seeking. Users who aren't as certain of what they're looking for will benefit from descriptive content components, such as a summary, ... to get a sense of what their search results are about."—Peter Morville and Lou Rosenfeld [169]

The essential content elements that search results should comprise must enable users to readily distinguish specific search results from one another *and* identify which results are relevant to them. Search results typically include the following content elements:

- **Thumbnail image**—Depending on the type of content an information space comprises and whether it features photos or videos, displaying a thumbnail image to the left of the textual information on the desktop—or above the text on mobile devices or in a gallery view of the search results on the desktop—offers stronger information scent and better differentiates search results from one another. Clicking the photo or video should display the linked page.

- **Title of the page**—A meaningful page title is the most salient element of a search result, particularly for users who know what information they're seeking. Place the title prominently at the top of the search result, left aligned, and emphasize it by rendering it in a larger font or in bold. The page title should be a link to specific content that is relevant to the user's search. According to Web conventions, the link text should be in blue by default; then once the user has visited the linked page, in purple, indicating that the user has already viewed the page. However, many information spaces no longer follow this convention, preferring to use their own branding colors to visually distinguish links from visited links that the user has already traversed.

- **Web address**—Optionally display the Uniform Resource Locator (URL) for the page to which the search result links. The Web address should also be a link. This element provides information about the source and topic of the search result for a more technically savvy audience.

- **Publication date**—This date is typically a run-in head preceding the content summary.

- **Content snippet**—This brief summary, description, or excerpt from the content on the linked page is a key element of each search result, especially for users who are uncertain about what information they need. Providing a meaningful snippet as part of each search result offers the user stronger information scent, effectively conveys what the search result is about, and better differentiates search results from one another. Within each snippet, highlight any text strings that appear in the user's query in bold.

- **Document type**—If a search result would not display a Web page, indicate the type of document it would display—for example, a PDF document or spreadsheet.

Additional content elements that users might find useful include the following:

- **Author**—Optionally providing the name of a page's author or authors can be helpful if the user is either seeking content by a particular author or values knowing that a page is from an authoritative source.

- **Key content**—Structured content elements are often available for *all* search results of a given type—for example, an organization's or person's address or phone number or a product's price. Including such content within the search results often obviates the user's need to navigate to linked pages of content.

Ecommerce search results typically include the following content elements:

- **Product photo**—Show a thumbnail image of the product to the left of the textual product information on the desktop—or above the text on mobile devices or in a gallery view of the search results on the desktop—to provide stronger information scent and better differentiate search results from one another. The photo should be a link to a product-details page.

- **Product name**—Display the product name prominently at the top of the search result, left aligned, and emphasize it by rendering it in a larger font or in bold. The product name should be a link to a product-details page.

- **Rating**—Provide a user-generated rating of the product.

- **Price**—Display the product's price.

- **Delivery data**—Display the product's delivery information, as well as its availability.

- **Other product-specific information**—Product descriptions include certain information that pertains only to a specific type of product—for example, a book's author; for clothing, colors and sizes; or for furniture, colors.

- **Add to Cart** button—For users who know exactly what they want, this button lets users purchase the product without navigating to the product-details page.

However, including too much information about each result would limit the number of search results that would be visible on a search-results page without scrolling, increasing the potential for the user to miss seeing some valuable results. [170, 171]

Search-results pages

"The true challenges of search are in understanding why people search in the first place, how do use the results, what types of results to show, what information to include in them, and how to handle each possible type of search outcome."—Peter Morville and Lou Rosenfeld [173]

Searching is an iterative process during which users might click various search results in an attempt to fulfill their information need. Thus, a search-results page presents a list of search results that collectively provide a response to the user's search query. [174] Search-results pages must convey both the overall scope of the results set and present the individual search results in the set at the appropriate level of detail to avoid users' pogosticking as much as possible. [175] Figure 13.13 shows an example of a traditional search-results page.

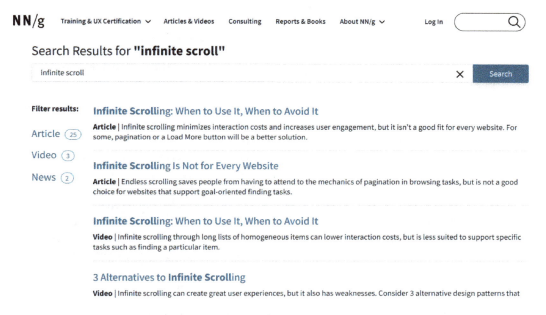

Figure 13.13—A search-results page on the *NN/g* Web site

The elements of a search-results page

A search-results page typically includes the following interactive features on a search bar:

- **Search bar**—This feature should reside at the top of the search-results page, display the user's search query in the search box, and let the user refine the query manually, if necessary, as described in the section "The search bar," earlier in this chapter; as well as the sections "Displaying the search query on results pages" and "Refining search queries on results pages," later in this chapter. The search box might incorporate an autocomplete or autosuggest capability, as described in the sections "Autocomplete" and "Autosuggest," respectively, earlier in this chapter.

- **Automated query–refinement features**—These include a spelling checker and autocorrection—as described in the section "By using a spelling checker," earlier in this chapter—as well as a *Did you mean?* feature, which is described in the section "Using *Did you mean?* to automatically refine search results," later in this chapter.

- **Scope drop-down list**—On an information space that organizes its content into categories or comprises sections *and* supports scoped searches, this list would reside on the search bar and might be labeled **Categories** or **Sections** or have no label at all. Selecting an item in the list would define the scope of the user's search. The section "Constraining the scope of searches," earlier in this chapter, describes this feature.

- **Save Query button**—This optional button would let users save their search queries, as described in the section "Saving search queries," earlier in this chapter.

A search-results page typically comprises the following informational elements:

- **Page title**—The title **Search Results** should reside at the top of the main content area of the search-results page.

- **Number of results**—This text should reside at the top of the main content area of the search-results page, immediately above the search results list. It shows the number of results that currently appear on the search-results page and the approximate number of results the search system has found for the user's query—for example, *Results [1–10] of about [###,###,###] results.*

    Note—Values within brackets indicate variables.

- **Search results list**—Depending on the concision and compactness of the search results, this list might comprise ten to around 25 individual search results per page. Highlight the user's query terms in the list of search results. For details about the content elements in each list item, refer to the section "The content elements search results comprise," earlier in this chapter.

Interactive features for navigating search results include the following:

- **Pagination controls**—If a query produces multiple pages of search results, provide pagination controls. These can take different forms, depending on the interaction model for pagination, as described in the section "Navigating search results," later in this chapter.

- **Filtering controls**—If the ability to filter search results would be useful to users and appropriate for the information space's content, provide a means of filtering the search results, as described in the sections "By filtering the search results" and "Filtering search results," later in this chapter.

- **Sorting controls**—If users might want to sort search results and the search results share one or more characteristics by which they are capable of being sorted—such as sorting by date, which would be useful for sorting news—provide a means of sorting the search results, as described in the section "Automatically sorting search results," earlier in this chapter, and the section "Sorting search results," later in this chapter. [176, 177, 178, 179]

Displaying the search query on results pages

"Working memory is very limited. ... People viewing search results often do not remember the search terms they just typed. ... Show the search terms that generated the results..., reducing the burden on users' working memory."—Jeff Johnson [180]

Displaying the user's initial search query in the search box at the top of all search-results pages provides a reminder of exactly what the user searched for. However, the way in which the search system has interpreted the query could differ from the query that the user typed—for example, if it has automatically corrected the query's spelling or has applied a *Did you mean?* suggestion. [181]

Refining the search query on results pages

If the user's search query returns either no results or too many results that do not adequately meet the user's specific information need, the user might want to refine or completely revise the query. Displaying a search bar at the top of all search-results pages with the user's initial search query in the search box facilitates the user's being able to refine or revise the search query, if necessary, rather than needing to retype it. Although it's sometimes better for the user to just start from scratch and type a new query.

If the user has conducted a scoped search, be sure to display the category drop-down list box in proximity to the search box at the top of all search-results pages, showing the selected category in the list box. The search user interface on the results page should clearly indicate any settings that constrained the user's original search such as filters, as well as any sorting controls. [182]

By using *Did you mean?* to automatically refine the search results

A search algorithm can automatically check the spelling of the words in a query. Then, if the user's intent remains at all unclear, offer an alternative spelling in the form of a *Did you mean?* suggestion above the results set for a search that used the alternative spelling. Use *Did you mean?* if an alternative spelling of a brief query would return a more useful results set.

To determine whether to simply correct a misspelling or offer a *Did you mean?* suggestion, follow this rule: If the search produced no results, automatically correct the misspelling. If the search produced fewer than 10 results, offer a *Did you mean?* suggestion. [183]

By filtering the search results

Consider providing one or more filters below the search box, enabling the user to narrow the search results to a specific category or search zone that has actually produced results. Such filters both clearly indicate what content the user searched and enable the user to broaden or narrow the scope of any revised search. For example, you might display the filter's options in a **Limit results search to** drop-down list box that comprises individual categories or zones.

In cases where the user might need to select multiple categories, either display an array of checkboxes below the search box *or* a drop-down list box with a checkbox for each option. Figure 13.14 provides an example of such a search-results page, which includes a **You searched for** box and two filters: a **Content type** drop-down list and a **Choose your priority** drop-down list. [184, 185]

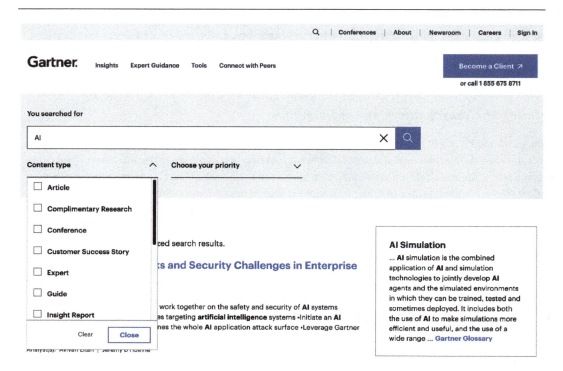

Figure 13.14—Gartner's search-refinement user interface

By returning partial matches to avoid getting no results

In cases where the user has overconstrained a search by typing too many keywords and, thus, found no useful results, rather than simply autocorrecting the query or using *Did you mean?* to automatically improve the results, consider the keywords in the query individually and return search results that match some of them.

In lieu of showing the number of search results at the top of the main content area on a search-results page, display a message that communicates the following: *Your search for "[the user's search query]" found no results. Either try searching again using fewer keywords or check out some of the following partial matches.*

Then display one or more variations of the query that do provide results. Create a group for each such variation of the query, comprising the following:

- **Group label**—This is a variation of the query with the keywords that the system did *not* use in conducting the search stricken—for example, ~~1961~~ *Gibson Les Paul guitar*.

- **A link to all the matching results**—Following the group label, provide a link to all the search results that match the partial query—for example, **View all [###] results**.

- **Partially matching search results**—Provide a list of search results that partially match this variation of the query, listing those that match more keywords first—that is, using *quorum-level ranking*. Display a diverse selection of results, avoiding any duplication.

Thus, the search system can guide the user in reformulating the original query in a way that produces potentially useful results. [186]

Displaying search results

Most search queries produce too many search results to display on a single page. To some extent, the scope of the overall results set determines the number of search-results pages or screenfuls of results across which the search results should appear.

Once the user conducts a search and a search-results page appears, the initial set of search results loads. The user can scan those results and click any links that are of interest to view their destination pages. If the user's information need has not yet been fully satisfied, the user must take an action to view additional search results. A particular information space might support any of the following ways of doing this:

- **Using pagination controls to view more search results**—The user can click a **Next** button or a link representing a range of search results to view another search-results page that displays additional search results.

- **Incrementally loading more search results**—The user can click a **More Results** button to load additional search results at the bottom of the same search-results page.

- **On-demand loading of more search results**—The user can scroll down to the bottom of the same search-results page to automatically load additional search results there. With infinite scrolling, the search-results page continues to load additional search results each time the user scrolls to the bottom of the page.

We'll look at each of these approaches in detail in the next sections.

Implementing any of these approaches requires determining how many additional results to display at one time. The right number of search results to display at once depends on what content elements and overall volume of information each search result comprises, as well as the screen resolution and connectivity of the device on which the user is viewing the search results. Overly detailed search results take up too much valuable screen real estate, while search results that lack sufficient detail might omit essential information that users need to determine what results would satisfy their information need.

To show a results set's overall scope, you could optionally display the total number of search results that the search system has retrieved, along with the approximate range of search results that the user is currently viewing—for example, *Viewing results [1–10] of about [###,###,###] results.*

The order in which the search results should optimally appear on a search-results page depends on the users' information-seeking needs and goals in searching for information. Users expect the search results that are most relevant to their search query to appear among the first few results on a search-results page. [187, 188]

By paginating search results

When a results set comprises more results than can fit on a single search-results page, paginating the search results lets the user view them across several search-results pages.

When using paginated search results, the user must click a **Next** button or numbered pagination link to navigate to a new search-results page on which additional search results appear. This requires a disruptive page load each time the user wants to view more search results, so displaying paginated search results can be quite a laborious process for the user. Plus, paginated pages provide no information scent.

When designing pagination controls, provide a **Previous** and a **Next** button at the bottom of each search-results page, as well as numbered links to the other search-results pages. For fairly limited results sets, it might also be worthwhile to provide **First** and **Last** buttons. Ensure that the numbered pagination links are sufficiently large touch targets. Highlight the numbered pagination link corresponding to the current search-results page. When the user is currently viewing the first page of search results, the **Previous** button and any **First** button should appear dimmed and be unavailable—that is, not be a link. When the user is currently viewing the last page of results, the **Next** button and any **Last** button should appear dimmed and not be a link.

Nevertheless, the pagination of search results provides several benefits, as follows:

- Pagination controls make a results set easy to navigate.

- Users can view and return to a specific page of search results.

- Chunking search results minimizes page-loading times for search-results pages.

- Displaying numbered links representing the search-results pages that the user has already viewed, as well as those that remain to be viewed, indicates the level of effort that viewing all of the remaining search-results pages would require. However, this is true *only* if the number of search-results pages is sufficiently limited that links to all of them would fit within the pagination control, which is often not the case. When there are too many search-results pages, the user can't view links to all of them at once. Figure 13.16 shows how Amazon handles this by inserting an ellipsis to represent missing numbered links. [190, 191, 192, 193]

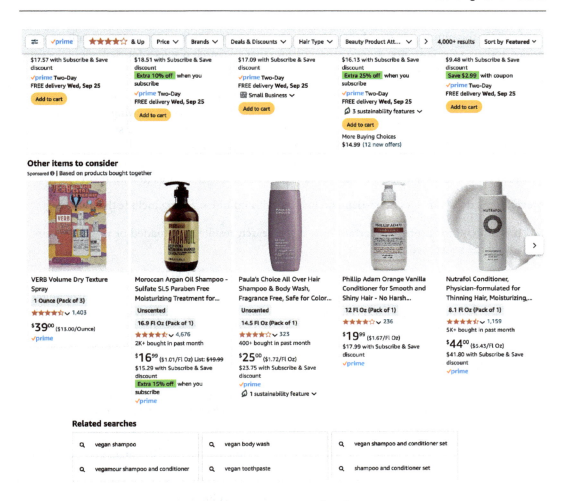

Figure 13.15—Pagination control on Amazon.com

By incrementally loading more search results

Various organizations have been experimenting with different ways of incrementally loading additional search results onto a search-results page.

The most usable way of incrementally loading more search results is providing a **More Results** button at the bottom of the search-results page. The user can click the **More Results** button to quickly load additional search results on demand at the bottom of the same search-results page—typically, ten more results at a time. Appending the search results to the same search-results page maintains the user's context, so there is no disorientation or confusion about the user's current place on the page. Plus, the user can easily scroll back up to view the earlier search results. Rendering links that the user has previously visited in purple also aids the user in keeping track of the search results that have already been scanned or visited. [194, 195]

To make clear the number of search results that are currently visible on the search-results page, consider updating the indicator showing the current range of search results—for example, *Viewing results [1–50] of about [###,###] results.*

While on-demand loading of search results with infinite scrolling provides a seamless experience for savvy users, this approach is better suited to less goal-oriented tasks such as viewing a social-media feed rather than seeking specific content. On-demand loading initially displays a specified number of search results on the search-results page, then additional search results dynamically load onto the page each time the user scrolls to the bottom of the page. This approach is more efficient because the search-results page loads and renders more results only when the user wants to view them.

However, some disadvantages of on-demand loading with infinite scrolling include the following:

- It is difficult for users to ascertain how many search results have loaded or where they are within the results set.

- Returning to a specific result earlier in the list of search results can be challenging.

- Users who are accustomed to using pagination controls and want that level of control might find the absence of a pagination control disconcerting, especially if they need to get to a search result that's buried deeply in a results set. [196, 197, 198, 199, 200, 201]

No-results pages

"The effort and ingenuity a product team invests in the no search results page is indicative of its overall dedication to customer success. Ignoring this special kind of search results page virtually guarantees a mediocre search experience...."—Greg Nudelman [202]

Greg Nudelman. Designing Search: UX Strategies for eCommerce Success. Indianapolis, IN: Wiley Publishing, Inc., 2011.

Despite using spelling autocorrection and **Did you mean?** to automatically improve search results and applying techniques for dealing with overconstrained searches such as returning partial matches, there still may be occasions when a user's search query returns no results. An information space simply might not include any content that matches the query—or any relevant content—or the search system might have encountered an error. In such cases, it's essential that the search system provide a helpful no-results page.

As on all search-results pages, the search bar should reside at the top of the page and contain the user's search query, allowing the user to revise the query. If possible, also include **Did you mean?**.

Of course, such a page must inform the user that the search produced no results by displaying a message such as the following at the top of the main content area of the search-results page:

Your search for [query text] found no matches.

Also, provide some tips for improving the user's search outcome such as the following:

To improve your search outcome, try the following:

- *Try using different words with similar meanings.*

- *Try using more general words.*

- *Try using more than one word.*

- *Try using fewer words.*

- *Check your spelling.*

- *Try searching all categories.*

- *Try browsing instead.*

Personalize this list of alternative approaches to reflect what the user has actually done. For example, if the user hasn't conducted a scoped search, don't include the following text: *Try searching all categories.*

Consider providing a **Popular Searches** list or a **Featured Products** list at the bottom of the page. [203, 204, 205, 206]

Filtering search results

Filtering search results temporarily removes certain results from the search-results set that appears on the search-results page and provides a highly interactive search experience that is very different from simply viewing or sorting the entire results set. Once the user has conducted a search, a set of filtering controls appears at the top or left side of the search-results page. Be sure the filters are noticeable. The user can employ these filtering controls to select, then access subsets of the results set. Thus, filtering is an iterative process that provides different views of the results set and lets the user narrow the results to view only those that would best meet the user's needs, while retaining the user's original search query.

Although filtering provides an efficient way of accessing particular search results, a filtered list of results does *not* give the user a clear conception of the full scope of the search results in a results set. Plus, this means of perusing a results set makes it more difficult for users to discern whether or when they've fully explored a set of search results. Because filtering search results requires more complex interactions, filtering is also a less accessible way of exploring search results.

The most common use of filtering is for faceted search, in which filters correspond to *facets*—independent attributes of the content or products in a list of search results. For more detailed information about faceted search, refer to the section "Faceted search," later in this chapter. Many ecommerce sites have implemented faceted search, enabling their customers to filter search results in myriad ways. Figure 13.17 provides an example of filtering faceted search results on the REI site, which lets users filter faceted search results by gender, size, brand, usage, features, color, price, ratings, and many product-specific facets.

Figure 13.16—Filtering search results on REI

Display meaningful headings for groups of filters on the search-sesults page, which should be collapsible and expandable. Expand the filters that users would most likely need to modify by default. Consider displaying the currently selected values for collapsed filters in addition to their headings. Individual filtering controls typically take the form of drop-down list boxes, drop-down list boxes with a checkbox for each option, groups of checkboxes, groups of option buttons, or sliders for setting ranges of values. The user changes the settings for the filters within the context of the search-results page, updating the list of search results accordingly, in real time.

Filtering search results risks overconstraining the user's search and producing no search results. For example, customers on an ecommerce site often want to filter products of a particular type by setting a price range. To support this capability, either provide a **Price range** slider that defines the lowest and highest prices of available products and lets the user select the desired range *or* **Lowest price** and **Highest price** drop-down list boxes that define the possible values for the lowest and highest prices. Do *not* provide text boxes for this purpose. Leaving it to the user to define possible values for the price range would likely overconstrain the user's search and produce no search results. [207, 208, 209, 210, 211]

Sorting search results

Sorting search results changes the order in which the results appear on a search-results page and sometimes provides a more useful results set than filtering the results would. Because most users limit their perusal of search results to those on the first page or screenful of results—or even just the first few of *those* results—sorting controls appear to act as filters by removing particular search results from view.

Enabling users to sort a list of search results—either alphabetically, chronologically, numerically, or in any other way—could help users complete their tasks. However, sorting by certain attributes could be more meaningful than sorting by others. For example, when searching for up-to-date news on a topic, users might want to sort search results chronologically—for example, using a **Sort by date** drop-down list box with the options **Most recent to oldest** or **Oldest to most recent** or using a set of toggle buttons comprising **Most Recent** and **Oldest** buttons. Toggle buttons make the options more visible than those in a drop-down list box.

Customers shopping on an ecommerce site and making purchasing decisions might want to compare products of a particular type and sort them by price or by ratings. Customers shopping for books or video or audio media might want to sort products by their publication or release date, respectively.

To enable users to sort a results set by a single, mutually exclusive sorting option, provide one of the following types of sorting controls at the top of the search-results page:

- Provide a **Sort by** drop-down list box comprising the available options and display the selected option in the list box. For example, to sort by date, provide a **Sort by date** drop-down list box with the options **Most recent to oldest** or **Oldest to most recent**; to sort by price, provide a **Sort by price** drop-down list with the options **Lowest to highest** or **Highest to lowest**; or to sort by ratings, provide a **Sort by ratings** drop-down list with the options **Best to worst** or **Worst to best**.

- Provide a set of toggle buttons whose labels present the available options. The selected toggle button should appear depressed. For example, to sort by date, provide **Most Recent** and **Oldest** toggle buttons; to sort by price, **Lowest Priced** and **Highest Priced** toggle buttons; or to sort by ratings, **Best Rated** and **Worst Rated** toggle buttons.

UXmatters uses a set of toggle buttons comprising **All**, **Authors**, and **Columnists** buttons to filter the content on the *Authors* page, as shown in Figure 13.18.

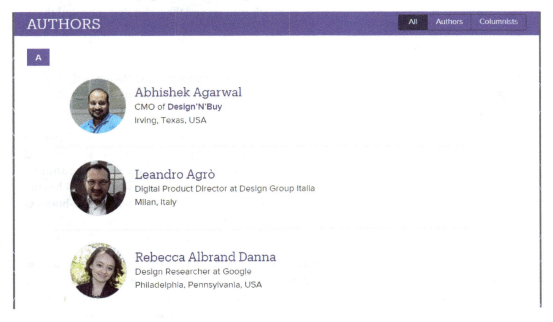

Figure 13.17—A set of toggle buttons on *UXmatters*

For detailed information about automatically sorting search results alphabetically, chronologically, or numerically, in addition to ranking them by relevance, recency, popularity, or document type, refer to the section "Automatically sorting search results," earlier in this chapter. [212, 213, 214]

Combining sorting and filtering

When searching a large volume of information, the main role of sorting is to display the search results in a way that better serves the user's goals rather than to simply reorder them—especially on mobile devices where screen real estate is limited. Consider different ways of better serving the user's goals by providing more useful combinations of filtering and sorting controls.

Users can employ a combination of filtering and sorting controls to select, then access either all of the search results in a results set or a subset of those search results.

When combining sorting and filtering controls on a search-results page, lay out the controls in close proximity to one another to ensure that the user can comprehend their collective impact on the user's view of the results set. For example, a scoping drop-down list box might reside immediately to the right of the search box, with a **Sort by** drop-down list box immediately to its right. [215, 216]

Some specialized types of search systems

Now, let's take a look at several specialized types of search systems and the technologies behind them.

Generative AI search

"GenAI will transform enterprise content search, discovery, and exploration, surfacing insights from various sources via a powerful natural–language user experience."—Annette Jump, Anthony Bradley, Eric Goodness, and Radu Miclaus [217]

In addition to the use of artificial intelligence's machine-learning (ML) algorithms and natural-language processing (NLP) in indexing and categorizing content to improve the precision and especially the recall of relevant search results, internal search systems can also leverage the capabilities of generative artificial intelligence (GenAI) to enhance the search experience and outcomes in various ways. GenAI has the potential to transform internal search systems across diverse domains, including business-to-business (B2B) manufacturing and commerce, ecommerce, customer service and support, healthcare, medical and scientific research, financial services, law firms, and knowledge-management systems. [218, 219, 220, 221]

According to *Gartner*, one key business use case on which GenAI will have the greatest transformative impact focuses on content discovery, including search, analysis, and knowledge management. With the proliferation of digital content across enterprises, the discovery of relevant, verifiable content has become more challenging for users. GenAI's ability to synthesize relevant responses to users' search queries or prompts from multiple sources of structured and unstructured data can support the following capabilities and more:

- Internal search and knowledge discovery
- Video and image search

- Business analytics and intelligence

- Analysis of data from diverse sources

- Summarization or visual representation of content from search results

- Analysis of the trustworthiness or accuracy of content from search results

- Conversational search [222]

GenAI search can understand natural language, as well as the holistic context of a search query or prompt, which often takes the form of a question, then quickly provide the most relevant search results, or answers. Plus, GenAI can use large language models (LLMs) to automatically generate content. So, once GenAI search analyzes a user's query or prompt, it can retrieve relevant information from one or more sources, then immediately generate a synthesized response. Ideally, a GenAI search system should cite the sources that informed its responses. [223]

According to research from Lucidworks, "Search practitioners expect generative models to transform the discovery experience in several key ways, including:

- Conversational search and browsing that can respond to simple requests ... with detailed results

- Semantic search that becomes smarter by leveraging LLMs to produce stronger training data

- Contextual document summaries that relate to the user's story

- Personalized product-detail descriptions that highlight unique product attributes relevant to the user's recent interaction" [224]

GenAI search, in combination with LLMs, supports conversational discovery and can provide personalized, information-dense responses to users' queries that can include summaries of content; suggest filters that let users narrow their search results; and offer personalized recommendations, which can drive revenues. [225]

However, while the integration of GenAI search holds great promise, several issues currently exist with this technology, as follows:

- **Obscuring information sources**—The most egregious and fundamental issue with GenAI search is its inability to consistently provide verifiable, accurate, relevant content. Responses to users' queries frequently include unsupported content and inaccurate citations. Citations fully support only 51.5% of generated sentences. Exacerbating issues regarding this lack of provenance, many search systems are now indexing generated content.

- **Making stuff up, or hallucinations**—The LLMs that power GenAI search generate coherent natural-language responses to users' questions—basically by predicting what the next words should be statistically. However, because LLMs cannot reason, the new content that they generate might lack essential context and, thus, fail to adhere to the actual meaning of their extensive training data, leading to unpredictable or undesirable results.

- **Failing to indicate uncertainty**—LLMs might guess what the response to a user's question might be rather than admitting their lack of knowledge—the equivalent of a conventional search system's returning no results. Thus, in cases where no relevant results exist, LLMs might hallucinate answers and even fake citations.

- **Reinforcing biases**—The training data for LLMs might reflect societal biases that result from the inordinate influence of a dominant group, including its social, cultural, ideological, and economic influences. The LLMs then learn, express, amplify, and reinforce these biases, which might include biases that are based on gender, race, nationality, religion, or class. Because LLMs tend to produce content that sounds authoritative, the existence of these biases might not immediately be apparent to users of GenAI search, who might trust the content and unquestioningly adopt these biases.

- **Trading off reliability for efficiency**—One goal of GenAI search systems is to reduce the effort of searching. However, in pursuing speed and efficiency, GenAI search might be reducing the breadth, depth, and accuracy of search results. When using a traditional search system, users are responsible for assessing the reliability and relevance of individual search results and can explore and verify the accuracy of a variety of results. But, when synthesizing a response to a user's query, a GenAI search system derives that response from particular information sources, reducing the diversity and depth of the information it provides, obscuring its provenance, increasing its bias, and reducing its reliability and verifiability. Unfortunately, users may misplace their trust in GenAI's plausible responses.

Different models of GenAI search systems are evolving, including the following:

- **Question-and-answer systems**—These conversational systems employ a chatbot user interface similar to that of Perplexity, shown in Figure 13.19, rather than the user-interface elements and interaction model of traditional search systems.

- **Hybrid systems**—These systems, which include Google and Bing, combine the user-interface elements and interaction models of traditional search systems with those of question-and-answer systems. Some of these systems prioritize the question-and-answer user experience. Figure 13.20 shows that of Google. [226]

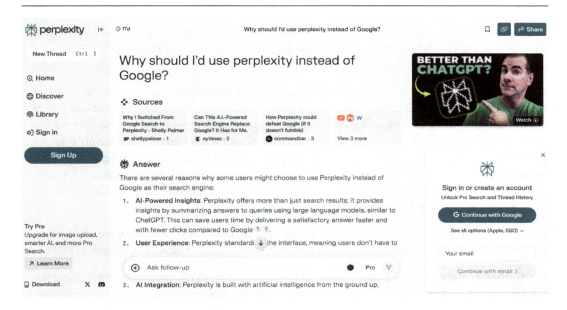

Figure 13.18—Perplexity's conversational user interface

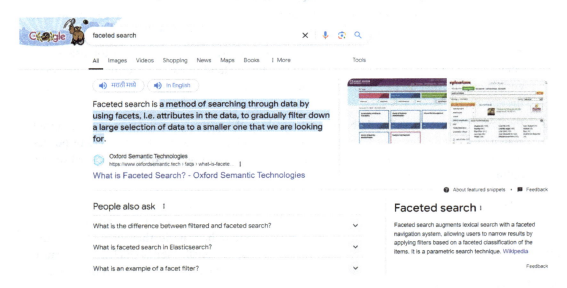

Figure 13.19—Google's hybrid user interface

Personalized search

"Generative AI is revolutionizing search by delivering personalized and context-aware search experiences."—Shelf [227]

"Personalization tailors search results to an individual user's preferences, behaviors, and geographical location, increasing the likelihood of delivering content that meets their expectations and is relevant and meaningful to them."—Elastic [228]

GenAI-based, personalized search leverages the analysis of comprehensive user data to generate precisely tailored search results that meet the user's unique needs. The personalization of search results can transform the search experience by enabling user journeys that provide the precise information the user needs to make optimal decisions. By generating search results that are specifically relevant to the user, personalized search provides a more satisfying and engaging user experience. The more relevant the results, the more efficient and effective the search experience is for the user. However, while personalized search increases the relevance of search results, it could also potentially limit the user's exposure to different perspectives or products. [229]

The data processing at the foundation of personalization comprises three distinct steps—each of which is essential to successful personalization—and consists of three corresponding components, as follows:

1. **Gathering data regarding the user**—This phase involves the collection of information at the necessary level of granularity for the system's analysis. This information pertains to a specific user or customer or a particular persona, as well as that user's context. The system collects historical and current data regarding the user's interactions, behaviors, preferences, and situational contexts and typically integrates user or customer information from other sources, as well as trend data sets. For example, the system captures usage data such as queries and clicks.

2. **Analyzing the data**—This phase consists of the comprehensive analysis of the usage and contextual data that the system has gathered to understand the user, customer, or persona, and determine that user's preferences. The analysis of this data occurs within the context of the specific discovery use case.

3. **Presenting the results**—Finally, based on the analysis, the system generates a set of personalized search results, tailoring their presentation to the user's information needs. The user's responses to these results can act as inputs to data analytics, providing a feedback loop that drives optimization. [230, 231]

Short head queries are those for which users most frequently search and with which the majority of users engage. Typically, short head queries correspond to a large part of all searches on an information space and drive the most revenue. Because these queries tend to be generic, they would benefit the most from personalization. But, for most information spaces, personalization is challenging because users sign in so infrequently. For example, 70–95% of users visit a given Web site less than twice a year. Therefore, the availability of the user data that drives personalization tends to be rather limited. As a consequence, companies should deploy AI-driven approaches to personalized search that can deliver personalized search results even when few user-data points are available. [232, 233]

Benefits of implementing a personalized-search capability include the following:

- **Enhancing the relevance of search results**—Personalization enhances the relevance of search results by leveraging contextual relevance in tailoring search results to a specific user based on geographical location, language, and device type; as well as the user's preferences, behaviors, feedback, and browsing, search, and purchase histories. Location awareness can play an important role in personalization and contextual search on mobile phones. Along with the user's selection of an application or submission of a search query constituting the entry point for the user's conducting a search, location data can help the search system infer the user's intent. Knowing the user's previous locations and current location—where the user came from and the path to where the user is now—could also inform user intent—what the user wants or needs now. Nevertheless, the clearest indication of user intent is the user's search query. Inferring user intent can also help the search system prioritize the display of certain search results to better meet the user's needs. [234, 235]

- **Expanding content discovery**—GenAI-powered personalized search uses natural-language processing (NLP) and large language models (LLMs) to interpret users' complex queries, enabling them to more easily access useful information that is deeply buried—for example, in extensive product catalogs or repositories of research data. Thus, personalized search can reduce users' search time and increase their productivity.

- **Anticipating the needs and expectations of users**—Personalized search uses machine learning (ML) and NLP to more accurately infer users' intent when searching for specific content and, thus, delivers more relevant search results that increase user satisfaction. ML uses dense-embedding representations to understand user intent through context and by employing *nearest-neighbor algorithms (NNAs)* such as the k-nearest neighbor (kNN) algorithm. These supervised ML classification algorithms use proximity in classifying or predicting the nearest neighbors to specific data points. The section "Using AI-powered search to achieve greater precision," earlier in this chapter, defines dense-embedding representations. Plus, personalization is a key element of predictive search, which leverages users' past searches in anticipating their future needs.

- **Increasing customer loyalty**—When users feel that a company understands their needs, their trust in and loyalty to that brand increases, making it more likely that they'll become long-term customers. Loyal customers contribute significantly to a company's value. Within ecommerce contexts, they make repeat purchases, increasing the company's revenues and adding directly to its bottom line.

- **Improving accessibility**—Supporting visual and voice search accommodates users' diverse preferences and abilities and, thus, facilitates users' employing their natural modes of communication and interaction when using a search system. For example, visual search lets users search for products without any need to specify precise keywords. Instead, users can identify photos of the products they want. This simplifies product discovery and decision-making for people who speak different languages and for dyslexics who find reading text and typing queries challenging. Voice search enables users with visual impairments to easily communicate what they need. Accessible search user experiences broaden a company's potential customer base.

- **Generating personalized recommendations**—When customers receive personalized recommendations for products and media such as movies and music, this often prompts them to buy more or engage more, increasing the company's revenues.

- **Personalizing product details**—Personalized descriptions of products can highlight particular attributes of products that are relevant to specific users by leveraging the context of their recent interactions.

- **Handling abandoned carts**—When customers abandon their cart on an ecommerce store, the company can send them personalized email messages, reminding them about the products they were considering purchasing or leverage their recent search and browsing history to recommend related products.

- **Incorporating facets and filters**—Facets, or smart filters, allow users to narrow their searches to particular relevant facets such as topics, categories, locations, price ranges, or date ranges. Providing filtering controls for facets is most useful on information spaces with large volumes of content. Some search systems can automatically detect filters. For more information about faceted search, see the section "Faceted search," later in this chapter; for more on filtering search results, see the section "Filtering search results," earlier in this chapter. [236, 237, 238, 239, 240]

Semantic search

"Semantic search is a search-engine technology that interprets the meaning of words and phrases. The results of a semantic search will return content matching the meaning of a query, as opposed to content that literally matches words in the query."—Elastic [241]

Semantic search uses machine learning (ML) and natural-language processing (NLP) to accurately interpret the meaning and determine the broader context of natural-language queries *and* the information space's content and to assess searchers' intent, as follows:

- **Searcher intent**—Because semantic search can capture and accurately interpret searchers' intent and, thus, understand their needs, it can identify the most relevant search results and return content that meets the searchers' needs. Semantic search ranks results in the order of their relevance based on both a searcher's query and its broader context. What was the searcher's purpose in conducting a search? Was the searcher seeking information? Purchasing a product?

- **Context**—Semantic search can interpret the contextual meaning of the searcher's query and analyze its context in responding to that searcher's intent. The context of the search also refers to any available information beyond the text string that the search query comprises, including the searcher's search history, the geographical location of the searcher, the context of the words in the query, the relationships between the words in the query, and any other contextual information that is available for analysis. By leveraging the context of the search, semantic search can determine the nuanced meanings of the words in the query and across an information space's entire corpus. Plus, semantic search can identify other words that searchers might use within similar contexts, broadening the scope of the search and making it more effective. Understanding the query's contextual meaning and, thus, the relevance of particular content significantly impacts the ranking of search results.

Semantic search ranks content, then delivers matching search results based on their semantic meaning, contextual relevance, and relevance to the user's needs. Thus, semantic search provides a superior experience to keyword search. [242, 243, 244, 245] Semantic search leverages large language models (LLMs) to produce stronger training data and, thus, improve semantic models. [246]

Vector search powers semantic search. Vector space models represent pages' content *and* queries as vectors in dimensional space, enabling a search system to compare and rank them. *Vector search* uses machine learning (ML) to perform the following steps:

1. When indexing an information space's existing pages, a vector-based search system captures the meaning and context of unstructured, natural-language data and images, transforming that data into embeddings—that is, numeric representations of searchable data and their related contexts—then stores them as vectors.

2. When the searcher submits a query, the search system transforms the query into embeddings, then stores them as vectors.

3. The search system uses the k-nearest neighbor (kNN) algorithm to compare the vectors that represent text on existing pages and match those vectors to the query vector to determine which vectors are most relevant to the query vector. Vector search can more accurately process ambiguous language in both content and queries.

4. The search system then generates results and ranks them according to their conceptual relevance. Vector search executes queries more quickly and delivers more relevant search results than keyword search. [247, 248, 249]

Some capabilities of semantic search include the following:

*   More accurately interpreting and effectively answering questions that searchers pose in natural language

*   Supporting voice search, which is wholly reliant on the ability to understand natural-language queries. Among people who are between the ages of 25 and 34, 50% use voice search daily, particularly in ecommerce and streaming TV contexts.

*   Supporting conversational search queries and answers

*   Considering the context of the search to distinguish the appropriate meaning among a word's different meanings

- Providing better targeted personalization and answering queries in the searcher's own language

- Delivering more comprehensive search results by identifying and searching for synonyms and variations of the words in searchers' queries

Disadvantages of semantic search include a lack of transparency, which makes it more difficult for searchers to understand why search queries return specific results. Plus, semantic search requires a great deal of computational power. [250]

Faceted search

"Facets offer an alternative to hierarchies. ... Strict hierarchical organization is ... limiting: there is only one way to locate a piece of information. ... With facets, [categories determine] the location of an item.... This offers multiple points of access [and] provides greater flexibility in locating information. ... Facets [are] mutually exclusive categories that describe the properties or dimensions of an item. ... Multiple values from a given facet can be assigned to a single item."—Jim Kalbach [251]

In comparison to keyword search, faceted search supports discovery and exploratory search more effectively. Faceted search can also facilitate conversational interfaces by using natural-language processing (NLP) to determine the meaning of users' questions, then providing answers to them.

Faceted classification is an approach to knowledge representation that provides the foundation for faceted search. *Faceted navigation* lets the searcher progressively refine or elaborate on a search query, observing the impacts of selecting certain properties or values for one facet on the properties or values that are available for other facets, then submit the query. Because selecting a property or value of one facet constrains the properties or values of the other facets, eliminating those that are not possible, facets and their properties or values are interdependent, as faceted navigation on the ecommerce pages of the Rosenfeld Media Web site demonstrates, as shown in Figure 13.21.

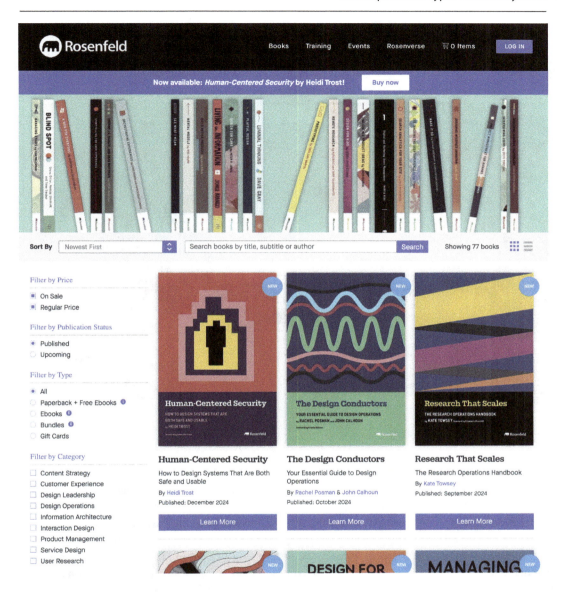

Figure 13.21—Faceted navigation

A *faceted search* user interface integrates faceted navigation with the ability to search a combination of both unstructured text and *metadata*—that is, structured attributes corresponding to a faceted classification system. The faceted navigation system searches the structured metadata, while a text-search capability searches the information space's content, which is unstructured text. The user first types a search query; then refines the query by selecting one or more facets, or filters, that narrow the scope of the search; then submits the search query.

In addition to providing multiple access points to an information space, faceted classification is scalable. Thus, faceted search is particularly useful for searching large-scale information spaces with extensive content such as Web sites, intranets, product catalogs, and ecommerce stores. However, with scale come the challenges of complexity and efficiency.

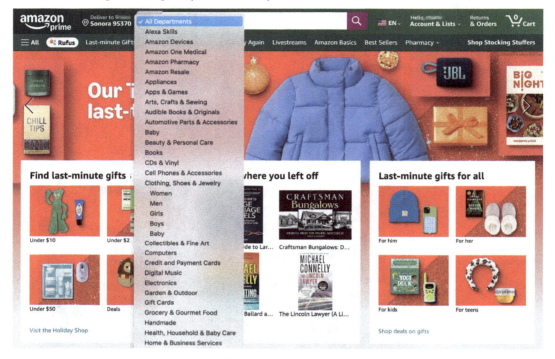

Figure 13.22—Amazon's faceted search

Managing scale must consider factors such as the number of pages or documents that an information space comprises, the number of facets that were assigned to each page, and the amount of searchable text on each page, as well as the storage requirements that derive from implementation of the faceted search system. The data structures that a faceted search system employs typically consist of two data tables: an *inverted index* that maps both the properties or values of facets and the instances of searchable text to particular pages *and* a *document table* that maps specific pages or documents to the properties or values of facets. To determine the scale of a faceted search system, add the number of facet properties or values and the number of instances of searchable text to determine the *average width* of each page, then multiply the average width by the number of pages on the information space.

With faceted search, processing a query is computationally intensive and follows this process:

1. The search system identifies the pages that meet the query's constraints and retrieves the results set from the inverted index.

2. The search system displays the facets and properties or values that let the user refine the results set.

Because of limitations in both screen real estate and the user's ability to handle complexity, displaying all of an information space's potential facets—the possible classifications for pages—and their many properties or values could be impractical. Therefore, determining the most valuable facets and properties or values to display is essential. Some guidelines for choosing these facets and their properties or values include the following:

- Try to limit both the number of facets and the numbers of properties or values under each of the facets.

- Provide facet labels that convey sufficient information scent *and* adequately differentiate the facets, which should be mutually exclusive.

- Ensure that the available facets provide comprehensive coverage of the results set.

- Prioritize facets for which there are properties or values assigned to all or most pages in the results set over those that apply to only a few.

- Choose facets for which the distribution of their properties or values is fairly consistent across the entire results set rather than concentrated in only one or a few properties or values.

- Merge facets that share common properties or values, creating a single facet from them and avoiding duplication.

For facets that have large numbers of properties or values, consider creating a hierarchy of properties or values, using progressive disclosure to display only those for a specific facet. *If* no value in the hierarchy has a large number of immediately subordinate values, there would be no need to display large numbers of values at once. However, usability problems could arise if this results in a deep hierarchy. If creating a hierarchy of facet values is impractical, some facets might necessarily have large numbers of values. To display facets with large numbers of values, consider doing one of the following:

- Initially display only the values that occur with the greatest frequency in the results set.

- Initially display only the values that occur more frequently in the results set than on an information space's overall collection of pages.

- Create a hierarchy of values by grouping values under ranges of letters—for example, A–C, D–F, and so on—or ranges of numbers.

- Create a hierarchy that groups similar values by inferring correlations among the values in either the results set or the overall collection of pages.

To increase facets' information scent, consider providing concise previews of the content that is included in the results set for the selected facet and properties or values.

To improve faceted search for collections of documents that comprise multiple entity types—for example, documents and authors—and support relationships between entities of different types, consider enabling users to conduct faceted searches on multiple entity types simultaneously. [252]

Challenges of designing faceted search

"Facets ... solve the problem of a user confronted with an unfamiliar search space. Facets structure the search space and make the user get a sense of the type of information she can expect to find there."—Raluca Budiu [253]

By providing controls such as drop-down lists, option buttons, and checkboxes that have natural-language labels, you can enable the user to narrow down large sets of search results to smaller, more relevant sets of results. [254] But designing a faceted search user interface poses some challenges, as follows:

- **Organizing and presenting facets**—Faceted search responds to the user's query by returning *both* a set of matching search results and a set of facets, or filters, that can help the user refine them. Laying out both the results and a large grouping of filters side-by-side on the search-results page is ideal, but can present challenges: users sometimes overlook the filters. While placing a vertical list of filters on the left side of the page is common practice, you must visually emphasize the filters to call users' attention to them.

- **Organizing and presenting facets' values**—The complexity of faceted search can overwhelm users. So, include only essential facets and values and present them using a consistent layout across searches. Display the most universally useful facets and most common values nearest to the top of the page. Then display facets that are specific to a particular type of document or product. Organize related facets into clearly labeled groups. If appropriate, arrange facet values in a clear hierarchy.

- **Supporting the selection of multiple values**—Typically, the user selects a single value for each facet. But it should be possible for the user to select multiple values for a given facet, using either disjunctive (OR) selection or conjunctive (AND) selection. The user-interface elements the user employs in selecting values must make the type of selection clear.

- **Providing a search bar**—Typically, the user submits a search query, then refines the results set using the facets. Always display the search bar on the search-results page, and display the user's original query in the search box, indicating the search that generated the results set. The available facets should reflect the entire results set.

- **Laying out search-results pages on mobile**—Although the concept of faceted search is mobile friendly, it is difficult to lay out a page comprising both filters and search results on a small screen. Typically, on mobile, the facets reside in a panel that overlays the search results. Alternatively, a faceted-search system could employ AI to analyze the ways in which the products in an ecommerce system's database correspond to a particular user's data, including the user's account information and behaviors, then display just a few filters, or *discovery tags*, that coincide with that user's needs. [255, 256, 257]

Advanced search

With the availability of modern search systems, especially those that support personalization and faceted search, most users now have no need for advanced-search capabilities. However, if an information space's audience includes such advanced users as academic researchers, doctoral students, medical researchers, attorneys who conduct patent searches or legal searches, or librarians, you could consider designing an Advanced Search page that would let them express complex queries using a query language, Boolean operators, or an Advanced Search form that uses parametric search or faceted search. If you decide to provide an Advanced Search page, should it be an easy-to-use query builder for novice searchers or a powerful tool for expert searchers?

Advanced search can help users learn a domain's vocabulary and become familiar with an information space's metadata. The fields of an Advanced Search form can facilitate greater precision in specifying searches.

Provide a link to the Advanced Search page near the search bar at the top of the search-results page, where users who are disappointed in their search results and want to refine or completely revise their query will readily find it.

Consider including one or more filters that let users narrow a search to a specific category or search zone. Display a filter's options in a **Limit search to** or a similarly labeled drop-down list box that comprises individual categories or search zones. Such filters clearly communicate what content the user is searching for and enable the user to broaden or narrow a revised search.

Be sure to provide a *Help* page that explains, in detail, how to use the advanced-search capabilities to create more precise queries. [259, 260]

Parametric search

In parametric search, the user selects parameters by using an extended form to select values for an array of properties using drop-down lists. The Library of Congress Catalog shown in Figure 13.23 provides an example of an advanced-search system that employs parametric search that comprises multiple search boxes and drop-down lists and incorporates Boolean logic.

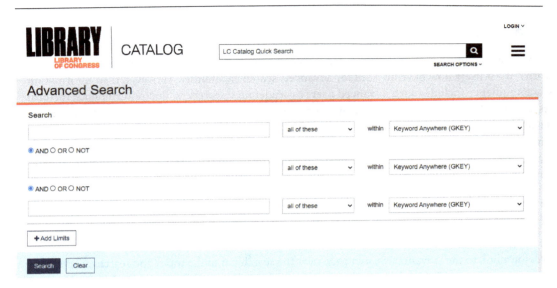

Figure 13.23—Library of Congress Catalog's Advanced Search

A disadvantage of parametric search is that it impedes exploration. [258]

Summary

Search actually does work as a user strategy in many cases…. … If you have a well-designed search facility and users are looking for a specific item with a well-defined name, they'll probably be successful."—Jakob Nielsen [261]

In this chapter, you learned about some challenges that users encounter in searching information spaces. You also learned how to assess the need to create an internal search system for a particular information space and the value that implementing an internal search system would provide.

The design of search systems is currently at an inflection point, with many powerful, new technologies radically changing the search landscape. You've discovered some of the technologies that are presenting new possibilities for optimizing the quality of search results and the overall search experience. You also learned how to design a usable internal search system that meets users' needs by using common search user-interface design patterns.

Finally, this chapter provided in-depth information about some specialized types of internal search systems that an information space might incorporate—alone or in combination with others.

References

To make it easy for readers to follow links to the references for this chapter, we've made them available on the Web: `https://github.com/PacktPublishing/Designing-Information-Architecture/tree/main/Chapter13`

Index

packtpub.com

Subscribe to our online digital library for full access to over 7,000 books and videos, as well as industry leading tools to help you plan your personal development and advance your career. For more information, please visit our website.

Why subscribe?

- Spend less time learning and more time coding with practical eBooks and Videos from over 4,000 industry professionals

- Improve your learning with Skill Plans built especially for you

- Get a free eBook or video every month

- Fully searchable for easy access to vital information

- Copy and paste, print, and bookmark content

Did you know that Packt offers eBook versions of every book published, with PDF and ePub files available? You can upgrade to the eBook version at packtpub.com and as a print book customer, you are entitled to a discount on the eBook copy. Get in touch with us at customercare@packtpub.com for more details.

At www.packtpub.com, you can also read a collection of free technical articles, sign up for a range of free newsletters, and receive exclusive discounts and offers on Packt books and eBooks.

Other Books You May Enjoy

If you enjoyed this book, you may be interested in these other books by Packt:

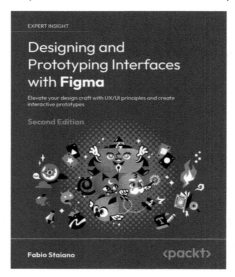

Designing and Prototyping Interfaces with Figma

Fabio Staiano

ISBN: 978-1-83546-460-1

- Create high-quality designs that cater to your users' needs, providing an outstanding experience
- Mastering mobile-first design and responsive design concepts
- Integrate AI capabilities into your design workflow to boost productivity and explore design innovation
- Craft immersive prototypes with conditional prototyping and variables
- Communicate effectively to technical and non-technical audiences
- Develop creative solutions for complex design challenges
- Gather and apply user feedback through interactive prototypes

101 UX Principles – 2nd edition

Will Grant

ISBN: 978-1-80323-488-5

- Work with user expectations, not against them
- Make interactive elements obvious and discoverable
- Optimize your interface for mobile
- Streamline creating and entering passwords
- Use animation with care in user interfaces
- How to handle destructive user actions

Packt is searching for authors like you

If you're interested in becoming an author for Packt, please visit `authors.packtpub.com` and apply today. We have worked with thousands of developers and tech professionals, just like you, to help them share their insight with the global tech community. You can make a general application, apply for a specific hot topic that we are recruiting an author for, or submit your own idea.

Share Your Thoughts

Now you've finished *Designing Information Architecture*, we'd love to hear your thoughts! Scan the QR code below to go straight to the Amazon review page for this book and share your feedback or leave a review on the site that you purchased it from.

`https://packt.link/r/1-838-82719-6`

Your review is important to us and the tech community and will help us make sure we're delivering excellent quality content.

www.ingramcontent.com/pod-product-compliance
Lightning Source LLC
Chambersburg PA
CBHW060110090326
40690CB00064B/4424